P9-CER-566

THE UNWANTED

THE UNWANTED

European Refugees in the Twentieth Century

Michael R. Marrus

New York   Oxford
OXFORD UNIVERSITY PRESS
1985

Oxford University Press

Oxford   New York   Toronto
Delhi Bombay   Calcutta   Madras   Karachi
Kuala Lumpur   Singapore   Hong Kong   Tokyo
Nairobi   Dar es Salaam   Cape Town
Melbourne   Auckland

and associated companies in
Beirut   Berlin   Ibadan   Mexico City   Nicosia

Copyright © 1985 by Oxford University Press, Inc.

Published by Oxford University Press, Inc.,
200 Madison Avenue, New York, New York 10016

All rights reserved. No part of this publication may be reproduced,
stored in a retrieval system, or transmitted, in any form or by any
means, electronic, mechanical, photocopying, recording, or otherwise,
without the prior permission of Oxford University Press.

Library of Congress Cataloging-in-Publication Data
Marrus, Michael Robert.
The unwanted : European refugees in the twentieth
century.
Bibliography: p.
Includes index.
1. Refugees—Europe—History—20th century.
2. Europe—Emigration and immigration—History—20th century.
I. Title.
JV7590.M37   1985   325'.21'094   85-15305
ISBN 0-19-503615-8

Printing (last digit): 9 8 7 6 5 4 3 2 1

Printed in the United States of America

For Carol Randi Marrus

Preface

SURVEYING the history of refugee movements in Europe since the 1880s, this book borrows and distills the work of several generations of scholars, statesmen, administrators, travelers, and others. Without their investigation and commentary I could not have contemplated a study such as this, let alone have completed it. I hope I have treated their work with respect, while maintaining a critical posture; in any event, I have done my best to be fair.

I am very grateful for support I received from many quarters. Work began at St. Antony's College, Oxford, in 1979, and was completed at the Institute for Advanced Studies of the Hebrew University of Jerusalem five years later. I want to thank Warden Raymond Carr and the fellows of St. Antony's, and Professor Aryeh Dvoretzky and his staff at the Institute, for providing ideal scholarly conditions as well as the good fellowship for which both institutions are famous. Ivon Asquith and Nancy Lane of Oxford University Press and Alain Oulman and Roger Errera of Editions Calmann-Lévy showed confidence in the project from an early stage, when it seemed only a frail scaffolding of ideas. Financial assistance from the Social Sciences and Humanities Research Council of Canada, the Connaught Foundation and the Office of Research Administration of the University of Toronto, and the John Simon Guggenheim Memorial Foundation enabled me to read and write in academic luxury.

Several colleagues offered help and encouragement along the way, particularly Raul Hilberg, George Mosse, Robert Paxton, and Eugen Weber. For their valuable suggestions on points of detail I want especially to thank James Barros, Yehuda Bauer, Chris-

topher Browning, John Cairns, Richard Cohen, Paul Robert Ma-
gocsi, Andrew Rossos, and Joseph Shatzmiller. My good friend
Edward Shorter read the entire manuscript while in the throes
of his own composition, saving me from many errors and infe-
licities. An able research assistant, Thomas Mengel, plunged
willingly into unfamiliar terrain, often with only fragmentary direction. Kate Hamilton typed the manuscript with exemplary accuracy and dispatch. Finally, Wendy Warren Keebler and Cecil
Golann at Oxford University Press helped see the manuscript
through to its published form.

   Throughout, my greatest and most lasting debt is to my wife,
Carol Randi Marrus—wise and discerning in human relations, the
essential background for European refugees.

Jerusalem   *M.R.M.*
April 1985

Contents

LIST OF MAPS   xiii

INTRODUCTION   3

1 / TOWARD A MASS MOVEMENT   14

The Nineteenth Century   14

The Jewish Exodus from Eastern Europe   27

The Balkans and the "Unmixing of Peoples"   40

2 / THE NANSEN ERA   51

The Great War and Upheaval in Eastern Europe   52

Refugees and the Collapse of the Tsarist Empire   53

Jewish Refugees in Eastern Europe   61

Empires in Ruin: Refugees and the Peace Settlements   68

Armenian Refugees   74

International Projects   81

Emergency Relief in the East   82

Fridtjof Nansen and the League of Nations High Commission for Refugees   86

Ministering to the Homeless   91

Greeks, Turks, and Bulgarians, 1919–28 96

  The Greco-Turkish War and the Convention of Lausanne,
  1923 97

  The Greco-Bulgarian Exchange 106

Stabilization, 1924–30 109

  International Planning in the Locarno Era 110

  Resettlement Schemes in the 1920s: Jews and Armenians 114

3 / **IN FLIGHT FROM FASCISM** 122

Fascism and Its Enemies Before 1938 123

  Refugees from Italy 124

  Refugees from Germany 128

  Refugee Movements in Europe, 1933–38 131

Refuge in the Depression Era 135

  International Climate 135

  New Threats Against the Jews in Eastern Europe 141

  Closing the Doors: Some Examples 145

    FRANCE 145

    GREAT BRITAIN 149

    SWITZERLAND 154

The Failure of International Organization 158

  The League of Nations 158

  The Crisis Year, 1938 166

  The Crisis Deepens, 1939 177

  Resettlement Schemes for Jews 183

Last Chance, 1939–41 189

  The Collapse of Republican Spain 190

  Refugees from Poland and Eastern Europe 194

  Flight in the West, 1940 200

  Prospects for Escape in the Years 1939–41 203

4 / UNDER THE HEEL OF NAZISM 208

Nazi Policy, 1933–44 209

Jews in Central and Western Europe, 1933–41 209
Lebensraum in Eastern Europe, 1939–41 219
Jewish Refugees and the Final Solution 227
Rise and Fall of Lebensraum 234

Wartime Escape Routes 240

The Soviet Union and Eastern Europe, 1941–44 241
Neutrals 252
SWITZERLAND 252
SPAIN AND PORTUGAL 258
THE VATICAN 265
SWEDEN AND TURKEY 270
Palestine 274
Italian Sanctuaries for Jews 278

Rescue Efforts 282

New Prospects in 1943 283
Rescue by Negotiations 289

5 / THE POSTWAR ERA 296

The European Picture, 1945–47 298

The Allies and the Refugees, 1944–45 299
Military Control in the West, 1944–45 308
Forcible Repatriation to the Soviet Union 313
UNRRA 317

Refugees on the Move 324

Germans 325
Jews 331

The IRO and the "Last Million" 340

EPILOGUE: CONTEMPORARY EUROPE **347**

Refugees and the Origins of the Cold War *348*

Shaping a UN Agency: UNHCR *354*

Cold War Refugees *358*

Diminished Importance of Europe 364

Settling Old Business 365
New Refugees 367

NOTES **373**

INDEX **401**

Maps

The Balkans, 1878–81 (43)

The Balkans, 1912–13 (47)

European Boundary Changes After World War I (73)

The Caucasus, 1918–21 (79)

The Balkans After World War I (99)

Poland Under Nazi Occupation (222)

Europe Under Nazi Occupation *(Before 22 June 1941)* (244)

Territorial Changes in Eastern Europe, 1947 (303)

THE UNWANTED

Introduction

REFUGEES, people obliged by war or persecution to leave their dwellings and seek refuge abroad, have tramped across the European continent since time immemorial. Yet only in the twentieth century have European refugees become an important problem of international politics, seriously affecting relations between states. "Never before has history seen such a universal upheaval," wrote one European observer in 1948, "tearing people loose from their homes and daily lives."[1]

I begin with the changes that the modern period saw in the character of refugee movements and their significance for the European state system. First, in sheer size the waves of forcibly displaced persons were vastly greater than the world had ever seen before. Sounding an alarm in 1938, the American journalist Dorothy Thompson told her readers that over the previous two decades some four million persons had had to leave their homes under political pressure. "A whole nation of people, although they come from many nations, wanders the world, homeless except for refuges which may at any moment prove to be temporary."[2] European population movement in the preceding century drained the countryside into towns and cities and generated the great waves of transoceanic settlement and colonization. But all that was dwarfed in the post-1918 period by the new phenomenon of refugees. "The history of international migration in the past thirty years has been largely the history of refugees," observed two researchers in 1944. "Ours may truly be called the era of refugees."[3] Contemplating the stunning refugee catastrophes of the Second World War, Malcolm Proudfoot noted how the numbers

3

had risen to unprecedented heights. During that conflict, he calculated, sixty million European civilians had been forced to move—more than ten times the number of refugees generated by the First World War and its aftermath. And the flow had not ceased. Writing in 1955, Joseph Schechtman said that postwar European population movements involved some twenty million people expelled, transferred, or exchanged.[4] Since then the flood has receded, and it is the Third World that is awash in refugees. But we might recall that this was not always so; Europe, not Africa or Asia, was once the continent of most of the world's homeless.

Second, along with the great rise in the numbers of refugees has come a radically new form of homelessness. The term "refugees" normally refers to persons who have crossed some international frontier, who have been forced to leave the state in which they once lived. Before the growth and consolidation of modern nation states there were certainly impoverished refugees, but they seldom appeared different from vagabonds or the itinerant poor who traveled from place to place in every premodern society. The latter were similarly outsiders, strangers, and hardly less vulnerable to the vicissitudes of weather, aggression, or the force of law than refugees who had been tossed up by political conflict. Since the eighteenth century the state has moved massively into many areas formerly reserved to the church or local institutions such as the village, the municipality, or assemblies of notables. National governments have increasingly defined rights and obligations of people, provided for their welfare, and certified their citizenship. Unlike vagabonds or the wandering poor, who at least were seen as part of society, refugees often found themselves entirely outside the web of national community. The late political philosopher Hannah Arendt, one of the first to define the singular predicament of refugees as it emerged after the First World War, described how they were reduced to a lonely, savage existence, hounded from place to place by national governments that alone accorded to people elementary rights. Outside the state from which they had come, refugees could not work, could not live unmolested, could usually not remain at liberty for any length of time. At the mercy of the state, they had become "the most symptomatic group in contemporary politics" in states that Arendt saw moving inexorably toward totalitarianism.[5] Modern refugees, in this way, differed from those of earlier times because their homelessness removed them so dramatically and so uniquely from civil society.

Third, modern refugee movements diverged from those of earlier times because of the extraordinary duration of the refugees' displacement. Refugees sometimes wandered for years through the interstices of the European state system, and many indeed passed on their anomalous status to a second generation. Partly this was so because some refugee agencies counted people as refugees until they satisfactorily resolved the question of their nationality—a technical procedure of often unending complexity. But in part modern refugees have remained homeless so long simply because of the great numbers involved. In 1951 the Council of Europe sponsored a committee of experts to study refugee problems. It estimated that since the end of the Second World War eleven million refugees of all sorts had poured into Western Europe, and, of these, 4.5 million had "neither finally nor temporarily been absorbed."[6] When the International Refugee Organization completed its task of resettling refugees at the end of that year hundreds of thousands of refugees remained unsettled in Europe, many of whom still languished in camps.

In contrast, refugee situations before the nineteenth century could scarcely endure for years at a time, aggravating international relations. Of course, some refugees could support themselves one way or another, often indefinitely. Others found refuge thanks to the charity of the clergy, princes, or local notables. Still others merged with the poor of the host society and survived by begging, stealing, or occasional labor. All these people might continue to think of themselves as outsiders in some sense. But premodern times knew no camps where masses of civilians could be interned for lengthy periods and needed no special category to suspend them outside the framework of the civilized community. Thrust unexpectedly on a society usually indifferent to outsiders of any sort, many refugees would quickly succumb to hunger, disease, or exposure. Large masses of people simply could not move from place to place supported by meager social services. Winters, generally, would finish them off.

In order to put modern refugee movements into perspective, it may help to scrutinize some earlier experiences of forced population movement in Europe. Beginning with the first expulsions of Jews from the Iberian Peninsula in 1492, one scholar estimates that there were over a million refugees in Europe during the next two centuries until the tempo slowed in the eighteenth century.[7] Generally, refugees expelled during this period were religious minorities, held to constitute some challenge to existing political authority. In the interests of consolidating their states, rendering them more homogeneous politically and culturally, rulers sometimes turned on cohesive groups whose loyalties were deemed suspect. In addition to the Jews driven out of the Spanish kingdom, followed by the Moriscos a short time later, Protestants from France and the Spanish Netherlands were similarly exiled, along with Protestants or Catholics from states and principalities in Central Europe during the Reformation era. Other refugees included Serbs fleeing Ottoman rule from the seventeenth century on, crossing the Danube into Hungarian lands; English, Irish, and Scottish Catholics following their Stuart monarch into exile after 1688; and later various Protestant sects leaving Scandinavia and Central Europe in the eighteenth century, seeking free religious expression. In addition, there were refugees caused by wars— populations temporarily displaced by the Thirty Years War, the campaigns of Louis XIV, and the wars of the eighteenth century.

It is impossible to generalize about the reception that awaited such very different groups of migrants, but one can say that they moved within a Europe in which the central organs of government usually considered that adding to the working population was an asset, not a liability. Long before the impact of the humanistic ideas of the Enlightenment, which may possibly have softened the impact on innocent refugees of man-made disasters, rulers tended to favor the controlled movement of people into their jurisdictions. New inhabitants, to whom neither the state nor localities by definition had any particular obligation, were potential generators of wealth and contributors to the strength of the society. According to mercantilist doctrine, immigration was to be encouraged. Guilds or municipalities often exercised controls fiercely, suspicious of outsiders and generally disposed to protect themselves from both the competition and the obligations that newcomers might impose. But these local authorities were sometimes overruled by central governments, and in any event localities were hardly inclined to distinguish between refugees and other outsiders. They mounted guard against both, with little distinction.

Central governments pursued their own interests by facilitating immigration and discouraging or even forbidding emigration. Whether to be taxed, to contribute to the growth of manufactures and commerce, to offer specialized knowledge, or to join the military, talented or affluent foreigners were frequently deemed useful to society and welcomed with open arms by European monarchs or municipalities. After the Spanish-enforced expulsion of Jews from the kingdom of Naples in 1511, for example, some cities successfully appealed to the king to allow their return, claiming that they could not pay their taxes without the services the Jews provided. For related reasons, governments also tried to block those who went elsewhere. Colbert, for example, tightened laws prohibiting people from leaving France until the death penalty was prescribed for those attempting to do so illegally. Widely implemented in the late seventeenth and eighteenth centuries, such policies prompted rulers to welcome refugees who were not judged a political menace. Thus Frederick William of Prussia, the Great Elector, invited Protestants to settle in his kingdom in 1685, the year of their expulsion from France by the revocation of the Edict of Nantes. Subsequent Prussian edicts extended into the eighteenth century invitations to a variety of other expellees. Peter the Great and Catherine the Great similarly tried to encourage immigrants to the tsarist empire.

For more humble folk, with less to offer the Great Elector or the Russian tsars, material factors limited the degree to which they could be uprooted. Throughout the premodern period armies were heavily dependent on local populations for supplies of all sorts, particularly food and shelter. Commanders, therefore, wanted the locals to stay put whenever this was possible, to help feed the troops and to be kept in line as well. It made no sense to drive away villagers before the harvest, to see the country stripped of their labor and means of transportation, and to have the conquerors left alone with their conquered territory. Devastation and expulsions did occur, of course, despite the injunctions of some strategists. Relying in part, for example, on a system of military magazines and waterborne supply, the armies of Louvois ravaged the Palatinate in 1689, making a good part of it uninhabitable. But even here, too many refugees could hurt. Particularly in the era of siege warfare, the late seventeenth and eighteenth centuries, when large armies remained relatively immobile, it did not pay to turn the occupied population into refugees.[8]

In such circumstances it is easy to see why refugees seldom were a bone of contention among states and why they rarely preoccupied people in positions of authority. As there was no generally accepted obligation to protect and succor strangers who arrived from afar, few people worried about the particular economic burdens refugees might impose. Because refugees only rarely threatened public order, they did not constitute any unusual danger at home. As central governments favored population growth and as military commanders wanted to avoid the flight of local inhabitants, neither of these powerful agents in society looked with equanimity on a massive flight into exile, and both sometimes welcomed refugees from elsewhere. Finally, given that the majority in these agricultural societies lived perilously close to subsistence in the best of circumstances, it was simply not possible for masses of them to survive for long as refugees. The affluent could enjoy the luxury of seeking refuge, but their numbers were limited. For the rest, many simply perished before even becoming refugees and before commanding the attention of anyone who mattered.

All this may help to explain why there were not even more refugees in Europe before the nineteenth century and why the existing refugees hardly troubled relations between states. The uprooted suffered as they have always—torn loose from their culture, belongings, and often their only means of subsistence. But Europeans did not view their calamity as a special kind of victimization, different from many other forms of oppression by the great and powerful. And as for the refugees themselves, the same was probably true. People were used to being ruled by foreigners of one sort or another in the days when the next valley or a nearby town was "foreign" in most respects. Assuming refugees survived to find refuge, their trials merged with the history of Jews, Moriscos, Protestants and Catholics, other resident aliens, or whomever; they would not have understood our modern preoccupation with a condition they shared—that of the refugee.

One sign that refugees as a category did not impinge on the European consciousness is the absence of a general term to designate them until the nineteenth century, the starting point for this study. Before this time, "refugees" almost exclusively denoted the Protestants driven from the French kingdom at the end of the seventeenth century. The third edition of the *Encyclopaedia Britannica*, published in 1796, first marked a change: "refugees," it said, was a term originally applied to the expelled French Protestants, but had since "been extended to all such as leave their country in times of distress, and hence, since the revolt of the British colonies in America, we have frequently heard of American refugees." Yet there are few indications that the shift in usage noted in 1796 was widely adopted. Well into the 1800s, French and English dictionaries refered to "refugees" as the victims of the revocation of the Edict of Nantes. Those who quit France at the time of the French Revolution, the "joyous emigration" of monarchists loyal to Louis XVI, preferred the term *"émigrés"* and hardly considered their decision to leave France akin to the expulsion of the French Calvinists a century before. Until the second quarter of the nineteenth century there was no mention of refugees in international treaties, and states made no distinction between those fleeing criminal prosecution and those escaping political repression. In German there was no term for refugees until well into the nineteenth century. German dictionaries included the French word *"réfugiés,"* repeating the generally understood definition applying to French Protestants. *Flüchtling,* the modern term for refugee, was noted in 1691 as designating a fugitive or a "flighty" person—"profugus, homo inconstans, fluctuans, vagus, instabilis."[9] *Heimatlos* or *staatenlos* began to designate certain categories of stateless refugees after 1870, but only following the First World War did the word *Flüchtling* denominate them all.

One of the purposes of this book is to trace the emerging consciousness of a refugee phenomenon since the 1880s. As we shall see, the nineteenth century was mainly preoccupied with exiles—individuals who left their native country for political reasons, usually after having engaged in revolutionary activity. These activists frequently appeared to their host societies as troublemakers—like the sinister character in George Eliot's novel *Daniel Deronda* (1876) who was "a Pole, a Czech, or something of that fermenting sort, in a state of political refugeeism." Then came the Jewish exodus from tsarist Russia and elsewhere in Eastern Europe that began in the quarter century before the First World War, a mass movement eventually sweeping up some 2.5 million impoverished and oppressed Jews. Sometimes called refugees and sometimes not, these emigrants helped condition Europeans to the phenomenon, and with them we can observe the first European reactions to what would become an international refugee problem.

Much discussion will concern the flood of refugees released by the crumbling of empires—Ottoman, Romanov, Hapsburg, and Wilhelminian. Before 1914 the slow retreat of the Ottoman Turk from the European continent and the accompanying series of wars that disrupted the Balkans generated hundreds of thousands of refugees. The historian C. A. Macartney once reported that there were seventeen different migratory movements in Macedonia alone between 1912 and 1925. To these refugees were added the former subjects of the tsar, fleeing revolution at home; the displaced civilians of the First World War; and the victims of the great Armenian tragedy, which continued into the 1920s. Following the First World War, Europeans were faced with a refugee problem of vast proportions: great masses of people wandered about the wastelands of Transcaucasia, languished in the slums of Constantinople, or pressed their case on the embassies of Western European countries. In response, the League of Nations established a High Commission for Refugees under the direction of the Norwegian polar explorer Fridtjof Nansen. This was the first occasion in which Europeans seriously addressed together a refugee crisis of international dimensions.

The remainder of this book will examine how international responses formulated in the 1920s changed and evolved under the impact of subsequent refugee problems. Best known are the refugee crises wrought by fascism, especially that caused by the Nazi persecution and expulsion of Jews from Central Europe. Here we meet the peculiar characteristic of the refugees in the twentieth century—the fact that they have nowhere to go. Awkward, confused, powerless, and often utterly demoralized—these refugees presented the international community with the by-now stock figure of the unwanted suppliant. Although Europeans spawned a variety of agencies and proposed a series of expedients to deal with the refugees, many continued to wander, homeless and rejected. Many more never even became refugees, for by refusing to receive some who had escaped, Western European countries effectively denied to others the possibility to leave. The 1930s thus witnessed a major retreat from refugee policies and principles articulated a decade before.

Succeeding waves of European refugees saw new definitions of the refugee and new approaches to the problem, building on preceding experience. After the Second World War the number of refugees climbed astronomically; not only were the displaced of a dozen nationalities adrift in Central Europe, but there were also massive new expulsions from the east—mainly *Volksdeutsche*, ethnic Germans settled throughout Eastern Europe. Chapter V will trace the eventual resolution of this massive human dislocation—the early repatriation of millions under the auspices of the Allied armed forces and the United Nations Relief and Rehabilitation Administration (UNRRA), settlement of the Germans by the West German government, and relocation abroad by the International Refugee Organization (IRO). Finally, we shall look at European refugees in recent decades, principally those seeking asylum from the Soviet bloc. The most important international agency concerned with refugees today, the United Nations High Commission for Refugees (UNHCR), developed its widely used definition of refugees with an eye to some of the earliest victims of the cold war—East Europeans fleeing their homes and unwilling to return to countries under Soviet domination.

A word now on the scope and focus of this book: my central concern is the impact of refugee movements on the international community in Europe, principally with reference to the relations between states. Important to this is a basic description of the refugees in question: who they were, why they became refugees, and what happened to them when they did. Vital also is an analysis of the variety of agencies that worked alongside national governments and ultimately became themselves actors on the international stage. I refer here to private relief societies like the Red Cross and to various international bodies such as the League of Nations and the United Nations. Left out of consideration is a cluster of issues associated with the "refugee experience"—psychological problems of refugees or questions of social integration and assimilation—except insofar as these determined the flow of refugees or affected policy on the national or international levels.

Given the evolution of the term "refugee," applying to ever wider groups of fugitives, it makes sense to include rather than exclude marginal cases in a study attempting to be comprehensive. Many refugees considered in these pages would not meet the current United Nations definition as

> any person who . . . owing to well-founded fear of being persecuted for reasons of race, religion, nationality, membership of a particular social group or political opinion, is outside the country of his nationality and is unable or, owing to such fear, is unwilling to avail himself of the protection of that country; or who, not having a nationality and being outside the country of his former habitual residence . . . is unable or, owing to such fear, is unwilling to return to it.

More simply, my discussion concerns people who, dispersed by persecution or man-made catastrophes like war or civil strife, have sought sanctuary and protection in Europe. The emphasis is on broad masses rather than on famous exiles or revolutionaries, many of whom were free to choose the path of exile or reject it. This focus does not make as much of international frontiers as the UN definition must. It seems reasonable to include many refugees displaced during the civil war in Russia even when they did not cross recognized borders and similarly to consider the millions who fled the advance of the German armies in France in 1940. At the same time, it does not seem wise to dwell, for example, on war evacuees in England during the Second World War. The simple test is the extent and duration of such displacements, together with their gravity in human terms and with respect to the relations between states. These refugees are people who suffer drastic upheavals in their personal lives and whose fate impinges substantially on the societies in which they live.

Generally, this study excludes people in captivity such as forced laborers, internal exiles, long-term internees, and concentration camp inmates. However ghastly the circumstances of their confinement, the latter meant at least that their situation was not "indeterminate"; that they were not dispersed in the sense understood here. After liberation or with the collapse of the authority that held them, such people often became refugees, and we meet some specific cases in the pages that follow.

The refugees studied here are moving within, out of, and into the European continent. Although we normally leave them after their departure from Europe, it seems impossible to avoid a few exceptions—for example, Jews banging on the gates of Palestine in the 1930s and 1940s or Polish refugees from Hitler who left the Soviet Union for destinations in the Middle East or Africa. Some developments in Transcaucasia, between the Black and Caspian seas, will have to be mentioned—because so many of its refugees, such as the Armenians, found their way to Russia and the West. Similarly, Muslim refugees from the shrinking Ottoman-controlled areas of the Balkans have a place in the story although Muslim refugees moving within Anatolia clearly do not.

Finally, a word of caution regarding numbers. Throughout, I will necessarily refer to the size of refugee movements and will cite statistics from a very wide variety of sources. The reader, like the author, must use these with care. Almost by definition refugees are among the forgotten, "people who have fallen into the cracks of history," as one writer puts it.[10] Census takers could hardly enumerate the desperate, panic-stricken civilians pouring across frontiers or wandering from place to place in crisis situations. Only after some dust had settled might the counting begin, assuming anyone cared to do so. And then, of course, the problems were legion. How were the refugees to be defined? When do people cease being refugees? How many refugees perished or were settled since the onset of crisis? And how does the historian evaluate the often unspoken definitions and procedures of times past? Complicating these technical issues are political considerations: some agencies inflated refugee statistics in pursuit of particular goals, and others deflated them; some considered certain groups to be refugees, while others denounced them as bands of traitors and quislings. In the end, we as historians must pick our way through these issues as best we can, armed with healthy skepticism and a variety of sources. No one should ignore, however, that substantial margins of error are possible and that refugee statistics can be among the most unreliable of historical data.

Numbers, however, do not form the backbone of the following account. More often than not, the refugees came in so great a torrent as to overwhelm the few who tried to keep track. Rather, the central theme is the emergence of a new variety of collective alienation, one of the hallmarks of our time. This is the story to which we now turn.

1

Toward a Mass Movement

DURING the century before 1914, Europe was little troubled by the kind of refugee problems that preoccupy us in this book. No masses of fugitives wandered across the continent or appealed to the international community for assistance. Contemporaries had likely heard of famous exiles—men like Mazzini, Bakunin, or Marx—whose refuge abroad sometimes provoked controversy or diplomatic activity. They may occasionally have seen East European Jews, huddling for shelter in railway stations or at dockside, waiting to continue their journey west. They may even have known about Balkan refugees—fleeing obscure conflicts ascribed to Turkish misrule. Yet none of these cases worried European political leaders or seriously affected relations between states. None of them prompted concerted action by the great powers, and none of them shifted consciousness about refugees appreciably. Only after the First World War did these things occur. But when refugees finally did emerge into general view, decision-makers were burdened with the perceptual baggage of earlier times. Men remembered the exiles, the Jews, and the Balkan refugees of time past. And when they acted, they sometimes had these earlier experiences in mind. It is, therefore, useful to begin this book by looking at these outcasts as a prelude to the mass refugee movements of the twentieth century.

THE NINETEENTH CENTURY

Refugees, writes Bernard Porter, were almost never seen en masse in mid-nineteenth-century London except at funerals.[1] Although

London was probably the refugee capital of Europe after the revolutions of 1848 and although exiles of every nationality congregated there, only those mournful, ritualistic celebrations of lost causes brought the faithful together. On such occasions, for example, they would parade a deceased rebel through the streets accompanied by flags and other paraphernalia of émigré politics; graveside ceremonies attracted throngs of exiles to celebrate their aging cause as well as their departed comrade. Funerals alone could bring the refugees together because practically all they had in common was the past, the imprint of their revolutionary struggles before they found refuge abroad. In London and elsewhere they were notorious individualists, impossible to unite around actual political objectives and forever bolting the many parties and factions they formed.

Behind this observation lies an essential fact about politically determined migration throughout most of the nineteenth century: it generally concerned *exiles*, individuals who had chosen their political path, rather than large masses of people torn loose from their society and driven to seek refuge. Until the last few decades of the nineteenth century, these exiles—and not the bewildered and helpless masses of later years—were the visible expression of refugee existence. And they were generally well received in liberal West European countries. Indeed, the generous treatment accorded them compares interestingly with the harassment and ostracism reserved for later refugee movements.

The first half of the nineteenth century has been called the age of revolutions—democratic, nationalist, and social. Each of the political upheavals of this era sent out its band of exiles; and virtually all of them, together with those who went abroad to escape political persecution in more normal times, found refuge in Europe if they chose to remain there.

Among the first political outcasts to define a distinct refugee identity for themselves were Polish nationalists, enemies of the absolutist tsar Nicholas I and proponents of Polish independence from the Russian empire. When Polish insurgents were finally overwhelmed by government forces in 1831, more than 5,000 left the country, the great majority going to France.[2] France was a natural haven for these battered champions of the Polish cause. Not only had the liberal French Revolution of 1830 been an inspiration for the rebels in Warsaw and elsewhere, and not only were there recent memories of the former emperor Napoleon as a protector of the Poles against the Russians, but the French had lavished sympathy on the insurrectionaries during the course of their struggle. If the French had failed to impede the progress of repression in Poland, they thought the least they could do was to welcome the Poles who had lost.

The Poles were hardly dangerous social radicals. Once the political elite of their country, they included an impressive array of high-ranking military officers, distinguished politicians, poets, intellectuals, and some glittering aristocrats—along with the wider pool of *szlachta*, members of the Polish ruling class with their legendary pretensions to aristocratic standing. In 1832, when the Poles arrived, the French government under King Louis Philippe wasted no time codifying in law the standing of these and other exiles. Together with the Polish nationalists, France had welcomed several thousand Spanish, Portuguese, Italian, and German dissidents; adherents of lost causes; and a variety of failed dynastic contenders—never more than 20,000 before 1848.[3] All these were formally subjected to regulations defined in 1832. Although the law empowered French officials to assemble the exiles in certain cities and keep them there and although there were also broad powers of expulsion, in practice the French were extremely lenient. The application of the threatening provisions of the 1832 law was regularly deferred until midcentury. Meanwhile, French authorities subsidized destitute refugees by paying them living allowances from the national treasury. In 1848 a law regularized such support, providing stipends that varied according to the military rank and social standing of the applicants.

Not all early nineteenth-century refugees had the gentility, real or proclaimed, of the Poles in France. The decades after the Napoleonic wars saw the emergence of various national movements in Italy, Ireland, Poland, and Germany that attempted to mobilize against outside domination. Known as Young Italy, Young Ireland, and so forth, these romantic rebels frequently challenged repressive local governments. In the German states, in particular, rebellious idealists ran into trouble. Young Germany, drawing heavily from the lower-middle-class and Jewish intellectuals, was highly vulnerable to police harassment and administrative sanctions, coordinated throughout the German Confederation by the Karlsbad Decrees of 1819. Inspired by the Austrian chancellor Klemens von Metternich, these regulations clamped down on radical agitation in universities, the press, and political assemblies throughout Central Europe. In their wake, many radicals went abroad—usually to France, relatively close to home.

The Young Italians, followers of the nationalist firebrand Giuseppe Mazzini, also contributed to the first waves of exiles in the nineteenth century. Notably following the suppression of a handful of uncoordinated and unsuccessful uprisings in 1820–21 and 1831, leaders of the movement fled to escape arrest and punishment. Some went to Florence or to Pisa, in the relatively liberal Grand Duchy of Tuscany. Some formed colonies of conspirators in France or in Switzerland, and others, like Mazzini himself, went to London.

The year 1848 saw an important new generation of refugees, now coming from across all Europe, as one uprising after another succumbed to counterrevolutionary repression. In the first flush of revolt, the French radicals swept away previous restrictions on the movement of exiles within France—excepting, however, Spanish Carlists, who were considered reactionaries. Exiles living in London or Paris quickly decamped, streaming home to join the upheaval. But many soon returned in disappointment as the fortunes of revolution sank. Before long, defeated leaders of the Paris June 1848 uprising were themselves driven underground or forced to seek asylum outside France. On their heels came hundreds of Germans, Austrians, Czechs, Hungarians, and Italians—victims of the forces of order that everywhere in Europe had gained the initiative by the summer of 1849. Switzerland probably received the greatest number of post-1848 political refugees—some 15,000, mainly from Germany and Italy, including 9,000 routed revolutionary troops from Baden, and the young Friedrich Engels, who had fought with them in several serious engagements.[4] About 4,000 Germans went to America, and a substantial number ended up in London, seen by the early 1850s as the great exile center of Europe.

The political refugees at midcentury might easily have been judged a more dangerous lot than their early nineteenth-century predecessors. After all, they practiced revolution as well as preached it; they had struck everywhere and had shown a measure of continentwide solidarity and even a degree of class consciousness that had not distinguished earlier bands of émigrés. The new communities, mainly in London and Geneva, included not only celebrated activists but also some noted theorists of revolution—one of the most famous of whom was Karl Marx, who crossed the Channel to England in the summer of 1849.

Yet despite the often ferocious reputation of these exiles, one is impressed from the vantage point of the twentieth century with the leniency shown to them in countries that had survived revolution or narrowly escaped it. Marx himself confronted the French government in July 1849, when he was briefly in Paris, hoping for a new wave of revolution. French leaders were by that time retreating from the radical experiments of the previous year. But the worst they would do to Marx was to order him to leave the capital and go to the department of the Morbihan in southern Brittany—a somewhat remote location for an agitated revolutionary. In correspondence with the Ministry of the Interior, Marx tried to persuade the latter that there had been a "misunderstanding" about his own activity. Later he learned that the Morbihan climate was dreadful; he concluded that the attempt at banishment was really a plot to murder him. Instead, he decided to take his family to London.[5]

Similarly in Switzerland, where the 1848 exiles were unpopular and always threatened to embroil the Confederation with its Austrian neighbor, there was no question of expulsion. Although the Swiss had emerged from 1848 sympathetic to liberals and supporters of the federal system, the international vulnerability of the Confederation made her a dubious haven for political fugitives from abroad. Yet even with more conservative men in charge, into the second half of the century, the Swiss managed to defeat foreign calls for a less hospitable policy. Geneva became what Alexander Herzen called "the Koblenz of the Revolution of 1848," reminiscent of the town on the Rhine that had hosted so many émigrés of the French Revolution.[6]

In England the exiles lived practically without restriction. Bernard Porter observes that throughout the nineteenth century not a single refugee was blocked at a port of entry or ordered to leave the country.[7] The refugees simply disembarked: after 1836 no one obliged them to notify the authorities, to register with the police, or to conform to special rules. British politicians, moreover, however they ruminated on the menacing presence of these insurrectionaries, never outlawed their agitation in the very center of their empire. When, for example, the Sicilian government asked the British in 1849 about a ship in the London docks suspected of being equipped for war by Italian revolutionaries, the foreign secretary at the time, Lord Palmerston, refused even to investigate, telling the Sicilians that both government policy and public opinion would not accept such partisan behavior. When the

Hungarian patriot Lajos Kossuth claimed in 1853 that he was being spied on by London police, there was a great public outcry, and meetings were held to protest this outrageous breach of tradition.[8]

In the 1850s and 1860s, political refugees were much less in evidence, their lot easing as one government after another offered amnesty to the former rebels living abroad. Gradually the exiles returned home. Polish revolutionaries burst once more on the scene after the failure of the 1863 uprising against the Russians. In London, Zurich, and Geneva, they mixed with Russian Populists, anti-Hapsburg rebels, and the leftovers of 1848; sometimes they went off to America, and sometimes they abandoned the life of exile to merge into English or Swiss society. But they had no difficulty finding asylum in Europe. It should be emphasized that receiving countries almost never assumed the refugees would abandon revolutionary politics. In exile, the newer brand of émigrés pursued the cause of revolution as conspicuously as their French aristocratic forebears had pursued counterrevolution. Under the liberal July Monarchy in France, Mazzini established Young Italy in Marseilles in 1831 and proclaimed to all the world the goals of an Italian republic, war against Austria, and the cause of republicanism throughout Europe—while enjoying the hospitality of the French king.

But the hosts were sometimes nervous. Even when the objectives of the exiles diverged profoundly, there was a tendency to lump together Polish nationalists, Rumanian patriots, or German Romantics, seeing behind them all the lurking presence of a revolutionary brotherhood. In 1834, Mazzini fed such fears with the establishment of a Young Europe society; its refugee adherents announced from Switzerland a "Holy Alliance of the Peoples" to carry on the fight for freedom everywhere. Such revolutionary fantasies could occasionally make people tremble, in part because the exiles lived the revolutionary roles Mazzini and others assigned to them—affronting society with a bohemian existence, dreaming continually of a reversal of revolutionary fortunes, with their trunks packed, as Herzen put it, waiting at every moment for the signal to join the insurrection at home. Indeed, both the refugees and the governments tended at times to exaggerate the political capacities of the exiles. The latter tried desperately to remain on the historical stage, to think of themselves and to be thought of as a continuing force in political life; receiving countries, by the same token, could make the same mistake, forgetting that the rebels' posture involved more show than substance.

Unlike their counterparts in the twentieth century, the refugees remained in some sense gentry. It may seem curious from our own standpoint on refugees that the world of political exiles was that of the relatively well-to-do or, at least, of the once well-to-do. Yet this is understandable given that national politics and a long-term investment in the business of revolution were not generally possible for ordinary Europeans and that it took some measure of affluence to flee abroad in the first place. Only through family money could young, Western-oriented anti-Turkish rebels from Moldavia or Walachia reach Paris in the 1830s and 1840s. And even the Russians who came in this period were once of high social standing—like the anarchist Michael Bakunin, grandson of a large estate owner and state counselor of Catherine II, and whose father was educated at the University of Padua. It is true that many refugees became a somewhat shabby lot. In London, *The Times* sniffily described a colony of Continental revolutionaries near Leicester Square, "wearing hats such as no one wears, and hair where none should be"—quite out of fashion and sometimes only barely making ends meet.[9] But most remained incorrigibly bourgeois, as Herzen himself noted, despite their bohemian preferences and their exaggerated calls for revolution. Their education, cultural affinities, modes of speech, and social habits set them off from the vast majority.

Finally, however dangerous the refugees were seen to be for neighboring states, they were seldom considered a threat in the countries where they made their new homes. The East Europeans appeared so exotic in their politics as to have very little relevance to countries like England, Belgium, France, or Switzerland. And there was often a real measure of sympathy for what was taken to be their cause—emulation of the liberal West. In liberal circles, opponents of tsarist, Hapsburg, or Ottoman autocracy often met an enthusiastic response. In provincial German towns, people poured into the streets to cheer Polish revolutionaries on their way to France in 1831. Chartists in London saw in Kossuth a Hungarian equivalent, and anti-Catholic Englishmen found it difficult to resist enthusiasm for rebels against the temporal powers of Pius IX. Mazzini described his passage through Switzerland in 1848 as a triumphal tour during which he was warmly greeted by the local population. When he arrived in Sweden in 1863, disguised as a Canadian, Bakunin enjoyed a sympathetic reception despite the bother he undoubtedly caused the government in Stockholm. Liberal newspapers defended to the hilt his right to pursue the cause of revolution then ablaze in Poland. Although the conservatives did not deny his right to asylum, they opposed unrestrained activity for refugees. So when Bakunin had an audience with the Swedish king, he apparently was advised to be politically discreet.

Generally, the exiles appeared so walled off in their fantasy world of emigration as to have little relevance to the political life of the host societies. Herzen certainly saw his fellow exiles this way: incapable of settling down where they were, yet cut off from the environment that had nourished them politically, they became increasingly disputatious, impractical, and out of touch. In addition, they seldom numbered more than a few thousand, however high their visibility. Serious political leaders usually understood these facts and ignored the refugees.

Occasionally, refugees impinged on the relations between governments although their fate was seldom considered important enough to provoke a serious crisis. Authorities in France complied when Turin officials asked them to expel Mazzini in 1832, and the Italian leader had to move to Switzerland. After 1848 the French and the Austrians repeatedly nagged the British about harboring refugees—all to no effect. Refugees indirectly sparked a confrontation of the great powers over the Eastern Question in 1849, when Turkey refused Russian requests to extradite Polish and Hungarian rebels; once the British and French moved their fleets, however, the tsar backed down. A cause célèbre occurred in 1858, when Felice Orsini, an Italian revolutionary living in England, tried to assassinate the French emperor Napoleon III and the empress as they arrived at the Paris Opéra. In the resulting explosion, eight people died, and many were injured. Immediately, the emperor took the diplomatic offensive, including several demarches concerning the reception of political refugees. Responding to serious French pressure, the Swiss agreed to remove certain Italian exiles from the Geneva district bordering on France. Napoleon was much less successful with the British, who rejected complaints that they had been too lenient toward characters like Orsini. (The latter had fabricated his bombs in Birmingham.) Lord Palmerston, then prime minister, was prepared essentially to meet the French demands. Opinion against this ran so strongly, however, that he was forced to resign. Although the affair turned upon many other political factors, sentiment against a harsher refugee policy was certainly among them;

the refugee community in England understood this and rejoiced in the result.

In reality, cracking down on refugees posed enormous technical problems for the tiny bureaucracies of Western countries. Police certainly could stop the open expression of seditious views by exiles living abroad. But preventing their entry into countries of refuge or tracking them down once they had arrived was practically impossible; even if they had wanted to eject foreigners, authorities would have been hard pressed to do so. Without fingerprints, photographic identification, efficient filing systems, and modern police communications, the refugees could cross frontiers with little difficulty and live virtually unmolested once they had arrived. In 1854, Mazzini, then in hiding, sent an open letter to the Swiss Federal Council, published in at least three Swiss newspapers, gently mocking his hosts' inability to track him down. "You are looking for me everywhere. You are wearing out your telegraph operators. . . . You do me the honor of buying my portraits." Continental officials complained continually to the British about Mazzini's finding refuge in London, but they were helpless to prevent his frequent return trips. Even the regulations that did exist were notoriously easy to evade. Few refugees carried identification papers or other documents. Alexander Herzen had a passport in 1847, when traveling in southern Italy; it was ponderously examined by a Neapolitan brigadier, but Herzen reported that the officer could not read.[10] In Geneva, removal of refugees after the Orsini affair was never thoroughly enforced, and the few, minor gestures in that direction were bitterly resented in the émigré community. The principal hardship, it turned out, was that refugees had to keep away from their favorite café on the Pont des Bergues.[11]

So long as the refugees remained few in number and so long as they seemed relatively innocuous guests in the handful of European countries that were willing to accept them, nothing changed. Refugees continued to trickle into exile after the revolutions of 1848, and they found asylum with little difficulty. Social Democrats, for example, fleeing Bismarck's antisocialist laws after 1878, or Young Turks, at the same time, evading the Ottoman sultan, Abdul Hamid, came without problems to Paris, London, or Geneva. Then, in the last third of the nineteenth century, new concerns began to arise, and with them came signs of growing impatience, even in liberal countries, with political refugees.

New kinds of refugees—and greater in numbers than before—were on the move as a result of the wars of German unification of 1864–71. The scale of mobilization of these conflicts, however short they were, was quite unprecedented in European terms. Vastly greater numbers of people than ever before heard the call to battle and were touched by the fighting. During the Franco-Prussian War the North German Confederation fielded 1.2 million men against France—twice the size of the army Napoleon had sent into Russia in 1812. Long-range artillery now sent shells crashing into cities many miles from the actual combat. Requisitions from local inhabitants also expanded greatly—partly, as one observer put it, "to crush the occupied portion of the country and make the inhabitants long for peace."[12] And on both sides of these contests, especially in the case of Germany and France, nationalistic sentiment, whipped up by a cheap and popular daily press, engaged increasingly large groups of civilians on behalf of their nation's cause.

Great masses of people were forced to move. The destructiveness of the fighting accounted for many refugees, of course—villages destroyed and fields ruined, the havoc wrought by the maneuvering of men and equipment in proportions hitherto unknown. In a single day in July 1866, for example, just before the crushing defeat of Austrian armies by the Prussians at Sadowa, in Bohemia, 13,000 terrified civilians abandoned the city of Prague, leaving all their possessions behind. This was a mere foretaste of what was to occur during the First World War. Moreover, the nationalistic focus of these wars also created migratory movements as conationals learned to abhor the pospect of being ruled by aliens. Between 1867 and 1871, an abnormally high porportion of Central European emigrants came from provinces newly annexed to Prussia—Hannover, Schleswig-Holstein, and Hesse-Nassau—presumably among them many who found life intolerable under the new Bismarckian order. In 1870, when hostilities began, France expelled some 80,000 Germans as part of a nationalistic campaign; and, with the annexation of Alsace-Lorraine to the newly formed German Empire, about 130,000 residents of the former French provinces left home to remain in their native country—a move permitted by the Treaty of Frankfurt of 1871.[13]

Nationalism created additional Polish refugees from the eastern provinces of the German Reich. Robert von Puttkamer, Bismarck's minister of the interior, raised the specter of an eventual independent Poland that would undermine the Prussian hold over a region where so many peasants were ethnically Polish. His response involved harsh anti-Polish policies for the eastern marches, including forcible removals—expulsions that continued into the 1890s. Thousands of Polish peasants were turned into refugees, sometimes sent to a tsarist Poland that they had never seen. Simultaneously, German refugees from the tsarist empire were sent to the Reich. Germans had lived in Russia since the days of Peter the Great and Catherine the Great, having originally come as colonists at the invitation of tsarist authorities. Faced with aggressive policies of Russification at the end of the nineteenth century, many left, sometimes under direct pressure to do so. Some 50,000 Volhynian Germans took this course between 1900 and 1914, and others quit the Baltic region. With the Revolution of 1905, violent outbreaks against the Germans occurred in the north, and warships had to be dispatched from Germany for evacuations.

Governments now found themselves host to indigent refugees quite different from the Romantic dilettantes of Young Germany or the Polish aristocrats of 1832. The Poles of 1863 had far fewer resources and much less social standing than their compatriots of a generation before; unlike their forebears, these Poles frequently blended into the indigenous working class. In Belgium and Switzerland they became strike leaders and socialist agitators; in Paris many surfaced during the Commune of 1871.[14] After the bloody suppression of that radical experiment, the Communards themselves were forced to flee. Some 20,000 Parisians died, and 35,000 were arrested in the immediate aftermath of the fighting; in addition, about 45,000 fled, many going abroad to England or Belgium.[15] These were certainly not a socially exclusive lot, including men and women from across the social spectrum. In London they established the Société des réfugiés de la Commune; riddled with spies and agents sent from Paris, the exiles' association did its best to defend the interests of a largely poor, demoralized, miserable community.

Tsarist exiles, perennially part of the émigré scene, were both more numerous and more dangerous to harbor in the decades before the First World War. The "Romantic exiles" of Herzen's vintage or the Populist idealists of the 1860s gave way, after the assassination of Tsar Alexander II in 1881, to a new wave of hardened leftists and violent desperadoes. Most spectacular, at the close of the century, were the terrorists and assassins, the anarchist devotees of "propaganda of the deed," and the defiant renegades prepared to wage war abroad with the Okhrana, the tsarist secret police. But these dangerous characters were merely the tip of the revolutionary iceberg. For every Russian anarchist there were dozens of intense political activists, organizing in exile the major political parties of the Russian revolutionary movement. There were politically involved students, often unable to return home because of their associations, living in abject poverty in European university towns. And beyond the students there were many others with less constant political preoccupations, driven abroad by the harsh Russification policies of Alexander III, escaping conscription into the tsarist army, or forced to flee in the wake of upheaval in 1905 and 1906.

Tsarist exiles, indeed, were everywhere; like the Russian empire itself, this political emigration outnumbered all the others. Galicia, under Austrian rule, welcomed the leaders of Ukrainian nationalism. Germany, traditionally home to Russian students, became a major focus of exile activity in the 1880s. The Reich had a common frontier with the tsarist empire, and with ample printing facilities Berlin and other German cities became leading centers of political agitation. Geneva may have remained the capital of the Russian revolutionary movement after 1905 as Angelica Balabanov said, but important provincial centers were London, Paris, and Brussels. Thus tsarist political refugees became the quintessential bearers of revolutionary culture in Western Europe; they were first thought of when the problem was raised to the level of international concern.

After the 1890s, conditions deteriorated further within the empire, culminating in the Revolution of 1905. Adding to the volatility of émigré politics was a flurry of anarchist violence that spread like an epidemic. From the early 1890s anarchists launched an international campaign marked by the "propaganda of the deed"—spectacular assassinations, political murders, bombs set of in cafés, and other extravagant gestures. Transfixed, the European public followed the story of the desperate perpetrators of these acts, and many first learned thereby about the wild men in their midst, many of whom were refugees from afar. Such fears mingled with another by-product of the age of nationalism—apprehension about spies. Foreigners of all sorts could come under suspicion as a result of spy mania, and the latter part of the nineteenth century witnessed a growing nervousness about refugees as a result.

In response, governments moved against the refugees, but in a distinctly limited way. There were a few expulsions. Immediately after the assassination of Alexander II, the Swiss responded to Russian demands to eject a quite harmless anarchist prince, Peter Kropotkin. (The latter simply moved across the border to a nearby French town, close to Geneva, where his wife was about to write examinations at the local university.) In another case, the French government refused asylum to "nihilist" émigrés after tsarist requests. Moving into an entente with the Russians at the time, Paris sought to propitiate the Russians on an issue of relatively little importance. Some states tried to distinguish between acts committed in the furtherance of revolts or revolutions, labeled political and hence not subject to extradition, and individual acts of violence like terrorist bombings, deemed criminal and not protected from extradition.[16] Governments everywhere after 1900 tended to favor extradition in doubtful cases. Increasingly, the more violent revolutionaries living abroad found themselves in danger of being sent home. In a celebrated case, Swiss courts did so to one member of the Russian Socialist Revolutionary party. France expelled more than 1,600 foreigners, mostly Italians, as anarchists between 1894 and 1906. English courts, on the other hand, continued to reflect the relatively benign British attitudes toward foreign revolutionaries. At the end of 1897, for example, Vladimir Burtsev, a Russian exile committed to revolutionary assassination, was arrested by Scotland Yard and convicted of having tried to organize the murder of Tsar Nicholas I. His sentence—eighteen months at hard labor.[17] In November 1898 the Italian government sponsored an international conference of police and other officials to discuss the problem of terrorists and anarchists. Deliberations went on in secret, but one result became known: Great Britain, Belgium, and Switzerland refused to alter their practices of granting asylum and would not surrender suspected anarchists to their native countries simply on demand. Liberal states, at least, stood firm on the question of individual rights for refugees. By the end of the century, however, another issue was just beginning to be posed: What was the responsibility of states in regard to large numbers of exiles—masses of refugees such as the European continent had begun to see in the decades before the First World War? Here the response, notably with the case of Jewish emigrants from tsarist Russia, was more problematic.

THE JEWISH EXODUS FROM EASTERN EUROPE

Between the early 1880s and the First World War about 2.5 million Jews left Eastern Europe for the West. In this tidal wave, perhaps the greatest population movement of postbiblical Jewish history, Jews joined the massive westward flow of peasants from Russia, Austria-Hungary, Italy, and the Balkans—where depressed agricultural conditions, overpopulation, and underdevelopment were writing the same story. Overwhelmingly, the Jewish emigrants came from the tsarist empire, in whose Polish provinces lived the greatest world concentrations of Jews. But Jews also poured out of two other great reservoirs, Austrian Galicia and Rumania. Was this a refugee movement? Clearly, conditions varied in the three areas under consideration; so also did the motives of the Jews who emigrated. Scrutiny of the circumstances under which the Jews left indicates why we must be cautious in applying the refugee label to their emigration.

The Russian empire contained the great majority of East European Jews in the century before the First World War. About 5.6 million Jews lived to the east of the German Reich in the 1870s—almost four million under the tsars, confined to the western part of the empire in what was known as the Pale of Settlement, over three-quarters of a million in the Hapsburg lands of Galicia and Bukovina, just under 700,000 in Hungary, and 200,000 in Rumania. Under Romanov rule, Jews had gone through a hopeful period with Alexander II, who assumed power in 1855. Early in his reign a series of improvements in the position of Jews within the empire augured well, many believed, for future progress. Then came a reimposition of Jewish disabilities after the suppression of the Polish uprising in 1863. This, combined with the growing pauperization of the Jewish masses, closed the door to successful integration into non-Jewish society.

While they shared the harsh living conditions of other subjects of the tsar, Jews also bore the particularly heavy weight of official hostility reserved for outsiders—in their case, a community apart—not only in religion, but also in language, culture, dress, and way of life. Jews walled themselves off with their rich religious tradition, which coexisted with a remarkable recent flowering of secular, Yiddish-language culture. Outside their community, however, pressures mounted. The last years under Alexander II saw deepening reaction, strengthened autocracy, and a freer hand to conservative Russian nationalism at the expense of subject minorities. Yet before the 1880s, Jewish emigration from this world was slight—an average of less than 5,400 a year from the tsarist empire between 1840 and 1880 according to the demographer Jacob Lestchinsky.[18]

The turning point was 1881, the moment when the Jews' migration history began to break with that of their non-Jewish countrymen. The year 1881 saw the first in a series of pogroms—collective attacks by the local population against Jews, usually abetted by the police, the military, and official representatives of government. These outbreaks followed quickly on the assassination of Tsar Alexander II and appeared linked to official policy. Early in 1882, the interior minister, Count Nikolai Ignatiev, declared: "The Western frontier is open to the Jews." Thereafter Jewish emigration soared. Through the 1880s an average of over 20,000 Jews per year traveled to America from the Russian empire, and the numbers continued to climb. There was a significant exodus after the Revolution of 1905, when anti-Jewish outbreaks were particularly intense and when many young Jews sought to escape military service for the war against Japan. Between 1906 and 1910 an average of over 82,000 Jews a year left for America. Throughout this period, moreover, the Jews' share of Russian immigration to America climbed far above their proportion of the population in the empire. In 1900, 408 emigrants out of every 1,000 from Russia were Jews; in 1906, when prewar Jewish emigration peaked, Jews constituted 581 out of every 1,000.[19]

To many contemporaries and to a substantial body of opinion ever since, the Jews were practically driven out of the Russian empire. The folklore of early migrants themselves sprouted with references to the persecutions. "People who saw such things never smiled any more, no matter how long they lived," wrote Mary Antin, describing her odyssey to America in 1894 following the outbreak of violence in White Russia. "[S]ometimes their hair turned white in a day, and sometimes people became insane on the spot." According to an editorial in the Hebrew periodical Ha-Melitz in 1882, the Jews had no previous desire to emigrate or seek a new life elsewhere. "It is as if we were sitting and waiting until the master of the house said to us—leave!; until they grabbed hold of our napes and threw us out into another land like a dish for which there is no longer any use."[20]

In the liberal West there was much dislike of tsarist autocracy and sympathy for the Jews. Throughout Western Europe, public protests denounced the pogroms as examples of tsarist barbarity. In France a Committee to Aid the Jews of Russia sprang up, headed by the aged Victor Hugo; among its members were distinguished politicians and the cardinal archbishop of Paris. In the Netherlands, Calvinists held prayer-meetings in churches for the salvation of "the folk of Israel." In England the *Jewish Chronicle* published a special supplement on "Darkest Russia" in 1891, including a stirring denunciation of anti-Jewish oppression penned by William Ewart Gladstone, who became prime minister for the fourth time the next year. A leading Radical M.P., Sir Charles Dilke, publicly drew the inevitable conclusion that every Jewish immigrant to Britain was a religious refugee.[21]

Sympathetic outsiders noted a sharp turn for the worse in the early 1880s, beginning with Alexander III. Following the first wave of pogroms came the Provisional Rules of 1882, imposing harsh new discriminatory measures against Jews. These regulations drastically limited Jewish rights to settle in the countryside or even to move about within the Pale of Settlement; they also established a numerus clausus in the Russian educational system, eliminating professional schooling for most Jews and thwarting their prospects for upward social mobility. More decrees followed: sharp cuts or elimination of Jews in various professions, restrictions on Jewish artisans, denial of state lands to Jewish mining enterprises, and so on. Beginning in the 1890s there were brutal evictions of thousands of Jews from Moscow and other cities and then a wave of new discriminatory orders and circulars that rained down through the decade. The objectives of this policy were supposedly set by Konstantin Pobedonostsev, the chief procurator of the Holy Synod or lay leader of the Orthodox Church, an architect of reactionary Russian nationalism before 1905: one-third of the Jews would convert, one-third would emigrate, and one-third would perish.

Even when viewed in the context of deepening reaction, Russification, and the persecution of other non-Russian minorities like Germans, Poles, Baltic peoples, Ukrainians, and others, one sees a particularly intense tsarist government assault against Jews. Rather than attempting to assimilate them, tsarist ministers one after another fashioned policies to close them off from the rest of society and to ensure their elimination from national life.

Unlike Jews from elsewhere in Eastern Europe, moreover, those from the Russian empire often left their homeland illegally and in circumstances that drew particular attention to the differences between their exodus and more usual emigration. The Russian bureaucracy took no official notice of Jewish emigration until 1891, when it expelled about 20,000 Jews from Moscow. Before that time, cumbersome procedures made departure extremely difficult for those who kept within the law; hungry for military conscripts and fearful of unexpected economic dislocations, the government resisted mass emigration. Then, for the next decade and a half, it removed some restrictions and allowed some emigration committees to function. But the obstacles could still be so formidable that emigrating Jews became refugees in the sheer manner of their departure. According to historian Hans Rogger, "most border crossings were accomplished illegally, under cover of darkness and with the connivance of frontier troops, but occasionally accompanied by their bullets."[22] Bribery and stealth were a normal part of the process; Jews going abroad were drained of their resources, hounded by local authorities, and harassed by the police. For the great majority of Jewish emigrants from Eastern Europe, therefore, the last encounters with Russian officialdom were part of a pattern of rule they had come to detest.

Along with these political factors, deteriorating economic conditions for Russian Jews helped explain the Jewish exodus. The German-Jewish demographer Arthur Ruppin drew attention to the strong natural increase in the Jewish population in this period, and the consequent difficulties Jews experienced in overcrowded traditional occupations. Jewish life in the tsarist empire at this time was mired in poverty and deprivation—to the point where by 1900 no province in the Pale of Settlement had less than 14 percent of its Jews on poor relief, and in most of the Ukraine the poor numbered at least 20 percent.[23] With millions of Jews reeling from impoverishment and social dislocation, emigration appeared as an attractive prospect. Also, a series of particular catastrophes spurred the Jews on their way: periodic outbreaks of cholera, the famine of 1891, and economic dislocation in industrial Russia following the Russo-Japanese war.

Then, too, forces of attraction were drawing Jews abroad: the lure of America, widely communicated in the 1880s and 1890s; the sudden lowering of transatlantic shipping rates by competing companies; the accelerating momentum of departure once the first wave of emigrants had settled abroad and started pulling on the chain of chain migration. Usually, the men went first—in a pattern common to other waves of migration from poor countries to lands of opportunity; then entire families followed. One

Jewish immigrant to England reported a typical case in 1889 in which there was no reference to pogroms or a sense of persecution:

> people began to leave our town, which is a small town, and began sending over money very often, and that made up my mind that I should go over there as well, and so I came here. Especially a man left our place—an old man with no trade at all. He was here only a few months, and he sent over £30. I made up my mind. "I am a mechanic. I believe when I go over there I shall be able to make more money than he can."[24]

Although persecution played its role in the Jewish exodus, there was certainly no concerted effort to force the Jews from Russia, however anti-Jewish the government officials. Tsarist ministers usually lacked a clear vision on the subject of Jews, and their measures were riddled with contradictions and hesitations. Some facilitated Jewish departures, believing the empire was best rid of the Jews; but others did precisely the opposite, loathe to relax the apparatus of strict, prisonlike controls. Some Jews responded to worsening conditions by emigration, but others clearly did not. Indeed, the overall Jewish population did not even decrease; with extremely low Jewish mortality in this period, apparently the exodus simply drained off the increased numbers. Neither in the persecution policy nor in the motivation for emigration do we find forcible uprooting in the usual sense. Jews from the tsarist empire seem to represent an intermediate case that we will meet again in this book—neither entirely refugees nor entirely voluntary emigrants, they included elements of both, sometimes to the confusion of outside observers.

The Jews exhibited another distinctive characteristic of modern emigration: the return of former emigrants to their countries of origin. Unlike normal patterns of mobility, which usually display substantial rates of return, refugee movements are distinguished by the impossibility of going back, at least so long as the causes of outflow persist. Jewish emigration from tsarist Russia, it has long been maintained, differed strikingly from non-Jewish emigration because the Jews could not return home. Unlike many other groups, Jews burned all their bridges behind them. Jacob Lestchinsky once wrote that from 1908 to 1925, 1,018,878 Jews came to the United States and only 52,585 departed—a return rate of just over 5 percent. In the same period the return rate for non-Jewish immigrants was about a third. But a more recent study concludes that substantial Jewish return *did* occur in the first two decades of mass migration—possibly between 15 and 20 percent in the 1880s and 1890s. We know that the rate was just over 15 percent from England during the period 1895–1902, for which we have some data.[25] As we shall see, some of the return was organized and encouraged by established Jews in Central and Western Europe. Many Jews had left in disappointment, having failed to adjust socially, to find work, or to prosper as they had originally intended. Others returned according to predetermined plans: in this view, the prosperity won in the West was to be enjoyed "at home." A long list of personal as well as structural reasons could be adduced to explain the return, complicating the picture of emigration as being propelled exclusively by persecution.

Jewish migration from Galicia, the Austrian province in Eastern Europe that sent a huge proportion of its Jews to America after 1881, was largely unrelated to persecution and anti-Jewish activity. After their emancipation in 1867, Galician Jews enjoyed full civil rights, were admitted to state schools, and achieved a measure of formal integration into the society. Jews felt the province to be well-ruled from Vienna; the Hapsburg emperor there was affectionately known to the Jews by a Yiddish name, "Froyim Yossel," and his birthday was celebrated in Jewish communities. An antisemitic Peasant party appeared in 1893, but this did not trouble matters greatly for the Jews; there were some anti-Jewish attacks in the summer of 1898, but these did not recur before the First World War.

Jews poured out of Galicia through the nineteenth and early twentieth centuries because of the extreme poverty of the region, not because of persecution. Galicia was probably the least developed of any Jewish settlement in Eastern Europe. "Only a few of these Jews," according to Arthur Ruppin, "had an approximately secure economic basis, the rest lived on petty trade, the sale of spirituous liquors, pawnbroking, on handicrafts carried on in the most primitive manner for the lowest remuneration, as commercial agents to the big landowners, or had to rely on casual occupations. Emigration alone offered them a chance to escape from this misery." The Jewish historian Raphael Mahler concluded that poverty was the driving force behind the Jewish exodus from Galicia, much more than was the case from Russia. It is true that the emigration of Jews from Galicia was proportionately greater than the emigration of non-Jews. But the latter emigrated to a much greater degree than did similar groups in the tsarist empire. All nationalities, it seems, had reason to leave. Moreover, it appears that economic conditions for Jews in particular deteriorated in the latter part of the century. The traditional Jewish occupational structure, with its heavy concentration in shopkeeping, artisanship, and peddling, made Jews particularly vulnerable to agricultural depression and to capitalism's penetration of the countryside.[26] The case of Galicia is a reminder that persecution was not always behind Jewish suffering in Eastern Europe and cannot always explain why Jewish activitity differs from that of non-Jews. It provides further evidence that not all Jewish emigrants from Eastern Europe were refugees.

Rumania, the third main source of Jewish emigration from Eastern Europe before the First World War, saw both extreme poverty and intense anti-Jewish activity. Here the causes of Jewish emigration became a matter of international debate. Rumania provides an interesting and rare example of international effort to prevent the expulsion of unwanted subjects and arrest the flow of refugees.

Having emerged from the Ottoman Danubian principalities of Moldavia and Walachia in the middle decades of the nineteenth century, Rumania achieved its independence by degrees—finally accepted by the great powers with the Treaty of Berlin in 1878. Caught between European countries who exercised their tutelage nervously, each with an eye on the other, the Rumanians experienced a peculiarly intense and insecure nationalism, which fastened on continued foreign domination. Jews became the symbols of this persistent and humiliating vassalage. Independent Rumania turned on the Jews in a "cold pogrom"—a deliberate and routine government policy that defined Jews as enemies of the nation, removed them from the state school system, and encouraged occasional violence. Large numbers were expelled from villages, concentrated in certain towns, or deported. Rumanian Jews, particularly the Hasidic communities of Moldavia, were also worn down by poverty. The great majority were bound by scores of laws that prevented them from earning a living—even from engaging in petty trades or commerce. Harassed and heavily taxed by officials, many of them survived only by virtue of handouts from the Alliance Israélite Universelle. By the end of the century, according to Howard Morley Sachar, "the sheer physical survival of Rumanian Jewry was becoming a matter of doubt."[27]

Periodically, the Western countries nagged the Rumanians to improve the lot of their Jews. The European powers maintained some leverage over Rumania as a result of several international agreements. From the 1856 treaty onward, they tried to force the Rumanians to enshrine constitutional rights for religious minorities. The Treaty of Berlin of 1878 insisted that Jews be granted full civil liberties. Despite pressures from the British, Germans, and Americans, however, authorities in Bucharest managed to withhold citizenship from approximately 200,000 Jews. Speaking to the West, Rumanian officials offered a mixture of denial, obfuscation, and procrastination. Jews had little recourse but emigration, which was prompted by economic circumstances as well. The result was a greater flow proportionately from Rumania than from anywhere else in Eastern Europe. Between 1871 and 1914 over 75,000 Jews, nearly 30 percent of Rumanian Jewry, emigrated to America.[28]

To outsiders, there was never much doubt that politics explained much of the Jewish exodus. Plenty of evidence existed. On the Rumanian side, authorities required departing Jews to sign statements indicating that economic distress alone had prompted them to leave. But Jews told a different story to their coreligionists abroad. The Bucharest Society of Jewish Artisans made the point forcefully around 1900: "We want to live like human beings, and if this is impossible here we shall live elsewhere. We want both, liberty and bread! That is the true cause of emigration of the Israelites from Rumania." The Alliance Israélite Universelle, a Franco-Jewish organization, repeatedly took up the Jewish cause at the highest diplomatic levels and continually presented evidence of persecution; so too did the Anglo-Jewish Association.

Publicity trained on the departure of these and other Rumanian Jews and exposed Rumanian contentions that the Jews were treated well. To capture attention, several bands of émigrés, involving a few thousand persons, walked all the way to Hamburg with the declared object of sailing for America. These *fusgeyers*, as they were known, aroused great excitement: in bizarre, sometimes exotic costume, they marched with homemade sandals, sometimes carrying wooden staffs; they camped outside towns and cities; everywhere, they prompted reports in the European press about the terrible oppression they had known.

In retrospect, the Rumanian-Jewish emigration seems more completely under the impact of persecution than either of the previous cases we have examined. One crude indication is how the Jews left Rumania to a vastly greater degree than did non-Jews who shared similar economic conditions; another is the way that Jews abandoned that country in a proportion to the total Rumanian emigration that considerably exceeds that for either the Russian or the Austro-Hungarian emigration. For the period 1900–1910, one investigator estimated that of the 59,467 Rumanian immigrants to the United States, Jews constituted 53,484—almost 90 percent. Comparable figures during the same period for Russia were 43 percent and for Austria-Hungary, just over 8 percent (of which Galician Jews made up a substantial proportion).[29]

Once they crossed the border, however, East European Jewish emigrants appeared remarkably alike. All spoke Yiddish, dressed in traditional garb, and appeared poverty-stricken and distressed. Masses of Jewish emigrants were a familiar sight in European capitals like Paris or Vienna, in the waterfront districts of ports like Hamburg or Bremen, at important railway junctions or frontier cities like Brody, in Austrian Galicia just at the Russian frontier. For many Europeans, these were the first "refugees" they ever saw. Some recorded the Jews' strange appearance, their stunned disarray, their exhaustion, the confusion that seemingly paralyzed them. Pierre Van Passen later recalled Jews encamped near the wharves at Amsterdam:

> Most of these people were in rags. Hunger and long years of destitution had left an ashen imprint on their faces. There was an air of hopeless impotence about their movements; a dumb defeatist resignation, almost unhuman. With their eyes they followed the visitors and the official who led us around. And into those same eyes crept that same look of fearful alertness of the hunted animal which I was to notice years later on the faces of natives of Somaliland, Nigeria and Ethopia, where the approach of a white man, any white man, inspires the blacks with dire apprehension.[30]

It did not take long before such uninvited guests became unpopular. By the end of the century, European opposition to the Jewish emigrants was widespread, nourished by popular antisemitism, xenophobia, fears about diseases carried by the Jews, and hosts of legends about their supposedly nefarious customs and habits. It was thus appropriate that Jewish organizations, like the emigrants themselves, wanted the tide to sweep westward to America.

Initially, most Jews crossed into the German empire—more than 700,000 in the period 1905–14 alone. Hostility toward them was widespread. To Otto von Bismarck, the Iron Chancellor, the Jews were "the unwanted element." To citizens of Hamburg in 1892, the Jews were responsible for an outbreak of cholera that raced through the port city that summer. To some strategists of the Central Powers, the massing of Jews on Germany's eastern frontier was a menacing move by the tsarist empire. Harold Frederic, London correspondent of *The New York Times*, had the fanciful idea that the predatory Russians would use the Jews as a moving human shield, behind which tsarist hordes would advance deep into Central Europe.[31] East European Jews in Germany, or *Ostjuden*, became a favorite target of a new generation of antisemitic rabble-rousers, who wildly exaggerated the number of Jewish emigrants remaining in the Reich, condemned their "infiltration" into society, and urged their immediate expulsion. And these campaigns had an effect: German authorities were alerted to the consequences of allowing too many Jews to stay.

In the end, the Germans never sealed the frontier with Russia. Instead, officials concentrated the Jews and encouraged some to leave; others, deemed economically useful, were permitted to stay. German shipping companies worked together with government and Jewish agencies to direct the Jews to German ports and German vessels. During the peak period, in the two years following the Revolution of 1905, 85,000 Jews sailed from Hamburg; and 50,000, from Bremen. Occasionally, when Jews remained too long or when local campaigns against these *Ostjuden* intensified, authorities simply expelled some of them back to Russia. But overwhelmingly, the Jews were dispatched to America. In 1910 there were just over 70,000 East European Jews left in Germany, and the proportion of *Ostjuden* in the German-Jewish community as a whole had scarcely increased at all in the period after 1900.[32]

Elsewhere the Jews had an easier time. There was no legal restraint on Jews moving within the Hapsburg empire. Large numbers came from Galicia, Bohemia, and Moravia to Lower Austria and especially the Imperial capital of Vienna, where the Jewish population rose from about 6,000 in 1857 to over 175,000 in 1910. Similarly, the French placed no obstacle before the relatively small number of East European Jews—about 30,000—who settled there before 1914. Belgium and Holland permitted thousands more to enter. But the most conspicuous open door, aside from America, was Great Britain, which received about 120,000

East European Jews in the three decades before the First World War.

During the first phase of Jewish emigration, until the peak period beginning in 1905, the British made no effort to regulate the admission or continued residence of destitute aliens. Unlike other European countries, which either controlled immigration or maintained some police supervision of newcomers, especially refugees, the British took no action whatever. European exiles, as we have seen, had flocked to London after the crushing of the 1848 uprisings. Russian revolutionaries entered freely, organized political parties, and published newspapers. And so, too, the Jews, who first appeared in large numbers as transmigrants, simply arrived at a British port and went about their business unrestrained. Later, many continued on to America in English ships.

Formally, this liberal Victorian policy ended in 1905, with the passage of the Alien Act. Responding to a variety of fears, notably about employment in the heavily settled East End of London, the statute set new limits to unwanted immigrants. Refugees, however, would still be granted asylum: No one could be refused admission to Britain, said the new law, "who proves that he is seeking admission to this country solely to avoid prosecution or punishment on religious or political grounds, or for an offence of a political character, or persecution involving danger of imprisonment or danger to life and limb." Yet the latter were still seen in terms of individual political activists rather than as masses of improverished and persecuted immigrants. To the lawmakers of 1905, refugees arrived in hundreds; but immigrants, in tens of thousands. To qualify, a refugee now had in theory to prove his or her case—a major break with the old policy of laissez-faire. Had a general question of the Jews' standing as refugees ever arisen, it seems very doubtful that they could have passed the test. By 1905, when substantial numbers had already settled in England, few, apart from conspicuous sympathizers, considered the Jews to be refugees. Nevertheless, Jews did not find themselves particularly singled out by the application of the new law. In practice, the new restrictions were only intermittently enforced, and the net effect was probably small. Although the law may have dissuaded some immigrants and even deterred some exiles from seeking asylum in England, Jews continued to arrive and settle. The Alien Act indicates rather a limited change in policy on immigration and a harbinger of future restrictions on refugees, who no longer had an automatic right to enter Britain. With it we see new apprehensions and distinctions; eventually, these would overwhelm what remained of the liberal policies of the nineteenth century.[33]

Everywhere the Jewish emigrants went in Europe, they encountered coreligionists nervous about the flood of Jewish travelers from the East. The 1880s saw the beginnings of a new wave of popular antisemitism, as we have seen, which harped on the threat to European societies caused by all Jews, but particularly by unassimilated, Yiddish-speaking *Ostjuden*. Spokesmen for established Jewish communities in the West worried anxiously about how the presence of these outsiders might undermine their newly acquired social status, might cast aspersions on their own acculturation. Confronted with this reminder of their own preemancipation identity, many Westernized Jews felt discomfort, even distaste. This disposition conflicted with a second sentiment—the urge to care for fellow Jews and protect those who were obviously in great need.

Somewhat ambivalently, the Jewish organizations began to assist the emigrants, but also to speed them on their journey westward. Jewish agencies stationed themselves along the paths of emigration from the first points of exit from tsarist Russia. Perhaps the best-known organization in the early days was the Paris-based Alliance Israélite Universelle, which prided itself on its early and energetic assistance to Jewish refugees who crowded into Brody in 1882. By the early 1890s German-Jewish relief societies deployed at border crossings in East Prussia and Upper Silesia, receiving long lines of Jewish migrants. Other Jewish committees from England, Belgium, Holland, and Austria were active in the field. To sympathetic outsiders, the work of these institutions was remarkable. Harold Frederic inspected a Jewish reception committee in Hamburg and reported, "I have never in my life witnessed more genuine, unostentatious, and intelligent philanthropy than I saw at work here."[34] At first, charitable societies were often ill-equipped and clumsy in handling the large numbers of people. In the spring of 1882, for example, over 24,000 Jews flocked to the Brody refuge station, overwhelming relief workers there. But before long, the various agencies were able to house, feed, and clothe the needy.

With few exceptions, these committees sponsored transatlantic migration, negotiating with steamship companies and government officials, subsidizing or purchasing tickets, and coordinating departures. Some investigated settlement possibilities abroad. They protected the Jews from swindlers and interceded with local authorities. Their correspondence betrayed at moments an impatient distaste for the *Ostjuden*—whom Charles Netter of the Alliance Israélite Universelle once called "the beggars of the Russian Empire." Along with the reasonable concerns to relieve the Jews' sorry plight and to provide them with the opportunity for a new life, one senses an almost visceral dislike by some philanthropists of traditional Jewish culture. Jewish relief workers attempted even to secure repatriation, especially in the 1880s and 1890s. Not long after the first pogroms, the Alliance made a determined effort to discourage Jews from leaving Russia; its officials felt that too much aid might prompt an even greater, totally unmanageable exodus. And the Alliance was not alone in such judgments. In Budapest and in Antwerp, local Jews mobilized to meet arriving emigrants and attempted to persuade the newcomers to return to the tsarist empire. The historian of the Jewish Board of Guardians, which coordinated London's Jewish charities, indicates that between 1880 and 1914 the board sent about 50,000 Jews back to Russia.[35]

But if the charitable impulse faltered at times, it would be wrong to deny it altogether. One of the important accomplishments of Jewish emigration from Eastern Europe, indeed, is the network of private agencies to deal with mass refugee movements. Following the spectacular Kishinev pogrom in 1903, a network of two dozen charitable agencies coordinated at an international conference in Frankfurt. A Central Office of Migration Affairs divided responsibility and made strategic decisions. This organizational scale towered above anything provided by governments at the time, and it assembled a multitude of representatives across an entire continent.

The Jewish mass migration from Eastern Europe had some attributes of a mass refugee movement. There was no refugee crisis, however, and European states seldom had to adjust their immigration policies because America was the great escape valve. East European Jews, therefore, did not force a revision of nineteenth-century notions about refugees. So long as America received virtually everyone who came, Europeans seldom had to define refugees, worry about their responsibilities, and ponder who qualified for aid and who did not.

THE BALKANS AND THE "UNMIXING OF PEOPLES"

Meanwhile, hundreds of thousands of refugees were roaming the Balkan peninsula in the decades before the First World War. In the background was the gradual decline of the Ottoman Empire after the seventeenth century and its territorial collapse in Europe in the decades before 1914. Caught in a dispute that became increasingly more damaging to Ottoman fortunes, refugees from both sides of the imperial divide sought to escape repression and the effects of war.

The slow disintegration of Ottoman power, which had once extended from the gates of Vienna south to the Mediterranean and east to the steppes of Russia, was one of the great dramas of European politics. Who would inherit the European possessions of the imperial giant? Rushing forward to make their claims during the nineteenth century, various national movements proposed independent nation states for the Balkans that would unite subject peoples whose national identities imperial domination had denied or left unrealized. At the same time, other European powers looked on jealously, anxious lest this "Eastern Question" be resolved against their own interests. Two empires, the Hapsburg and the Romanov, watched with particular apprehension and avidity—nervous at the prospect of so strategically vital an area on their doorstep falling into the wrong hands, yet eager to enhance their own, sometimes sagging imperial fortunes. Occasionally onlookers wanted to hasten the end of the Ottoman "Sick Man of Europe"; but sometimes they helped prop him up, knowing that the final collapse might bring about their own undoing. Southeastern Europe, as a result, was a focal point for political manipulation and intervention by outsiders—ensuring its own instability throughout the period.

Complicating this situation was the polyglot character of the Balkans, whose mountainous geography was the home of many different linguistic, ethnic, and religious groups. Under Ottoman rule these had lived side by side—frequently exercising their own religious, legal, and administrative jurisdiction under the supreme control of the central government. But as their autonomy grew and as independence loomed, they found themselves ever more at odds.

Broadly speaking, these conflicts produced refugees of two sorts, moving in opposite directions: Muslim populations turned south, retaining Turkish protection and participating in the diminishing advantages of imperial rule; Christian populations went north, escaping Ottoman control and eventually sorting themselves into states and regions that assumed particular national identities. This was what Lord Curzon referred to as "the unmixing of peoples"—a quip that obscured how much mixture still remained, how uncertain were the definitions so sweepingly applied, and how much remixing went on simultaneously, bedeviling efforts at simplification and peaceful resolution.

During the nineteenth century, national feeling gnawed at the structures of Ottoman rule, fatally weakening the remshackle institutions of the sultanate. Greece was the first country fully to escape Turkish domination. Assisted by the Russians and to a lesser degree by the British and French, the Greeks finally achieved full independence in 1832 after more than a decade of bloody warfare. About the same time came autonomy for Serbia and the Danubian principalities of Moldavia and Walachia. At the expense of the Turks, the Russians and Austrians increased their involvement at the fringes of the empire—in Bulgarian lands, Moldavia and Walachia, Montenegro, Bosnia, and Herzegovina. More conflict followed, eventually leading to a situation in which, by the outbreak of the First World War, the Turks were left with nothing but eastern Thrace of all their former European territories.

Accompanying this progressive shrinking of Turkish control was a continual uprooting of local populations. In 1822, for example, in one of the most famous episodes of the Greek war of independence, the Turks devastated the island of Chios in the Aegean, off the coast of Asia Minor: they massacred part of the local population, sold others into slavery, and sent thousands into exile. For much of the nineteenth century an identifiable Greek diaspora of Chians maintained distinct communities in London, Trieste, and Marseilles. During the 1860s and 1870s thousands of Bulgarian revolutionaries against the Turks and their Greek-dominated officialdom fled north across the Danube to Moldavia or Walachia, where Ottoman control was weak. Christian peasants from Bosnia and Herzegovina, the most westerly Ottoman territories, streamed over the borders to Austria as well as to Serbia and Montenegro when an extensive tax rebellion broke out against the Turks in 1875. As well as embittering such conflicts, national feeling helped to direct the flow of refugees: the presence of conationals in neighboring states prescribed ready-made havens, as well as centers for continuing sedition and the export of revolution.

We see the difficulty of sorting out the mixture in the Balkan crisis of 1875–78, a focal point in the history of independence movements in the peninsula. Following the harsh repression of uprisings against the Turks in Bosnia and Herzegovina and Bulgaria by the new sultan, Abdul Hamid, the Russians declared war on Turkey in 1877, ostensibly leading a Pan-Slavic crusade against Ottoman rule. After unexpectedly tough resistance, tsarist troops drove across the Balkans to reach Constantinople, and the fighting finished. At San Stefano, the Russians imposed extremely onerous terms on the Turks. In reaction, anti-Russian sentiment both in England and in Austria forced revision of these arrangements at the Congress of Berlin in 1878. The latter scaled down the San Stefano settlement with the Turks, providing for Hapsburg occupation and administration of Bosnia, formal independence and slight territorial gain for Serbia and Montenegro, and a division of San Stefano Bulgaria, still nominally Ottoman, into three parts—an autonomous principality, a semiautonomous province of Eastern Rumelia to the south, and Macedonia, which remained under Turkish control.

Responding to these momentous political changes, refugees tramped wearily across new jurisdictions in a sometimes desperate effort to avoid massacre. They surged to friendly sanctuaries in regions that had achieved national autonomy or independence, and they appealed to intervening powers who promised refuge; simultaneously former beneficiaries of Ottoman rule gravitated back to the heart of the empire. The Congress of Berlin and its sequel saw refugees meeting each of these descriptions.

For example, Muslim settlers or representatives of former Turkish institutions were immediately affected. These refugees abandoned territories annexed by Serbia in 1878 and the newly independent state of Montenegro. Thousands more quit Bosnia after the dramatic Turkish reversals there; among the most vigorous defenders of Ottoman rule, these thousands flocked to Constantinople, where they railed against the newly imposed Austrian authority. Similarly, Muslim refugees fled before the Russian armies as they advanced into Bulgaria, eager to avenge the Turkish slaughter of Bulgarians the year before. Muslim refugees hated the former Christian subjects who had just won a

THE BALKANS, 1878–81

After R. R. Sellman, *A Student's Atlas of Modern History* (London, 1952).

43

measure of national independence, despised the Russians who had smashed Turkish domination, and condemned the sultan who had yielded too easily. Such diehards doubled the population of Constantinople, contributing much to the instability of imperial government there.[36]

Meanwhile, many Christian subjects in lands that remained Ottoman also had to evacuate their homes. Macedonians and Bulgarians in Thrace who had suffered at Turkish hands during the preceding period fled north to Eastern Rumelia or the Bulgarian principality. After the 1877 war, the flow continued. As the Turks tightened their grip on territory they had so nearly lost, subject peoples began to move out. By the mid-1890s some 100,000 Macedonians had moved to Bulgaria, doing in Sofia roughly what their Islamic counterparts did in Constantinople—energizing radical politics, embroiling their hosts with their national enemies, and threatening to destabilize the existing regime. Macedonian exiles clamored for Bulgarian help to liberate their homeland and may well have been behind the assassination of the Bulgarian prime minister Stefan Stambolov in 1895. That year they founded a Macedonian revolutionary organization in Sofia, challenging the Turks across the frontier.[37]

Following the Berlin settlement, national conflicts continued in the bosom of the Ottoman Empire, together with disputes among the newly independent states. Each of these confrontations generated refugee movements. After a brief Greco-Turkish war in 1897, Muslim refugees crowded once again into Ottoman territory, permitted safe passage thanks to the insistence of the great powers who had saved the Greeks from defeat. Macedonian nationalists followed their predecessors on the road to exile in Bulgaria, notably after the anti-Turkish revolts of 1903. Ottoman repression intensified after the Young Turk Revolution in 1908. The latter began under the direction of the Turkish army officers stationed in Macedonia and for a time improved relations between Turks and their Christian subjects. But nationalist extremists soon gained the upper hand on the Ottoman side, and in the resulting terror a new Macedonian emigration ensued.

However violent, such refugee spasms were insignificant compared to the mighty waves produced during the Balkan conflicts of 1912–13 and after. Feeding this new conflagration was the accumulated nationalist debris of more than a century: the unresolved quarrel against the Turk, the still unsatisfied aspirations of various subject peoples, the fearful antinationalist reaction of the Hapsburg monarchy, and the nervous concerns of the Russians and other great powers. Here was a major threat to regional and European peace. Tensions exploded in 1912. In the first of two wars, the young Balkan states combined to fight the Turks, hoping to drive them finally from the peninsula. In the second, following quickly on the first, the Balkan allies quarreled over the Macedonian spoils. Greece and Serbia joined Turkey to defeat the Bulgarians, yielding a new settlement that frustrated Macedonian national feeling, outraged the Bulgarians, maintained Turkey in a small redoubt in eastern Thrace, and embittered relations among most of the other Balkan states.

Short though they may have been, these wars set new standards for cruelty, violence, and destructiveness in a part of the world not known for refinement in the observance of the rules of warfare. Only the outbreak of the First World War almost immediately after the settlement of these wars prevented them from marking a new standard in the scale of horrors associated with modern conflict. An international commission sponsored by the Carnegie Endowment concluded grimly in 1914: "From first to last, in both wars, the fighting was as desperate as though extermination were the end sought." Massacres occurred in every quarter: noncombatants suffered heavily, with the suffering of the Macedonian population perhaps heaviest of all. Refugees followed every course we have seen in the Balkans thus far and in devastating proportions. Close to 100,000 Turks attempted to escape the advance of the Balkan armies when they first attacked in the autumn of 1912; tens of thousands were evacuated from western Thrace when the guns fell silent the following year. In early September 1913, the Carnegie Commission heard that 135,000 Moslem refugees had passed through Salonika on their way to Anatolia—the survivors of widespread atrocities against the Turkish population begun the previous year. Arnold Toynbee reported that during the period 1912–13 over 177,000 Moslem refugees entered Turkey after the onslaught of the Balkan powers. During the second war, 15,000 Macedonians fell back from Macedonia, following the Bulgarian army in retreat; 70,000 Greeks quit western Thrace when it was taken and held by the Bulgarians. Civilians from eastern Thrace dispersed in all directions because of the fortunes of war: formerly Turkish, this territory was conquered by the Bulgarians in 1912 and reverted to Ottoman hands within a matter of months. As a result of these wars, Greece, Serbia, and Bulgaria carved up the disputed

Macedonian region and shared it among them. Finding their homes assigned to a partitioning state where they feared persecution in an alien linguistic, religious, or ethnic environment, Macedonian partisans of another partitioning power took to the few roads through the country to find refuge. Others simply sought clusters of relatives elsewhere or opportunities to emigrate abroad.[38]

During the postwar negotiations, several former combatants attempted to bring chaotic refugee movements within the framework of international agreements through the mechanism of population exchanges. Initiatives came from the Ottoman side, where since 1908 the Young Turks had been trying to hammer their sprawling empire into a more centralized, more homogeneously Muslim state. From 1910 a broad program of Ottomanization formed part of the Young Turks' efforts to modernize their country and secure its national independence. The Turks were acutely aware of how the presence of Christian minorities had created endless pretexts for foreign intervention in their empire; far better, it was now thought, to get rid of many of them in exchange for Muslims from elsewhere. On the Bulgarian side there was also reason to think in terms of orderly population exchange. Between the Congress of Berlin and 1912 more than 250,000 refugees flocked to Bulgaria, to be followed by the latest round of fugitives during the Balkan Wars. Some 120,000 crossed Bulgarian frontiers from 1912 to 1914, posing urgent problems for the state that had just been defeated by its Balkan neighbors.[39]

The Convention of Adrianople, concluded between Bulgaria and Turkey in November 1913, attempted to order the flow of refugees. It has been termed the first interstate treaty on the exchange of populations in modern history, but, in fact, most of the people affected had long since been uprooted: During the two most recent wars just under 50,000 Turks left Bulgarian territory, and about the same number of Bulgarians went in the other direction.[40] The agreement confirmed that fait accompli, but it also provided rules and procedures to regulate the consequences of what had happened. Referring to a voluntary exchange of persons living in a fifteen-kilometer strip along the mutual frontier, it permitted removal of property and established a mixed commission to settle outstanding property claims.[41]

Having succeeded with this project, the Turks turned immediately to Greece, hoping to negotiate a similar accord. According to the Carnegie Commission, Greece harbored 157,000 refu-

THE BALKANS, 1912–13

After R. R. Sellman, *A Student's Atlas of Modern History* (London, 1952).

47

gees when the fighting ceased—an extremely heavy burden for
an impoverished country with a population of 2.6 million. Ottoman authorities nudged the Greeks toward the negotiating table
by generating more refugees: the Turks began large-scale depor-
ations of 150,000 Greeks from the Aegean coast and forced another 50,000 to the interior of Anatolia. Obliged to settle, the
Greeks concluded a preliminary arrangement with Turkey in May
1914. This envisaged a much more ambitious transfer than that
with the Bulgarians in the previous year. Also voluntary and aided
by a mixed commission, the exchange was to occur over far more
extensive areas than in 1913, involving Macedonia, Epirus, Thrace,
and western Anatolia. Significantly, the eligible populations had
not necessarily been uprooted by war and indeed had often been
living peacefully, undisturbed by the recent fighting. The goal was
clearly to eliminate minorities and incorporate conationals in the
most orderly and expeditious fashion.

The outbreak of the First World War in 1914 aborted these
schemes before the mixed commissions could operate. For the
Balkans, this was the most devastating refugee experience yet.
Serbia practically became a nation of refugees, with about a million of its citizens, one-third of its entire population, violently
uprooted at one point or another before the battles ended. With
a respectable army of 350,000, the Serbs did quite well against
the Austrians in 1914, twice throwing back invading armies from
the north and west. Then came a series of catastrophes. In the
winter of that year, typhus spread unchecked through the
crowded, stinking cities, teeming with war evacuees. Within six
months it had killed over 150,000, including about a quarter of
the Serbian army. In the end, about a tenth of the population
died of the disease.[42] The year 1915 saw a renewed offensive by
the Central Powers, now joined by the Bulgarians, who advanced from the east. The Serbs were overwhelmed. In order to
escape what promised to be a ruthless occupation, hundreds of
thousands scattered.

The invading armies captured not only the Serbian soldiers, but
civilians as well. Some 10 percent of the population, it has been
estimated, was rounded up and dispatched to camps in Hungary or Bulgaria, often to be used as forced laborers.[43] The greatest concentration of refugees began a march to the Adriatic coast
in one of the extraordinary popular epics associated with the history of warfare. Ravaged by cold and illness and battered by their
enemies, Serbian troops and civilians dragged themselves through

rugged country in Montenegro and Albania. Progress was extremely slow, made difficult by the mountinous terrain, primitive roads, and the occasional harassment by Austrian aircraft. For weeks, thousands of campfires dotted the sides of Mount Jlyeb at the gateway to Albania while soldiers and refugees plodded into that country. As many as 500,000 civilians may have set out from Serbia; many returned, and perhaps 200,000 perished on the way.[44] From Durazzo and other towns on the Adriatic, survivors were eventually rescued by British and French warships. Some of the Serbs went to the Greek island of Corfu, and others took advantage of the popularity of the Serbian cause elsewhere; these ended up in France, Italy, or Switzerland.

Other refugees trudged across the Balkans during the war, though nowhere on such an enormous scale as in Serbia. Serious fighting occured in Rumania in 1916, when the latter joined the Triple Entente side. After the capture of Bucharest and the cession of Dobruja to Bulgaria, more refugees took to the roads. Toward the end, when the Serbs finally cleared their country of the occupation and when the Bulgarians were thrown back, many of those displaced in earlier refugee waves returned, but new groups were made homeless. Bulgaria choked with refugees from territories occupied by the victorious powers—Greece, the new state of Yugoslavia, and Rumania. An investigation by the International Labor Organization about this time showed about 300,000 refugees in Bulgaria who simply could not be absorbed.[45]

The war completed the movement toward independence of the various Balkan states; it left national aspirations of various groups unsatisfied, however, and can hardly be said to have resolved minority problems in the peninsula. Refugees were to flood the Balkans once again, spectacularly as we shall see, but not because of the establishment of new territorial entities. This last was the heritage of the century before 1914.

The Balkan refugee movements we have just scanned were by-products of the state-building process. One student of refugee problems has recently referred to "integration crises" as important precipitators of refugee flows in the modern period. By such convulsions he means the mobilization of new states to achieve the goals of state-makers—independence, national cohesion, and the achievement of such nationalistic projects as the absorption of coveted territory.[46] Nationalism in the Balkans fueled refugee movements in precisely this manner. Over the course of the nineteenth century, nationalism meant that the ethnicity of local populations would become badges of political allegiance. Then the wars of national liberation, the formation of new political entities, and reactions to these processes within the Ottoman Empire contributed significantly to the sense of incompatibility among various groups. The result, as we have seen, was hundreds of thousands of people cast adrift.

The Eastern Question preoccupied European diplomacy because the fortunes of the great powers were at stake; there was also a measure of sympathy for Christian victims of the Ottoman Empire. But to outsiders, often secure in their own long-standing national independence, the ferocity of Balkan peoples toward each other was simply beyond understanding. Despite his deep familiarity with the region, Arnold Toynbee refered to "the chaotic, unneighborly races of southeastern Europe," as though there existed some special genetic predisposition toward hatred of outsiders in that part of the world.[47] Europeans were inclined to write the entire region off as a sea of barbarism and saw southern Hungary as the frontier of Europe. No one expected high standards of behavior from the new Balkan states, and their refugee movements made little impact on the European consciousness. Apart from those who managed to leave as proper emigrants, the refugees themselves were contained within the peninsula and thus had no effect on the European state system. No one, therefore, felt prompted to staunch the flow of refugees, and no one understood that similar movements were about to flood the center of the European continent.

2

The Nansen Era

In practically every way we can imagine, the First World War imposed on contemporaries the awesome power of the nation-state. Modern machinery of government mobilized seventy-four million men and harnessed every innovation of bureaucracy and industry to the destructive enterprise—in the end killing between ten million and thirteen million combatants and wounding another twenty million. With such energy poured into the war effort, it was unlikely that civilians would be spared. Not only were technical means available to carry hostilities far beyond the confines of traditional battlefields, but it now also made more sense than ever to extend the range of the fighting to destroy the enemies' war-making potential by ruining their towns and cities, their industry and agriculture.[1]

So far in this book, I have argued that the growth of the modern nation-state implied not only the naming of certain people as enemies of the nation, but also the expulsion of significant groups for whom the state would not or could not assume responsibility. With the First World War, this process accelerated powerfully. The war itself schooled the new masters of the state apparatus: civilians could become dangerous enemies; fighting could not stop simply because they were there; on the contrary, it was best to eject unwanted or menacing groups when they threatened to weaken the beleaguered mation.

In 1918 huge masses of refugees appeared in Europe, victims of new-style nation-states—especially those consolidating their precarious existence in the postwar world. It was estimated in 1926 that there were no less than 9.5 million European refugees,

51

including 1.5 million forcibly exchanged between Greece and Turkey, 280,000 similarly exchanged between Greece and Bulgaria, two million Poles to be repatriated, over two million Russian and Ukrainian refugees, 250,000 Hungarians, and one million Germans expelled from various parts of Europe.[2] The bulk of the refugees concentrated in the east, where successor states were the most fragile and the most threatened by outside forces and where nationalities were so explosively intermingled, posing a challenge to statesmen and rival champions of national independence. Some of these refugees, such as the Jews or the Russians, fled direct persecution or the upheaval of revolution and civil war; others, such as the Greeks or Armenians, were deliberately cast out by rival and dominant national groups.

For the most part, these refugees had to fend for themselves. Occasionally, the peacemakers in Paris attempted to control this great human catastrophe, introducing provisions in their treaties to block some of the more egregious instances of expulsion or persecution. In addition, the League of Nations created a high commissioner for refugees, a post filled by the Norwegian Fridtjof Nansen until his death in 1930. Usually, these efforts were like using bedroom sheets to block a hurricane; but, as we shall see in the case of Greek resettlement, international organization could be remarkably effective. This chapter examines the refugee crisis in the years following the armistice of 1918 and its gradual resolution in the decade that followed.

THE GREAT WAR AND UPHEAVAL IN EASTERN EUROPE

For centuries four dynastic empires dominated Eastern Europe: the Ottoman, the Romanov, the Hapsburg, and the Hohenzollern. The First World War sent all four crashing into ruin. In their place emerged a series of states and an invigorated ideal of national self-determination, a model of what independent nationhood might achieve. Sometimes these nationalist energies ran against such forces as the dominant nationality of Rumania, unwilling to permit each group free choice of its future; or the revolutionary momentum of the Bolsheviks in Russia and Atatürk in Turkey; or the determination of the victorious Allies to prevent another war and to impose a lasting settlement in Europe. On the whole, refugees were the people left out of the settlements of 1919, those whom the treaties failed satisfactorily to ad-

dress, whom the Great Powers considered of secondary importance if they thought of them at all. As Eastern Europe saw the
greatest postwar political transformation, this was also the cradle
of the new refugee movements.

Refugees and the Collapse of the Tsarist Empire

The outbreak of war in Eastern Europe in 1914 started an avalanche of refugees that continued into the early 1920s, when the
new Soviet government finally brought the process under control. Shortly after the first shots were fired, Tsar Nicholas II consigned to military rule a huge portion of territory in the western
provinces of the empire, essentially everything west of a line
running from St. Petersburg to Smolensk and then south along
the course of the Dnieper River. Thereafter, military personnel
controlled civilian life in much of European Russia. Ill-prepared
for this task, the soldiers engaged in massive requisitions and often
brutal harassment of local populations. "Arbitrary meddling by
the military authorities in nearly every aspect of life," writes one
historian, "produced widespread chaos and a progressive dete-
rioration of the status of both institutions and individuals."[3] By
the summer of 1915, when the Russian army began a headlong
retreat before the German attack, people in the western part of
the empire were often desperately trying to avoid further
suffering.

Huge numbers of civilians had already fled. Hundreds of
thousands were displaced in Polish territory and in Galicia in the
first stage of the fighting, attempting spontaneously to escape the
lines of fire. The military authorities systematically forced civilians further to the rear when they were deemed of doubtful loyalty. Jews fell into this category, as we shall see, and so did ethnic Germans, an obvious target in the chauvinistic excitement of
war. Some 200,000 Germans from Volhynia in the northeastern
Ukraine were caught up in this net.

When the Russians fell back in mid-1915, however, the uprooting intensified. Terrified at the collapse of forward positions,
army commanders marched entire villages to the railway stations for deportation or simply forced them onto the roads at short
notice with the few possessions they could carry. Retreating troops
now practiced a policy of scorched earth: abandoned territory was
to be made uninhabitable in a desperate attempt to slow the Ger-

man advance. In the Volga districts, authorities fell savagely on the local German population; by this time it was simply assumed the latter were in league with the kaiser, and officials decided to deport practically all Germans to Siberia or Central Asia. Deportees were sealed in cattle cars and dispatched by rail to the east with the slightest preparations and the most meager facilities. "Now and then," according to a historian of the Volga Germans,

> an entire train with its human cargo got shunted onto a railway siding and was forgotten for days because of locomotive priorities, chaos, traffic mismanagement, or simply the indifference of slovenly Russian train crews who functioned best, or only, through bribery. When the doors of the wheeled cages were finally opened, they yielded the tortured living and the rigid dead who had succumbed to hunger, thirst, disease, freezing, or heat prostration.[4]

In December 1915 the Russians counted over 2.7 million refugees, and the tide continued to rise.[5] Practically nothing was done for the displaced civilians. The tsarist government, says one writer, "was slow to recognize the dimensions of the problem and failed miserably in its attempts to deal with the refugee situation and minimize its ill effects on the country."[6] Clashes broke out when desperate, exhausted fugitives from the war zones landed among settled populations in the interior. Antisemitic riots flared up often when Jews arrived in this way. By the beginning of winter the refugees began to jam major Russian cities. In Moscow crowds of starving fugitives gathered to be fed at the Alexandrovskii railway station in the bitter cold; in Kiev there was near panic in the face of cholera, typhus, and dysentery brought by the refugees.

An English nurse, Violetta Thurstan, reported from Russia that as many as five million refugees were adrift in early 1916, facing a catastrophic situation in the bitter cold. She described aristocratic ladies' charities that worked together with government committees and municipal organizations, providing a bare patchwork of assistance amid the general disorder. One Russian relief association headed by Countess Tolstoy established automobile patrols to rescue abandoned infants by the roadside. Countess Tolstoy's automobiles collected four hundred such babies by the beginning of 1916. In the Caucasus, Thurstan indicated, workers sheltered more than 5,000 lost children who did not know their own names or the names of their villages.[7] In the winter some refugees camped in summer buildings on the frozen banks of the river Volga flowing through Kazan Province. But the Volga flooded in the spring as the ice began to melt; once again the fugitives were on the move. Everything had to be improvised to save these refugees from freezing and starvation. Everywhere those in charge tried to send the tide farther east— "to the interior," as was often said. Once set in motion, there was a powerful momentum to this great and only partially controlled migration.

When the revolutionary turmoil of 1917 began, many of these refugees had still not been settled. The avalanche, however, was now sliding faster and with greater force than anyone could imagine. From every corner of the stricken empire fugitives of all descriptions began to move: deserters, sick, and wounded streaming back from the front after the Revolution of March 1917; Jews, driven from the Pale of Settlement; peasants, dragooned to work in war industries, now trying to return home. Then came the collapse of the Russian army, chaotic demobilisation, another burst of revolution in November, and a new round of warfare generated by counterrevolutionaries, foreign intervention, and rival national movements.

The Bolshevik coup in November 1917 opened a period of intense political struggle within the former tsarist empire, marked by the slow and difficult extension of Soviet control. With the March 1918 peace at Brest-Litovsk, in which Russia lost to Germany as much land as Hitler took by conquest in 1941–42, the Bolsheviks showed how helpless they were in the early stages and how little they were able to exercise dominion. It took time to put the new house in order. The process worked from the center outward. In region by region, jurisdiction by jurisdiction, nationality by nationality, Lenin's revolutionary party fought to establish its authority, smothering its opposition only by the early 1920s.

Enemies of the revolution and those made homeless in the struggle to secure it were thrown back by this conflict to the peripheries of the former empire. Although some affluent Russians went immediately abroad, the main thrust of the refugee movement in the Revolution and Civil War, until 1920 or 1921, was centrifugal within onetime tsarist territory. Moreover, even when émigrés fled to Finland, the Baltic countries, Poland, or the Ukraine, they generally saw their move as *within* the Russian Empire. Many considered that Bolshevik rule would soon collapse, and in any event their territorial perceptions remained those of the pre-1917 period. Participating in the exodus were not merely the partisans of the old order—aristocrats, former bureaucrats, high-ranking officers, clergy, and so forth—but also liberal and even socialist opponents of the new regime. As the Bolsheviks tightened their repressive apparatus and girded for the desperate struggle to keep their revolution alive, terror helped determine the contest for power. Some who accepted the revolution in the early days took flight as soon as repression intensified and the opposition to the Bolsheviks burrowed underground.

In the north, refugees poured into Finland and the former Baltic provinces, which won effective independence at the end of the First World War. These émigrés included natives of the region, often of German origin, freed by demobilization or simply seizing the opportunity provided by the collapse of central government to return to their native lands. Poland received not only refugees from the fighting on the Polish-Ukrainian frontiers in the period 1918–20 but also hundreds of thousands of former inhabitants who had fled or been driven eastward during the war, especially in 1915. To the south went Ukrainians hoping to join the independent government established there, as well as political émigrés hostile to the party of Lenin. Many of the latter congregated in Kiev, which for a time became the focus of the anti-Bolshevik emigration. Others moved further south to Odessa, the Crimea, and later the Don and Kuban regions.

Fighting between the Bolsheviks and their enemies generated many thousands of refugees. To conservatives, Leon Trotsky's advancing troops were the force of the Devil incarnate. Lenin declared his intention to "probe Europe with the bayonet of the Red Army," believing that the Soviet forces might stimulate revolutionary energies elsewhere in Europe.[8] But many also panicked before the Whites, the antirevolutionary armies aided by the British, French, and Americans. As, for example, General Anton Denikin moved westward for the counterrevolution in the Ukraine in 1918, he drove masses of civilians before him, many of them crossing paths with others fleeing the Bolsheviks and moving in the opposite direction. The Ukraine was particularly devastated for close to three years, being the scene of intense and sometimes three-way conflicts between Bolsheviks, nationalists, and counterrevolutionaries. During 1919 the region passed under Soviet domination, and in early 1920 practically the entire Ukrainian nationalist army under Simon Petlyura sought refuge in western Galicia, where they were interned by the Poles. That year saw the debacle of anti-Bolshevik forces in the south: the abandonment of Kiev and Odessa, followed by the withdrawal of counterrevolutionary armies under the command of Baron Peter Wrangel in November. In the north, General Nikolai Yudenich was turned back from St. Petersburg in October 1919, forcing many to seek shelter in Finland and the Baltic states. Counterrevolutionary forces evacuated Archangel shortly afterward, forcing many more to take flight. Battles also occurred in western Siberia, where Admiral Alexander Kolchak was swept from the field and captured by the Bolsheviks at the end of 1919.

Poland saw most of the sustained conflict and probably the greatest number of refugees in this period. In 1918, with the simultaneous collapse of the former partitioning powers—Russia, Austria, and Germany—the Poles achieved national independence. Poland's borders were only defined, however, after protracted fighting with her neighbors. Even before they formally declared a republic in Warsaw, the Poles went to war with the Ukrainians in Galicia. Polish troops also battled the Bolsheviks near Vilna and clashed with the new Czechoslovak state over the former Duchy of Teschen and other territories. After what appeared to be a settlement of the eastern frontier along the Curzon line, war with Russia resumed in the spring of 1920, as the Poles attempted to incorporate parts of Lithuania, White Russia, and the Ukraine. At first, the Polish side did poorly. By late summer the Bolsheviks reached the gates of Warsaw; then, after stubborn resistance, the Poles threw back the Red Army and came to terms. The Treaty of Riga, in March 1921, set the frontiers between the former combatants and enabled the passage of displaced persons from one country to the other.

Six years of warfare had thrown out of their homes a good proportion of the Polish nation. In 1914 the Germans and Austrians, like their Russian enemies, evacuated hundreds of thousands of Poles from zones of military operation. Once hostilities ended in 1920, these, together with those who had been removed on the Russian side, began to return. The Polish government set up agencies to assist repatriates and to settle them, generally in eastern Poland closest to the Soviet side from which most of them came. Crossing into what had been the scene of intense combat, the refugees met utter devastation. Parts of eastern Poland resembled a lunar landscape with scarcely a living thing visible. An Australian relief worker described hills covered with skeletons of Russian soldiers who had been killed years before and who still lay as they had fallen. Barbed wire and ruined fortifications traversed the countryside; the wanderers made homes and shelters in long-abandoned trenches, living in the natural graveyard of battles fought years before.[9] At the beginning of 1923, Polish authorities estimated that they had already repatriated 703,250 people and anticipated the arrival of 300,000 more refugees; in 1920 the Poles said that over one and a quarter million had returned to Polish territory.[10]

The Poles heard about refugee conditions in Russia from returnees who had crossed Soviet territory. "Recently," reported a repatriation commission in 1922,

> a train arrived at the Polish frontier which carried 1,948 persons at departure. . . . Of this number 1,299 people died of exhaustion, privation and infectious diseases; only 649 reached their homeland, that is to say a third.

This train had come all the way from Kazan on the Volga, a distance of over 1,700 kilometers; the repatriates had traveled at a snail's pace for three months, eating little but the scraps of black bread given to them by the Soviets. The dead were simply left unburied, thrown out at stations along the way.[11] This description matches many others by travelers to the Soviet Union during the terrible years of War Communism, when the new regime mobilized for its very existence. Trainloads of refugees crisscrossed the vast spaces of Russia, carrying starving fugitives from town to town, spreading malaria, cholera, and typhus from one end of the Soviet state to the other.

By 1921 this drama reached its culmination. To the chaos of civil war and political upheaval were now added the effects of forcible requisitions under War Communism, the paralysis of transport, and a severe drought. The result was a collapse of Russian agriculture and a spectacular famine, which has been called the worst in European history, killing over five million people.[12] Famine cast even more people adrift, adding to the millions of refugees. Outsiders tried to grasp the dreadful character of the Soviet disaster. "Was there ever a more awful spectacle in the whole history of the world than is unfolded by the history of Russia?" Winston Churchill asked rhetorically in 1920. In the *English Review*, Sir Arthur Shipley wrote of the "collapse of a nation": "We are watching the decay of one of the greatest nations of modern times, a nation surpassing in numbers almost all others, and covering a wider continuous territory than any other community of people." According to Shipley, Russia was a prey to infectious diseases, especially malaria, on a colossal scale resulting in millions of deaths. Fridtjof Nansen was especially struck by the orphans, said to number 1.5 million in famine areas. "Many hundreds of thousands of these poor little creatures are wandering about in the towns and in the country, in conditions of incredible suffering, want and destitution. Thousands of them die by the roadside or in the streets from hunger and disease . . ."[13]

Until the Soviets closed their frontiers to those in flight in the mid-1920s, come of the internal refugees managed to escape Russia and find asylum abroad. Individual émigrés had left at the first sight of political trouble; more proceeded to German-occupied territory thereafter and also to neighboring countries, as we have seen. Politically, the most important of these evacuations was in the autumn of 1920, when over 130,000 persons, the defeated soldiers of Wrangel's army, together with their dependents, sailed from the Crimea to Constantinople.

At the time the Allied troops were occupying the Turkish city and provided emergency relief. But the evacuees completely swamped existing facilities, creating an international scandal that the British politician Philip Noel-Baker called in 1927 "one of the black spots in the postwar history of Europe." Within Constantinople everything was lacking: food, shelter, clothing, medicines. British High Commissioner Sir Horace Rumbold arrived on the Bosporus about the same time as Wrangel's men and was apalled at what he found. He wrote to King George V that the streets were filled with starving and utterly demoralized Russians. Lady Rumbold helped organize a British Emergency Committee, which undertook to feed, clothe, and bathe 10,000 Russian women and children. Because conditions were so bad, the Allied High Command kept many refugees on ships, sailing to and fro on the Sea of Marmara. The greatest number finally disembarked at Constantinople; others eventually dispersed to Gallipoli, nearby Greek islands, Yugoslavia, Bulgaria, and North Africa.[14]

Wrangel wanted to keep his men together as a fighting unit; the Allies, beginning to see now that the anti-Bolshevik struggle was hopeless and staggered by the problems of relief, argued that the Russians were simple refugees and no longer an intact military force. They favored dispersion of the émigrés and even proposed repatriation for those desiring to return to Russia. Gradually, the British and French prevailed. But all this took time; meanwhile, cholera, typhus, and smallpox devastated the refugees. Sir Samuel Hoare, then deputy high commissioner of refugees for the League of Nations, attempted to coordinate the relief activities of a variety of agencies. Some 24,000 still remained in the city in March 1922. Evacuees from Russia continued to arrive until the frontiers were closed in mid-1923. That year the Barracks Hospital in Constantinople, made famous by Florence Nightingale during the Crimean War, housed 11,000 destitute refugees. In January 1924 the Turks handed over to the Soviets the Russian embassy and consulates, rendering the situation of anti-Bolshevik refugees extremely precarious. Thereafter the few remaining Russians melted into the local population or escaped as best they could.

Throughout Europe the Russian exiles made their way. In June 1924, Nansen reported to the League of Nations that there were over a million Russian refugees abroad. Emigrés were now abandoning their temporary locations on the frontiers and had started moving westward, tired of waiting for the Bolshevik collapse. Meanwhile, the Soviet government began to exile troublesome dissidents on its own; these included intellectuals and professional people, politicians and officials, all of whom were supposedly unwilling to adapt to the new order. From the beginning some of the exiles went to the Far East. Colonies of émigrés existed in Turkestan, Manchuria, Mongolia, and China. Principally, the Far Eastern emigration centered on Shanghai and Harbin. About 60,000, according to Nansen, were in China in mid-1924. Others went to the Middle East—to Syria, Palestine, and elsewhere. Nansen also estimated that there were about half a million refugees in Germany which had more of them than any other country. France came next with 400,000; Paris was undoubtedly the political capital of the refugees, with the greatest variety of émigré groups, the most intense degree of organization, and the most energetic anti-Bolshevik activity. According to Nansen, Rumania had about 80,000; Poland, 70,000; Yugoslavia, 45,000; and Czechoslovakia, 27,000 refugees.[15]

At the time, Nansen's estimates were considered extremely modest; close study by experts in the 1930s suggested, on the contrary, that his figures may have been too high. When the first estimations were made, everyone was overwhelmed by the scale of the Russian catastrophe and thus disposed to accept evaluations of two or even three million Russian émigrés.[16] Subsequent investigation indicated that many of the refugees were counted more than once as they moved from place to place. Research undertaken for Sir John Hope Simpson's authoritative survey of 1939 pointed to approximately three-quarters of a million Russian refugees in 1922; this figure did not, however, include those in the Far East. In 1926 the Refugee Service of the International Labor Organization also counted 750,000 Russian refugees although this excluded Germany, for which statistics were not available. Clearly, by this time, the numbers had started to fall: there was an excess of deaths over births, common in refugee communities; some émigrés migrated overseas or for other reasons were not counted as refugees; and some returned to the Soviet Union.[17] We may conclude that the refugees from Bolshevik Russia probably numbered close to a million at the highest point; the total fell quite substantially in the late 1920s as the situation stabilized and as the refugees accommodated themselves to the postwar situation.

Jewish Refugees in Eastern Europe

Although flight from the former tsarist empire involved every nationality, Eastern European Jews probably experienced greater upheaval than any other group. The war in the east had been fought, for the most part, over the Pale of Settlement, where Jews had been concentrated for over a century. On both sides, Jews had been singled out for deportation during the course of the fighting. Then came a series of pogroms in Poland, the Ukraine, Russia, and Hungary. Jews caught in this maelstrom faced the bloodiest disaster of modern Jewish history to that point. During its course hundreds of thousands were displaced, and when it was over, Jews resumed their exodus from Eastern Europe, which had been interrupted in 1914. Now, however, the Jewish disaster prompted a refugee crisis that engaged a good part of Eastern Europe.

The displacement of Jews began in the early stages of the war with their systematic persecution along the western borderlands of the tsarist empire. The Russian High Command believed that Jews living there constituted a particular security problem and had to be removed. Nourishing this evaluation were the powerful traditions of Russian antisemitism in the military and also the soldiers' nervousness at the prospect of a German breakthrough. High-ranking officers as well as local commanders feared that the

Jews would welcome a German victory. The Jews' Yiddish language, a German-Jewish dialect, prompted misunderstandings about their loyalty. Before long, rumors spread to the effect that Jews were spying on behalf of the Central Powers or were engaged in smuggling across the battle lines. The army began to deport Jews in 1914, and before long a policy of widespread eviction covered entire regions. By early 1915 soldiers were forcing Jews out of ghettos throughout White Russia, Lithuania, and the Ukraine. Evacuations were carried out abruptly and with stunning brutality. In May 1915, for example, the entire Jewish population of Kovno, about 40,000 people, was ordered to leave within forty-eight hours. Vandals accompanied their departure with arson and pillage. Within a few weeks all the Jews living in eastern Lithuania and the province of Courland were expelled. Some 600,000 Jews were uprooted; thousands were kept by soldiers as hostages, and many others were set on by rampaging troops.[18] Liberal deputies in the Duma protested loudly. The French ambassador in St. Petersburg, Maurice Paléologue, was horrified at the consequences of the expulsions of Jews in the winter of 1914–15: "Hundreds of thousands of unfortunates have been seen wandering across the snow," he wrote in his diary, "driven like cattle by squads of Cossacks, abandoned in distress in railway stations, parked in the open on the outskirts of cities, dying of hunger, of exhaustion, of cold."[19]

The military wanted to expel the Jews from their jurisdiction. At first deportations were simply to major nearby cities. At the end of 1914, for example, about 80,000 Jewish refugees jammed into Warsaw. Jews from the western Ukraine were dispatched to Kiev. Later, particularly with the great retreat in the spring of 1915, Jews were dispersed throughout the Russian interior, where official censuses lost track of them. Only after hostilities ended and with the collapse of the regime could Jewish relief agencies evaluate the situation. In 1918, considerably after the Brest Litovsk Treaty, the Central Committee for Relief of Jewish War Sufferers noted that over 211,000 Jewish refugees still remained in their places of exile within the Russian interior.[20]

Jews in the domains of the Central Powers also took to the roads. Galician Jews who lived under Austrian rule were terrified at the prospect of falling into Russian hands; between 200,000 and 300,000 fled westward at the beginning of the war. In the autumn of 1915 the Austrian Interior Ministry reported 137,000 refugees in Vienna, of whom about 60 percent were Galician Jews.

Most of these returned home after the Austrians had broken the Russian advance and had retaken most of Galicia. Despite the best efforts of the tsarist military, large numbers of Jews found themselves on the German side of the front after the 1915 German offensive. Eventually, the Germans engulfed Lodz, Vilna, Warsaw, and other major Jewish population centers. The Germans raided Jewish communities for laborers, forced loans, and requisitions. Some 35,000 Jewish farm and factory workers were deported to Germany from Poland and western Russia.[21]

The end of the war brought no relief. Rather, the collapse of the Central Powers opened an even grimmer chapter in the troubled history of the Jews in Eastern Europe, marked now by wild outbreaks of popular hostility—an epidemic of pogroms such as the Jews had not seen since the seventeenth century. Between 1917 and 1921, according to one scholar, there were more than 2,000 anti-Jewish riots in the affected regions. In Russia and the Ukraine alone, these made about half a million Jews homeless, destroyed 28 percent of Jewish homes, killed more than 30,000 people, and may have been indirectly responsible for five times that many deaths. Jewish existence was paralyzed, and it took over a decade for their economic life to recover. In the Ukraine, the focus of much of the destruction, the Jewish share of the total population fell from 8 to under 6 percent. Although the overall population increased by 36 percent between 1897 and 1926, the Jewish population actually declined by almost 5 percent.[22]

Attacks on Jews occurred in Poland, in Galicia, in Russia, as well as in the Ukraine. Polish troops who captured the city of Lvov from the Ukrainians in November 1918 fell on the local Jewish community, killing and looting. Elsewhere in Galicia, in contrast to the days of the Austrian kaiser, soldiers and civilians ran amok, murdered Jews in the streets, set Jewish property ablaze, and humiliated Jews in public. These assaults continued through 1919 and 1920. Polish nationalism suddenly appeared in close association with anti-Jewish hatred, and with the Russo-Polish War the tempo increased. Jewish officers who had volunteered for the Polish army were promptly interned by the Poles in concentration camps, the general assumption being that all Jews were traitors.[23] In Soviet territory the White armies often lashed out at Jews. We have already noted the intensity of rioting in the Ukraine. Sometimes counterrevolutionary groups hit Jews who had already been victimized by Ukrainian nationalists and the Soviets. Caught in a bitter civil war, Jewish casualties were even higher than in Poland, especially among those returning to their homes after wartime exile in Siberia. Although Bolshevik authorities denounced this violence as counterrevolutionary, they were unable to establish order.

In Hungary as well, counterrevolutionary forces took a heavy toll of Jews. Behind the disorders was the struggle to determine who would rule the reduced Hungarian state. Communists led by Béla Kun, a revolutionary of Jewish origin, imposed a communist regime and then were swept away by conservative nationalist elements who struck out at Hungarian Jews. Pogroms occurred in some fifty towns. Emerging victorious from the contest with communism, the Hungarian ruling class, led by Admiral Miklós Horthy, officially stamped the new order with antisemitism. Once supportive of the Jewish minority, the Hungarian leadership now embraced the anti-Jewish pattern of Eastern Europe. In the autumn of 1920 the new government passed a numerus clausus statute, the first anti-Jewish law of the interwar period.

Throughout Eastern Europe Jews were now on the move. Projecting an image shared by many outsiders, Violetta Thurstan wrote that "Jewish refugees do not suffer so acutely from the terrible homesickness that attacks the refugees of other countries; they are wanderers by nature, and are not so rooted to one particular soil as those with a heavier sense of nationality."[24] In October 1921, Nansen indicated that there were 200,000 Jewish refugees—and the tide was only just beginning.[25] From the Ukraine, Jews crossed the Dniester River to reach Bessarabia, now in Rumanian hands. Those who could continued on to the Regat, the heart of the expanded new Rumanian state. The liberal prime minister, Ion Brătianu, informed Woodrow Wilson in 1919 that the Jewish problem in his country was like the "Yellow Peril" in the United States—the result of excessive immigration of unassimilable foreigners. In August 1921 the Rumanian ambassador to Switzerland told a conference on refugees in Geneva that the total number of refugees in Rumania exceeded 100,000, of whom 95,000 were Jews. Although he probably exaggerated the Jewish proportion in order to press for their continued migration westward, there is no doubt that the Rumanians felt swamped. A year later, after continuous Jewish transmigration, a League of Nations agent in Bucharest still reported 45,000 Russian Jewish refugees in the country.[26] Poland received the greatest number of Jews, some repatriates returning to what they hoped would be their homes after years of wartime deportation, some in flight from anti-Jewish attacks further east, some fleeing Bolshevik rule, and some pursuing their goal of emigration for other reasons. In 1921, when the frontier between Poland and Russia reopened, some 100,000 Jewish refugees were reported in Poland.

Jews also entered the Free City of Danzig, an important way station for emigrants from Russia and Poland. Between 1920 and 1925, 60,000 Jews passed through its port heading westward. Jews came to Czechoslovakia, often directly from eastern Galicia, concentrating in Prague. A Jewish relief organization there, for example, noted 10,000 Jews in transit during a three-month period in mid-1921.[27] Germany received large numbers of Jews on their way to America, as did Belgium, Holland, and France. Agents of Western shipping companies sometimes assembled transports of Jews in Warsaw and Kovno, sending them directly to the docks at Hamburg, Bremen, Rotterdam, or Antwerp.

This lively traffic of Jewish refugees worried authorities in receiving countries. The American government, which had generally been imposing immigration restrictions at the time, closed off most transatlantic migration with the Johnson Act in 1924. A Canadian Order in Council of the previous year moved in the same direction, and other countries followed suit. The Jews were to remain, therefore, on the European continent. Arthur Ruppin reported that the flow of Jews overseas, dammed up during the war, reached 136,000 in 1921; thereafter, overseas migration dropped substantially, averaging 78,000 annually in the period 1922–25.[28]

Even before the effects of these restrictions were felt, the Polish and Rumanian governments moved against Jewish refugees. From Bucharest, authorities threatened summary expulsion for 10,000 fugitives who had reached Bessarabia, and the Rumanians began wholesale removals of Jews from border communities. Of 22,000 Jewish refugees in Rumania in 1921, only 4,000 were allowed to prolong their stay. Somehow the others were to move on.[29] In July 1921, eager to stop the influx from Russia, Poland closed its eastern frontiers to migrants. The government then notified the Polish Jewish community that unless the refugees then present in the country emigrated westward within a fixed period, they would be expelled eastward. Expulsions, indeed, were already under way. The Poles drove stateless Russian Jews, unacceptable to the Soviets, over the frontier to Danzig. Officials in the Free City similarly wanted no part of the Jews and sent them back. Jewish refugees in Poland clamored to leave the country: the American consulate in Warsaw, for example, was besieged on a daily basis by Jews hungry for visas. About 100,000 managed to emigrate from Poland during this period, but many remained behind, without papers and without the possibility of going elsewhere. At the end of 1922 the Poles once again threatened expulsion and designated 6,000 helpless and stranded Jews. League of Nations representatives intervened and secured a reprieve for one year. Meanwhile, Galician Jews who had reached Austria instead of Poland faced a similar situation. In 1920 the Austrians decided on expulsion to Poland. When the Poles complained and refused to receive the Jews, the League of Nations Council broke the log jam, permitting the Jews to remain. On a vastly smaller scale, the right-wing government of the German province of Bavaria joined the bandwagon in 1923: notices went out to eject some 180 families of Jews from Poland and Austria. Despite the objections of the German government and international protests, the Bavarians actually carried out their threats.[30]

In the early 1920s the Jews of Eastern Europe thus presented a refugee crisis of major proportions. The president of the Alliance Israélite Universelle, the French scholar Sylvain Lévi, appealed to the League in December 1920:

> From Odessa to Vilna, a multitude of people, maddened by their sufferings, are appealing for help, and in despair, are preparing to abandon their homes. The countries on the other side of the Atlantic are watching with alarm the arrivals of the first batches of immigrants. . . . The problem is fundamentally an international one. . . .[31]

But the League of Nations, still a fledgling institution, was hardly in a position to intervene forcefully. Sometimes successful in single, well-publicized confrontations between states, it was scarcely prepared to address the general conditions of homelessness and had no capacity to meet the refugees' material needs. Moreover, as we have seen, the Jewish refugee crisis coincided with a major exodus of dissidents from revolutionary and famine-stricken Russia. Jewish organizations alone were able to provide aid to Jews on an international scale.

As in the period following 1882, Jewish relief agencies on the spot offered rudimentary assistance and helped subsidize emigration abroad. Jewish communities in Rumania, Poland, Germany, Czechoslovakia, and Austria, as well as those in Western

Europe, mobilized for this kind of work. More than a dozen Jewish societies operated on an international level, including the Alliance Israélite Universelle, the Jewish Colonization Association, the Hilfsverein für Deutsche Juden, and others. Among the most important was the American Jewish Joint Distribution Committee, a philanthropic body established in the United States in 1914 precisely in order to succor the Jews of Eastern Europe. In 1920 the Joint, as it was called, launched its Department for Refugees and Repatriates under the direction of Bernard Kahn. Offices sprouted in Poland, Rumania, the new Baltic states, and Central Europe. In eastern Czechoslovakia, for example, where Jewish refugees from the Ukraine were in deep difficulties, the Joint set up its headquarters in the town of Munkacs in Subcarpathian Ruthenia, sponsoring 150 local committees in the surrounding region. Such committees not only funneled aid from America, but they also stimulated self-help projects and interceded with local authorities on behalf of their Jewish clients.[32] Between July 1921 and April 1923, the Joint assisted some 300,000 homeless East Europeans to emigrate, repatriate, or settle where they first found refuge.

The Jewish Colonization Association (ICA) also ministered to Jewish refugees. Established in 1891 by the Paris philanthropist Baron Maurice de Hirsch, this British-based body assisted the emigration of East European Jews and sponsored colonization schemes, mainly in Latin America. Galvanized by the emergency, the ICA assembled a special conference in Brussels in June 1921 of twenty leading Jewish emigration committees, securing their agreement to work together to assist refugees. At the same time, the organizers worked to limit the flow abroad and to prevent a general Jewish exodus from Eastern Europe. An ICA press release in mid-1921 made this important point clear: "The main task . . . must be to deal with urgent cases and . . . no steps should be taken which might be calculated to stimulate unnecessary migration."[33] As we saw in the case of pre-World War I Jewish emigration, philanthropists feared that their work might help provoke an emigration flood. The ICA conferences did, however, assure the emerging League of Nations High Commission for Refugees that it would assume all financial responsibility for Jewish refugees. That same year another conference, in Prague, gave rise to the United Committee for Jewish Emigration, known as Emigdirect. This body secured transit permission for Russian refugees from states bordering the Soviet Union and reached an agreement with the Soviet authorities regarding further emigration. Emigdirect cleared the way for overseas travel and explored settlement possibilities, notably in South America, Canada, Australia, and South Africa. The latter activity was especially important after the restrictions on immigration to the United States in 1924. Together with the American-based Hebrew Sheltering and Immigrant Aid Society (HIAS), Emigdirect and the ICA joined together in 1927 to form HICEM, taking its name from the initials of all three: this was an international Jewish body for pursuing immigration and settlement, with its headquarters in Paris. Six years later, with the rise of Nazism, this organization was to cope with a new flood of refugees.

Empires in Ruin: Refugees and the Peace Settlements

To the diplomats and statesmen assembled in Paris in 1919, perhaps the most important task was redrawing the map of Europe—a terribly complicated process, both in strategic conception and in tactical detail. Weighing on the draftsmen were not only the political, economic, and military considerations commonly associated with the negotiating process, but also a set of ideals. Broadly speaking, the peacemakers hoped to produce a just, stable, and lasting peace, consistent with the interests of nationalities. They believed that national rivalries and frustrated national aspirations had done much to generate prewar tensions and that the new arrangements had to be in the interests of the peoples concerned. Harold Nicolson, then a young member of the British delegation, remembered his own intoxication with such notions: "We were journeying to Paris, not merely to liquidate the war, but to found a new order in Europe. We were preparing not Peace only, but Eternal Peace. There was about us the halo of some divine mission. We must be alert, stern, righteous, and ascetic. For we were bent on doing great, permanent, and noble things."[34]

Notably, the leading thinkers of 1919 hoped to avoid some of the more egregious assaults on particular group interests and thus preclude the precipitous flight of refugees that sometimes occurred when territory changed hands. Geopolitical changes in general, it was assumed, would satisfy national aspirations to the greatest possible degree. Frontiers of states would run along the lines of nationality, and inhabitants would accept the jurisdiction of their compatriots. Inevitably, some people would find themselves on the wrong side of the ethnolinguistic divide; for such as these, strenuous efforts were required to prevent the kind of refugee situation that, for example, plagued the Balkan peninsula during the previous decades. To address this problem, the Paris treaties included a very old provision in international regulations permitting those living in affected areas to choose their subsequent national allegiance. If they identified with another state than that to which the territory in which they lived was assigned, individuals had twelve months in which to leave; and if they did so, they could retain their property. On the other hand, in some cases people were allowed to choose the nationality of the annexing country even if they were not living there at the time of annexation. Against the primitive claims of *raison d'état*, therefore—the contention that governments had the right to dispose of entire populations for political purposes—the peacemakers generally accommodated the claims of individuals, who could define their own national allegiance and choose where they wanted to live.[35]

For example, by the provisions of the Treaty of Saint-Germain, signed by Austria, a former Austrian subject living in land assigned to Italy could normally claim Austrian nationality, move to the new Austrian state, and assume full citizenship rights there. According to the Treaty of Versailles, people of Polish or Czech background from territory remaining part of Germany could, nevertheless, opt for Polish or Czechoslovak nationality. Germans in Poland or Czechoslovakia born within the new Polish or Czechoslovak republics, on the other hand, could choose the nationalities of those states irrespective of the wishes of those governments—which were understandably nervous about acquiring large German minorities. Across Europe, people now had choices to make to a greater extent than ever before according to elaborate rules written into treaties and given the force of law within signatory states.

In addition to these provisions, the postwar treaties directly addressed the issue of national minorities. Millions of people, it was recognized, would continue to live in a culturally, linguistically, or religiously alien environment. Of course, the treaties reduced their numbers. The migration expert Joseph Schechtman estimated that some sixty million Europeans were ruled by an alien jurisdiction before the war; after the peace settlements, he claimed, this number fell to between twenty million and twenty-five million.[36] In the past, the persecution of these minorities had caused many of them to seek refuge in neighboring states and had been a source of international instability; henceforth, it was hoped, special minorities treaties would guarantee elementary rights to such people and thus eliminate a historic injustice. Overriding vigorous objections from the governments concerned, the great powers obliged Poland, Czechoslovakia, Greece, Yugoslavia, and Rumania to sign treaties providing minorities with basic human rights and, in the case of Poland, specific protections for the Jewish minority. The newly established League of Nations took responsibility for supervising the enforcement of these treaties and was empowered to receive petitions about real or prospective violations.

The immediate postwar period, therefore, saw an international climate apparently conducive to the resolution of refugee problems. Long-range difficulties appeared, of course. The entire framework of guarantees to minorities was only as strong as the postwar settlements themselves; the enforcement machinery involved an experimental leap into internationalism, requiring for success a measure of agreement among the Great Powers. Nevertheless, in formal terms, the peacemakers seemed to have done well.

In practice, however, severe refugee problems arose from the beginning. Despite the careful attention paid to minorities and to the status of persons whose country was transferred to a new state, some people still found themselves made homeless by the settlements. Notably, with the dissolution of three multinational empires, there were some individuals who would not or could not assume the nationality of a successor state. If there were no other state to accept them, these became stateless persons, known as *apatrides* or *Heimatlosen*. Usually, such people had been buffeted about the continent in the course of recent upheavals and finished the war far from their place of birth. One such case reported to a Quaker relief worker in Spain was of a man born in Berlin, but of Polish origin because of his Polish parents, who was technically designated *apatride;* he claimed Ukrainian nationality, but was claimed in turn by the Russian government for repatriation and service in the Red Army.[37] According to Hannah Arendt, for whom the experience of people thrust outside the boundaries of legal protection was particularly ominous, these *Heimatlosen* were mostly Jews who "were unable or unwilling to place themselves under the new minority protection of their homelands." But there were many others. Polish workers in Belgium, to take another example, sometimes found themselves made stateless when they failed to acquire Polish nationality according to the terms of the 1921 Treaty of Riga, which settled the Polish-Soviet frontier. Such cases illustrate the appearance before the international community of large numbers of people who simply did not fit into the legal and political categories negotiated after the war. At any moment these stateless persons could become refugees, and many defined themselves as such from the moment the war ended, given the reluctance of any state to accept them.[38]

Throughout Central Europe, moreover, the postwar era saw a new generation of refugees resulting directly from territorial changes. The humiliated Weimar Republic, for example, received close to a million refugees. Germans poured across the new frontiers from Alsace-Lorraine, now reattached to France; from northern Schleswig, which went to Denmark; from Eupen and Malmédy, now joined to Belgium. In the east, nationalism and the turmoil of postwar politics forced about half a million Germans to move. Polish nationalism prompted a series of economic boycotts against German shops and businesses in the western part of the country. Ethnic Germans left the former provinces of Poznan and Pomerania, now part of the Polish Corridor, and the port of Danzig, also separated from Germany by the Treaty of Versailles. From Upper Silesia, where a Polish uprising and civil strife followed a 1921 plebiscite supporting a return to Germany, refugees moved westward to the Reich, and more followed when this area was partitioned in 1922, giving Poland a substantial portion. Other German refugees came from the new Baltic states—Lithuania, Latvia, and Estonia. Within the former Russian empire, ethnic Germans from these regions, some of whose families had settled there centuries before, had been among the fiercest opponents of self-rule by the local nationality; in 1920, when the independence of these states was assured, their nationalism was often accompanied by intense anti-German feeling. German aid to these refugees, known as *Flüchtlingsfürsorge*, was a mammoth undertaking and signaled official state involvement in refugee matters on a previously unimagined scale. The government maintained camps and placement offices, looking both to the initial care of the refugees and to their eventual resettlement in Germany. At the same time, the Germans faced massive problems associated with the repatriation of some two million Allied prisoners of war and hundreds of thousands of forced laborers who had been deported to Germany during the conflict.[39]

Nowhere were the issues posed by postwar refugees more acute than in the truncated Hungarian state that emerged from the Treaty of Trianon, signed in 1920. With the collapse of Hapsburg authority, the Austrian and Hapsburg components of the former empire went their own ways. Hungary emerged from the peacemaking process a shrunken, landlocked remnant of former glory, one-third of its previous size, with one-half of its prewar population. This imposed settlement triggered a massive exodus of Magyar-speaking loyalists from the sizable areas lost to Rumania, Yugoslavia, and Czechoslovakia. In Budapest, the Hungarians considered the new borders an unpardonable affront, and each Hungarian refugee a grievance to be nursed in the cause of revising the hated settlement. The National Administration for Refugees, created by Hungarian authorities in 1919, carefully kept track of the fugitives as they crossed the frontier.

In 1921 the Hungarians announced that about 234,000 had come: 139,390 from Rumania, 56,657 from Czechoslovakia, and 37,456 from Yugoslavia. The exodus from newly Rumanian Transylvania was particularly important and followed an active campaign by Bucharest to integrate this huge new province into the Rumanian kingdom. Sharp debates over this issue ensued in Geneva, where the League of Nations Council rang with accusations from both Hungarian and Rumanian sides. Meanwhile, as we have seen, refugees moved to the rhythms of civil war in Hungary itself. The advent of Béla Kun in the spring of 1919 prompted a wave of emigration escaping communist revolution; the crushing of that experiment sent a current of implicated commissars in the same direction, spurred on by the White Terror of the nationalist right.

In an effort to stem the tide and put pressure on their neighbors, the Hungarians closed their frontiers to new arrivals in 1921 and 1922. Battered by the war, Hungary labored to absorb tens of thousands of disgruntled refugees. These ranged from the cream of Magyar society—former Imperial officers, administrators, and estate owners—to far more humble fugitives, swept up in the bitter turmoil of the Hungarian nation. Hundreds of immigrants remained for months and even years in the railway stations where they arrived. Reports reached the capital of refugees stranded in frontier towns, preying on orderly communities and a heavy charge on charitable services. Three years after the ar-

EUROPEAN BOUNDARY CHANGES AFTER WORLD WAR I

After William R. Keylor, *The Twentieth Century World: An International History* (New York, 1984).

mistice of 1918, 18,000 refugees still camped in Budapest and its vicinity, unsettled after their original flight. Among these people embers of irredentism glowed brightly into the 1930s; demagogues continually addressed their grievances, and by the latter part of that decade many had found their political home in fascism, notably with the Arrow Cross movement of Ferenc Szálasi.[40]

These were the most important, but not the only refugee movements within Central Europe following the postwar treaties. In addition, the Austrian Republic received many thousands of displaced wanderers, often German-speaking former Imperial civil servants and military officers from various parts of what used to be the Hapsburg domain. Some came from Bukovina, once ruled from Vienna but now part of Rumania, or from Czechoslovakia and Yugoslavia. As in the case of Hungary, imperial notables frequently reached Austria penniless, with their political worlds destroyed and the social fabric of their existence torn apart.

Attentive as the treaties were to the needs and wishes of individuals caught in an alien geopolitical environment, they could not address every anomalous situation, every combination of circumstances. Inevitably, some people fell between the stools—born in the wrong village, speaking the wrong language, naturalized at the wrong date, finishing the war in the wrong part of Europe. Many such people became fugitives, joining those who found the carefully drafted political arrangments an intolerable affront to their national sensibilities: Yugoslav minorities, for example, who could not accept the centralist constitution of 1921 and its concomitant Serbian hegemony, or Germans from Czechoslovakia, who preferred Austria or the Weimar Republic to life under the Czech president Thomas Masaryk. But Central Europe was basically able to deal with these refugees, to feed, shelter, and resettle them, however great the difficulties in doing so. In the Balkans, on the other hand, or in the poverty-stricken Transcaucasian republics, refugee catastrophes of unprecedented magnitude were unfolding without any prospects of effective short-term relief.

Armenian Refugees

The end of the First World War brought the peacemakers face-to-face with the Armenian question, an issue that for Westerners had been singularly obscure. By 1919 hundreds of thousands of Armenian refugees wandered across territory whose disposition was uncertain and where Britain, France, the United States, and Russia maneuvered for a say in the future settlement. On one level the encounter was philanthropic: charitable agencies, especially from America, became intensely involved in the rescue of Armenians; and a few local governments had to cope with the arrival of thousands of utterly destitute immigrants. But the question was also political; in Paris, Armenian representatives eloquently presented the case for a national homeland for their people and demanded the rectification of historic injustices. For a short time, as we shall see, an independent Armenian republic came into existence. Then followed more disasters and more refugees. In 1924 the League of Nations finally assisted those who had managed to flee to Europe or the Middle East, perhaps 320,000 in number. By then the Armenians had suffered an unprecedented tragedy, and the chance to resolve their problem on a national basis had been lost.

The Armenians were a Christian minority within the Ottoman Empire whose lands had come entirely under Ottoman rule in the sixteenth century. Heirs of a proud tradition of cultural and national independence that flourished before the Christian era, the Armenians once formed a buffer between Rome and Persia and then between the Arabs and Byzantium. Over the centuries the Armenians maintained their identity and kept a large degree of cultural autonomy within the Ottoman Empire. During the last quarter of the nineteenth century there were between 1.5 million and 2 million Armenians scattered among some 35 million Ottoman subjects, but living mainly in Anatolia and what has been called the classic Armenian plateau in the eastern part of the empire, in a mountainous region dominated by the majestic Mount Ararat. Relatively prosperous, highly literate, and well-placed within the Ottoman administration, the Armenians were also evident in the Imperial capital of Constantinople, where their elite were closely integrated into the established order. But following the accession to the sultanate of Abdul Hamid II in 1876, the repressive apparatus of Ottoman rule clashed increasingly with Armenians, an inevitable result of the aggressive Ottomanization policies of an insecure and progressively decrepit central authority. Defeat by the Russians in 1877–78 seems to have turned the Turks against their hitherto "loyal people," as the Armenians were known; thereafter the Turks became obsessed with the campaign of an Armenian minority for national independence.

Massacres of Armenians occurred in the years 1894–96, part of the deepening quarrel between the beleaguered central authorities and an emergent Armenian nationalism. Thousands died afterward of starvation and destitution; others fled across the Ottoman frontiers to the tsarist empire or to Persia. Opinions hardened on both sides, and the confrontation intensified in the few years before the First World War. In 1907 Armenians hoisted the banner of revolt and called for liberation through a revolutionary party, the Dashnaktsutiun. The next year the Young Turks established themselves in the Ottoman capital, soon outdoing the old regime in nationalism and brutal repression. In 1909 about 30,000 Armenians were killed.

During the First World War attacks on Armenians assumed genocidal proportions. Massacres began in 1915, when the Turks, then fighting on the side of the Central Powers, accused the Armenians of aiding the Russian invasion in the Caucasus. The deathly drama began with the elimination or murder of Armenian intellectuals; then followed deportations and periodic slaughter designed to remove entire Armenian communities from Turkish Armenia and Asia Minor, sending them southward across arid wasteland from which they were not expected to return. Franz Werfel later described the significance of the deportations in his arresting novel *The Forty Days of Musa Dagh*:

> Though the government had always arranged such massacres it had never admitted having done so. They were born of disorder, and vanished in disorder again. But disorder had been the best part of such rascally business, and the worst to fear of it had been death. Banishment was a very different story. Anyone might think himself lucky who was released from it by death, even the most cruel. Banishment did not pass like an earthquake, which always spares a certain number of people and houses. Banishment would go on until the last Armenian had been slaughtered, died of hunger on the roads, of thirst in the desert, or been carried off by spotted fever, typhus, or cholera. This time it was not a case of unbridled, haphazard methods, of whipped-up blood lust, but of something far more terrible—an ordered attack.

We shall never know for certain how many perished in this cataclysm, which continued into 1916 and even beyond. In October 1915, when the operation was still under way, the British statesman Viscount Bryce estimated the dead at about 800,000; there seems to be some agreement that about two-thirds of the Ottoman Armenians perished, with serious dispute turning on the total Armenian population before the war. The number killed appears to be close to a million or even more. As early as August 1915, in a chilling observation that has become fixed in the history of genocide, the Turkish interior minister, Talaat Bey, declared: "The Armenian question no longer exists."[41]

Despite these horrors, many Armenians were able to flee. They followed the path that tens of thousands of Armenians had trodden before during the periodic massacres of the late ninteenth and early twentieth centuries. The greatest number moved northeastward to tsarist Russia, to the Caucasus between the Black and Caspian seas; others went south to Syria, and a few thousand straggled into Egypt. Some of the refugees reached Constantinople, the heart of the empire, where they received help from foreign charities. Relief workers telegraphed to the West shattering descriptions of their arrival in the Caucasus in 1915: "There seems to be no end to these solid columns moving forward in a cloud of dust. The majority are women and children, barefoot, exhausted, and starving." American missionaries and medical personnel were utterly overwhelmed by what they encountered:

> The stream of refugees still flows [went another cable a few days later] but with a slacker current. . . . The situation is extremely harrowing. . . . There is a shortage of bread. The majority of the refugees are ill. . . . In the Echmiadzin School, 3,500 children who have lost their parents are huddled together. . . . Yesterday evening I visited a building and in the big hall I counted 110 babies lying naked on the floor.[42]

Three years after the flood began, an official of the principal American charity, Near East Relief, reported that about 300,000 Armenians were on their lists. From Tiflis the U.S. consul told his superiors that the conditions of the refugees had reached the critical stage, for the fugitives were incapable of caring for themselves. Food supplies were exhausted, local government was without resources, and its authority in any case was in doubt. When the war ended in 1918, chaos prevailed throughout the regions where the refugees gathered; starvation and disease were everywhere, and the trials of many of the survivors were just beginning.

Sympathetic onlookers hoped that an independent Armenian entity, as part of a postwar political settlement, might be able to absorb and care for the hundreds of thousands of refugees. Circumstances seemed propitious. The refugees concentrated in an area in which a power vacuum suddenly appeared. Momentarily, neither of the neighboring imperial powers was capable of preventing a surge for national independence in the Caucasus, and three Transcaucasian nationalities moved quickly into the breach: the Georgians, the Azerbaijanis, and the Armenians. When the tsarist empire began its slide into chaos in 1917, the Caucasian peoples inhabiting lands to the south of the Great Caucasus mountain chain shook themselves loose from Russian control and briefly formed a Transcaucasian Republic, a federation of the three national groups. Attacked by the Turks from the west, this frail structure collapsed in the spring of 1918 and was succeeded by three separate states. Then Turkey, under whose control most of the Armenians had lived, went down to defeat with the Central Powers and signed an armistice with the Allies at Moudros at the end of October. Here at last was the moment for which Armenian nationalists had been waiting. With the Turks subject to Allied occupation and with Russia stunned by civil war, its Bolshevik government still weak and having difficulty asserting its authority, the prospects for the Armenians seemed better than at any time in the recent past.

Unhappily, the three Transcaucasian states had sharply different interests and conflicting claims. In addition, both the Turks and the Russians were soon able to reassert themselves and even to agree with each other over the future of the area. Neither was eager to see troublesome independent states on its periphery, and neither would countenance Allied meddling in so strategically important a region as the oil-rich Caucasus. Working together, a resurgent Turkey under Kemal Atatürk moved in from the west and the increasingly confident and securely established Bolsheviks descended from the north. In the brief period of its independence—from May 1918 to November 1920, when a Soviet Republic was proclaimed in Erevan—the Armenian state found itself at war with its Transcaucasian neighbors as well as with the Turks and Russians. Finally, in March 1921, Turkey and the Soviet Union divided the Armenian plateau by the Treaty of Moscow. The European powers felt in no position to challenge this fait accompli; the Armenians, therefore, watched while politics extinguished their small flame of independence.

For Armenian refugees these events were cataclysmic. Survivors of the murderous policies of the Turks, masses of Armenians packed into the Armenian Republic at the end of 1918. Es-

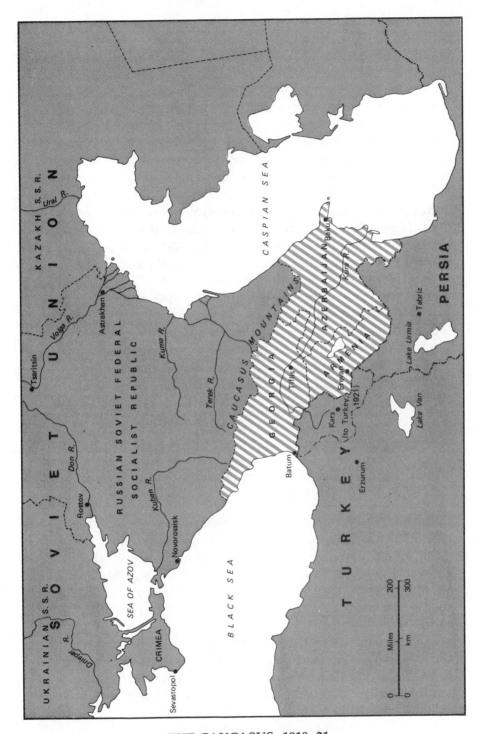

THE CAUCASUS, 1918–21

After Michael Pereira, *Across the Caucasus* (London, 1973).

timates of their number hover around half a million, about half the entire population of the new state. Armenia at the time, moreover, comprised only about 11,000 square kilometers of dry and war-ruined country; much of what Armenians considered historic Armenian land was still under Turkish control.[43] Everywhere but in their truncated state, the smallest of the Transcaucasian republics, the Armenians were pariahs. To the north, Georgia blocked Armenian refugees who sought entry; and in Azerbaijan to the west, Armenians were massacred in Baku, the oil town on the Caspian Sea. Some Armenians tried to return to their former homes in Turkey, but here, too, they were stopped. On the other hand, Armenian refugees continued to reach Armenia from Turkey, tramping into this disaster area at least until 1925. Old rivalries, bitter memories of recent national humiliations, and panic at the prospect of unlimited numbers of mendicants all combined to make the Armenians outcasts. Within their state, on the other hand, the refugees died by the tens of thousands. The winter of 1918–19 brought famine, cholera, typhus, and unspeakable suffering. An Armenian visitor relayed an anguished appeal to the United States during a visit in 1919:

> No bread anywhere. Government has not a pound. Forty-five thousand in Erivan without bread. Orphanages and troops all through Erivan in terrible condition. Not a dog, cat, horse, camel or any living thing in all Igdir region. Saw women stripping flesh from dead horses with their hands today. . . . Another week will score ten thousand lives lost. For heaven's sake hurry!

According to a leading Armenian historian, about "200,000 people, almost 20 percent of the Republic's population, had perished by midyear [1919]. . . . It was verily a land of death." In a melancholy observation at the time the British diplomat Robert Vansittart noted that "the Armenians are dying everywhere at such a rate that there soon will be none left for the future of Armenia."[44]

In Paris, Armenian nationalists urged the Allies to guarantee the infant state and to procure for it the historic homeland still occupied by the Turks and the other two Caucasian states. They also demanded Allied pressure on Turkey to permit Armenian refugees still wandering about the former Ottoman Empire to reclaim their homes and villages. Despite their initial good will, however, and despite their stated sympathy with the Armenian cause, the Allies were increasingly reluctant to maintain their involvement in this anarchic, distant land where both Russian and Turkish leverage could be so much more easily brought to bear. By the end of 1919, the Americans, who had never been at war with Turkey, showed that they were unwilling even to commit themselves to Europe, let alone the politics of this desolate, devastated region. The Armenians, therefore, were abandoned. The British evacuated the Caucasus in 1920, realizing that they had more important interests to the south in Egypt, Palestine, and Iraq. Having committed soldiers to the area in 1918, when the anti-Bolshevik forces were still performing well, the British were disinclined to invest more energy there as the White armies faltered and the costs of such deployment became more difficult to bear. Outside help continued to come mainly from the United States—a meager enough supply, although "sufficient to nourish the parched root of Armenian independence," in the words of a sympathetic observer.[45]

Survivors of the Armenian debacle began to appear throughout Europe in the early 1920s, having managed by luck or persistence to escape the fate of their compatriots. As we shall see, the Turks expelled some 90,000 Armenians to Greece during and after the fighting between the two countries, and many thousands were massacred in the Greek collapse at Smyrna in 1922. Some Armenians left the Soviet Republic, set up at the end of 1920, as revolutionary rule settled on the once-independent state. Armenian patriots organized an uprising against the new order in 1921, but the Red Army retook Erevan before long. Although many Armenian refugees ended their homelessness by emigrating to the Soviet Union in the later 1920s, others proceeded westward. Nearly every major city in interwar Europe seems to have had some Armenian refugees. League of Nations officials became particularly preoccupied with them because of their stateless condition, finally addressed by means of the Nansen passport. Throughout this period close to 225,000 remained outside of traditional Armenian lands, about half of these in Europe. France had the greatest proportion, about 63,000 in the 1930s.[46]

INTERNATIONAL PROJECTS

Immediately following the collapse of German armies in November 1918, Europe braced before the prospect of economic ruin and revolutionary upheaval. Leaders in the victorious states first directed their efforts to the relief of Western Europe and stabilization in Germany, threatened with Communist revolution. Massive shipments of American aid began almost immediately to reach France and Belgium; political and administrative energies focused on peacemaking, demobilization, and the rebuilding of shattered economies. Western statesmen worried about the exportation of Bolshevism and looked to a cordon sanitaire of states to seal the region off from the West. Under these circumstances little governmental concern remained for refugees in Russia, Asia Minor, or the Caucasus. Catastrophes in these distant lands preoccupied philanthropists, of course, but such involvement was unofficial and even eccentric. It therefore took several years for the international community to address the refugee problem and to construct an institutional framework for dealing with it.

Emergency Relief in the East

Before the establishment of the League of Nations High Commission for Refugees in 1921, private agencies not only shouldered the principal burden of refugee aid but coordinated international effort as well. Charitable societies based throughout Western Europe and the United States negotiated with consular officials, politicians, or bureaucracies, moving in where governments feared to tread. Much of the assistance came in traditional philanthropic forms—soup kitchens, emergency medical treatment stations, and orphanages—organized from headquarters in Washington, D.C., New York, London, Paris, Brussels, or Geneva. The Red Cross, the Quakers, the Save the Children Fund, and a cluster of special organizations addressing Russian émigrés were all busy at work in Eastern Europe, as were the Jewish societies whose work we have examined briefly. Hundreds of thousands were kept alive by these efforts, rescued from starvation and disease, particularly in the dreadful winter of 1918–19. The scale of relief operations was unprecedented, involving unheard of sums of money and extraordinary private initiatives on the diplomatic front. But the results were paradoxical. Having kept so many refugees alive during the critical postwar period, the private organizations helped to maintain the pressure of refugee crises. In the long run, this activity helped to elicit a response from governments and from the international agencies set in place after the First World War.

The most important single association devoted to this work was a federation of Christian missionary groups and other philanthropic societies operating in the Balkans and the eastern Mediterranean. At its origin was the call for help from the American ambassador in Constantinople, Henry Morgenthau, Sr., one of the first to grasp the scope of the Armenian massacres. While the slaughter was still under way, Morgenthau called for a private fund-raising effort to save the Armenians who still remained. Responding, a core of respected and influential Americans launched an Armenian Relief Committee, which soon coordinated other private charities and missionary societies both in the United States and in Turkey. At first the committee operated through Morgenthau's office in Constantinople; then, when the Americans broke off relations with Turkey in the last year of the war, the new organization took wing, dispensing relief under its own auspices. As the fighting ended, this organization became known as the American Committee for Relief in the Near East (ACRNE) and was practically alone in the field, dispensing millions of dollars' worth of assistance.

Near East Relief, as it became known, soon operated in twelve different countries in the Balkans, Asia Minor, and the Middle East. ACRNE brought under one administrative roof the disparate efforts of the United States government, the British army, the London-based Lord Mayor's Fund, the Red Cross, a Canadian fund, and a few others. It raised the extraordinary sum of nearly 20 million dollars during 1919 and dispatched research teams throughout Turkey, the Caucasus, Palestine, Persia, and Syria in an effort to evaluate the refugee situation. In the stricken Armenian Republic, according to James Barton, president of ACRNE, the organization supplied 338 towns and villages with flour from September 1919 to April 1920 and fed an average of 332,716 people daily. ACRNE furnished practically the entire hospital services for Armenia—at one time thirty-nine institutions. Between January 1919 and the summer of 1920 (the end of Armenia's independent existence), Near East Relief distributed supplies worth over 28 million dollars to destitute Armenians, of which ACRNE itself contributed over 11 million dollars. ACRNE became especially concerned with orphans and abandoned children: the agency took charge of about 75,000 of these and had

over 30,000 living in its institutions, making Armenia, as Barton
put it, "the largest orphanage center in the world."[47]

Delivering this aid throughout Asia Minor and the Middle East
proved extraordinarily difficult. Authorities in Turkey remained
hostile to the Armenians and deeply suspicious of Christian missionaries, fearing, for example, that Near East Relief orphanages
would be used to Christianize Turkish youth. Travel throughout
the region was hazardous, roads nonexistent or in poor repair,
often little more than animal tracks. The Caucasus remained a
wasteland torn by war, and the geographic situation of Armenia
required that supplies be transported through hostile territory in
Georgia. Food and supplies went by ship to the port of Batum
and then south by rail to Armenia, a perilous trip under the best
of circumstances and complicated even further when the two
countries were locked in battle.

Western governments became much more interested in refugees from revolutionary Russia, partly because they were sympathetic to the defeated counterrevolutionaries. Since 1918 the
Allies had provided munitions, men, and money for each of the
major anti-Bolshevik efforts fighting the Red Army, movements
associated with Kornilov, Yudenich, Denikin, Wrangel, and Kolchak. Counterrevolution finally ran into the sand by the autumn
of 1920, when Wrangel's troops evacuated the Crimea. Yet even
at this point the Allies recognized some obligation to their erstwhile protégés. Another reason for Allied interest was the cultural affinity Westerners generally felt for Russians, whom they
knew something about through literature and occasional travel.
Near Eastern peoples, by contrast, even when Christian, were
largely unknown in popular terms and did not engage European
sympathies to nearly the same degree.

During 1919 and early 1920, the Allies remained preoccupied
with the Civil War. Everything hung on its outcome. Sir Samuel
Hoare, later to become deputy high commissioner of the League
of Nations for the care of Russian refugees, wrote to Winston
Churchill in May 1919 that "for the last six months I have been
convinced that the whole future of Europe, and indeed the whole
world, depends upon a Russian settlement and the destruction
of Bolshevism."[48] Thus, the West considered the first flights from
Russia and the first evacuations of White armies as tactical re-
treats. It took some time to realize that these were not temporary
withdrawals but rather the beginnings of a permanent migration. In Constantinople, occupied by the Allies and a major em-

igration center in the closing phase of the Civil War, British, French, and United States representatives doled out some help directly and also organized rudimentary assistance through the American Red Cross. The Jewish Colonization Association was also on hand. But governments had no concerted policy until 1921, when the League of Nations put some muscle into resettlement and repatriation programs.

Acting on their own, the Americans launched an ambitious scheme for emergency relief to Europe as the war drew to a close and as reports reached the West of massive starvation in Russia. American President Woodrow Wilson asked Herbert Hoover, one of his key advisers and wartime director of the Belgian Relief Commission, to take charge of what soon became "the most massive relief program in history"—the delivery of nearly one billion dollars' worth of goods to twenty-two countries in the nine months following the 1918 armistice. Hoover became director general of the American Relief Administration (ARA), with a staff of four thousand persons, supposed to continue this work in the immediate postwar period.

Early in 1919, Hoover proposed to extend relief efforts to the embattled Bolshevik state, which at that very moment was fighting for its life against Allied-backed counterrevolutionaries and a naval blockade. Taking the model of relief to Belgium during the Great War, Hoover wanted a neutral relief commission to organize aid, headed by the Norwegian polar explorer Fridtjof Nansen and backed by various governments uninvolved in the recent war, supported by the Western Allies. Hoover envisioned that, in exchange for the relief activity, the Bolsheviks would cease hostilities against their enemies and let the Allies distribute aid. Behind the plan lay a broad liberal belief that Bolshevik ascendency depended on popular suffering: when the latter diminished, the regime would weaken and crumble. The key for Russia and for Europe was social betterment, to be achieved through American-style economic development. Nansen enthusiastically put his name to this scheme, which not unexpectedly ran into a blank wall of Soviet opposition. By early May, when the Russians replied, their own fortunes had brightened notably in the fighting against the counterrevolutionaries. Speaking for the regime, Georgi Chicherin held the door open to humanitarian aid but indicated there were too many strings attached to the existing proposal. No international relief would be possible until the Civil War had been resolved.[49]

The following year, when events signaled an eventual Bolshevik victory, Hoover renewed his proposals. Talks between the ARA and the Soviets began in the summer of 1920 and finally led to agreement in August 1921. Nansen agreed once again to take charge of a vast lifesaving effort. For the next twenty-two months, with great stretches of Russia affected by famine, an International Committee for Russian Relief beamed aid to the Soviet Union from a variety of sources in Europe and America. The bulk of the food came from the United States. According to the ARA, this effort sustained some ten million people threatened by famine; the Soviets did not dispute this evaluation, publicly acknowledging how important the supplies had been.[50]

Politics thus put strict limits on the timing and character of help that was extended to the war-ravaged colossus of Eastern Europe, the source of most of the postwar European refugees. Political considerations blocked an agreement to send government help to Russia until 1921 and precluded international action before then on behalf of refugees. Aid still flowed at this time to nominally independent Transcaucasia, as we have seen, although the failure to underwrite the Armenian state politically condemned many of its inhabitants to continued refugee status. ARA support did reach Poland, together with that of the Joint Distribution Committee, the American Red Cross, and other private groups. There were plenty of philanthropic organizations in the field. These could hardly staunch the flow of refugees, however, or invent European havens where they could be settled. Only in 1921, when the postwar international order stabilized somewhat and when the refugee problem appeared as great as ever, could the Allies mobilize international action on the refugees' behalf.

Fridtjof Nansen and the League of Nations High Commission for Refugees

International help finally arrived under the auspices of the League of Nations and Fridtjof Nansen, a Norwegian explorer, scientist, and public figure. Fifty-eight years old in 1919, Nansen was one of the great celebrities of the age. His fame derived from the two prewar decades and the worldwide public fascination with hero-adventurers—men whose achievements were broadcast wherever there were mass circulation newspapers and magazines. In

1893, following a dangerous expedition to cross the Greenland ice fields several years before, Nansen set out for the North Pole. Accompanied by a single fellow explorer, Nansen finally reached the highest latitude then achieved by men—soon described in widely read books, *Farthest North* and *In Night and Ice,* both published in 1897. Following this expedition, Nansen continued scientific and academic work, but also assumed a political role, participating as a Norwegian patriot in the separation of Norway and Sweden in 1905, serving as Norwegian ambassador to Britain in the years 1906–08, and then leading the Norwegian Association for the League of Nations after the war. Tall, athletic, with snow-white hair, clear blue eyes, and a drooping mustache, Nansen of the North played his part well. He was universally admired and respected, believed to be "above politics." At innumerable international conferences in the immediate postwar period, he became a regular fixture, an idealist and visionary, a man of proven personal courage, standing out in a crowd with his famous broad-brimmed hat—always tilted to one side.

Nansen's personal standing made him the ideal candidate to help solve one of the most technically difficult postwar problems—the repatriation of prisoners and internees from Russia, Germany, and the successor states of Austria-Hungary. In Russia there had once been 1.25 million prisoners of war from the Central Powers, of whom more than half perished in the chaotic wartime and revolutionary conditions. Some 330,000 interned civilians also remained in Russia, citizens of the Central Powers living in the tsarist empire in 1914, when they were sent to Siberia. At the same time there were more than a million Russians in Germany in the spring of 1918, men whom the German army once wanted to mobilize against the new Soviet regime. During the first part of 1919, Allied leaders took up this former German project, forbidding the return of prisoners and hoping to launch them against the Red Army after combining them with the White forces operating in the Baltic and southern Russia. Eventually, the Allies decided that the plan was practically unsound, and so, toward the end of 1919, their policy shifted, permitting the resumption of repatriation. The following March, Nansen was called in to negotiate and to organize the process, made awkward by the persistence of the Allied blockade and intervention in Russia and by the reluctance of Western countries to recognize the Bolshevik regime.

Building his own organization based in Berlin, known as Nansen Help, the Norwegian explorer brought the parties to an understanding in 1920. Despite the scarcity of transportation, the suspicions of the Soviets, and continued warfare in Russia and Poland, Nansen won the confidence of the negotiators and carried out the exchange. During the next two years over 427,000 men, belonging to more than twenty nationalities, returned to their homes. Nansen carefully insisted that no one would be sent to the Soviet Union against his will, and many former subjects of the tsar refused repatriation. Establishing a pattern that would become familiar in the interwar period, Nansen drew on the League of Nations only for its prestige and administrative assistance; the actual operation was supported by private agencies. Thus, it was the International Red Cross that organized assembly camps, provided food, set up disinfecting stations, and supplied transport for the repatriates. Utilizing his League of Nations title of high commissioner, Nansen himself raised a great deal of the money needed to finance the operation.[51]

As we have noted, Nansen was simultaneously negotiating with Herbert Hoover and the Americans over famine relief in Russia. By 1920 he was recruiting men for this task and was preparing to implement extensive assistance schemes. (It was in this effort that a footnote in refugee history was earned by Vidkun Quisling, in 1921 one of Nansen's trusted deputies as a relief worker in Russia and later to become the pro-Nazi prime minister of Norway under German occupation in the Second World War.) Throughout this period of tentative humanitarian demarches, Nansen became an almost indispensable mediator between the Soviets and the League of Nations countries.

Given his success in the delicate task of exchanging prisoners, the Norwegian leader then became an obvious candidate for a related project, settling refugees from revolutionary Russia. This emerged as an especially difficult problem for the private agencies that had largely carried the postwar burden of assisting these émigrés. By early 1921, Russians were clogging the streets of Constantinople in desperate straits and were presenting severe problems in Poland, where they swamped existing aid facilities. Many of them clamored to enter or to remain in other European states, and this appeal was broadcast by émigré spokesmen in Paris. But the former subjects of the tsar were often without identity papers, belonged to suspect political groups, and threatened to arrive destitute on European doorsteps without possibility of reemigration. Moreover, the problem seemed likely to worsen with the winding down of civil war and the terrible famine in Russia and the Ukraine. At the end of 1920, some of the philanthropic societies and a few governments believed that the League of Nations, which had not previously been considered a proper agency to treat refugee matters, should become involved. Early in the following year the International Red Cross pressed the appeal of several private agencies on the Council of the League, calling for action. Speaking for the Red Cross, Gustave Ador, then president of the Swiss Confederation, noted that there were 800,000 Russian refugees scattered throughout Europe, bereft of legal protection or representation. The Red Cross envisaged a special League of Nations commission to deal with the Russian refugees, to define their legal status, to assist them where possible, and also to secure their repatriation or employment outside Russia.

Although opinions differed sharply about the implications of such involvement, the Council of the League agreed to act. Having failed to persuade Ador himself, the League of Nations turned to Nansen and asked him in mid-1921 to head its organization to assist the refugees. Only the Russians were really at issue at this moment, as indicated by Nansen's cumbersome title, "High Commissioner on Behalf of the League in Connection with the Problem of Russian Refugees in Europe." As with prisoner-of-war repatriation, this office was to receive administrative support from the League of Nations but no funds for refugee assistance; it was expected that the latter would continue to come from private agencies. The latter were to send representatives to an advisory committee to assist the high commissioner. Principally, Nansen was to negotiate with governments to resettle refugees and also to permit their unimpeded return to the Soviet Union. No one bothered to define "refugees" because it was assumed that only one group was being addressed and that in any case the problem would be solved before too long. The projection was ten years, after which there would be no more need for the High Commission. This last is an important basis for understanding the structures set up at the time. Particularly in his fund-raising efforts, Nansen stressed that a permanent solution was being sought that would eventually involve many refugees returning home.

The League of Nations accepted the new High Commission with some reluctance, member states having in 1921 far higher priorities than an international agency to succor refugees. Nansen's very idealism, indeed, especially some of his grandiose plans for Russian relief, made some members nervous. Prisoner-of-war repatriation and fighting epidemics were acceptable because such projects included no long-term commitments; refugee assistance, however, threatened to drain resources badly needed for postwar reconstruction and implied an obligation to receive unwanted immigrants. F. P. Walters, the historian of the League suggests that the High Commission was unpopular among most member states. The Americans, of course, had rejected the League entirely by the time Nansen's agency was born; the Italians became openly hostile in the following year, when the advent of Fascism prompted an emigration of bitter opponents of Mussolini's new order. Great Britain and the Dominions felt unwilling to receive refugees and feared the High Commission might pressure member states to do so. Only the French and Scandinavian countries offered warm support. Nansen's initial fund-raising efforts were not a great success, and he was unable to persuade member states that they should invest in permanent solutions to refugee problesm. Each year, as if to underscore the lack of League enthusiasm for the refugee agency, the assembly reminded Nansen's administration that it had only a few more years to live.[52]

Despite Nansen's personal prestige, moreover, he had difficulty persuading critics that refugee assistance was not a veiled pro-Bolshevik enterprise. Suspicions arose partly because of Nansen's reputed sympathy for repatriation just when some émigré diehards were contemplating further military adventures. Repatriation of prisoners of war had been the principal diplomatic achievement of the new high commissioner, and Nansen certainly favored return of some of the refugees, as did many governments. The Czech leadership, for example, tied its refugee assistance to a training scheme, hoping eventually to dispatch the newly schooled refugees to rebuild Russia. Nansen supported such approaches, noting simply that the Russian government should be persuaded to receive the trainees at the appropriate moment.[53] The London *Times* attacked this Nansen strategy, noting ominously in 1923 that the Norwegian was on friendly terms with the Bolshevik government and had been made an honorary member of the Moscow Soviet.[54] Russian exile associations in London and Paris took up this charge and even launched an anti-Nansen petition calling for his replacement. In response, the high commissioner stressed that he favored only voluntary return of refugees and was merely acting in the best interests of the émigrés themselves. He emphasized that repatriation was part of a long-term solution of the refugee problem and not a tactical move on the diplomatic chessboard. Tens of thousands of refugees eventually did return to Russia, of course, including large numbers of Jews who ended up in Central Europe after the war.[55]

Nansen's famine relief in the Soviet Union in the early 1920s similarly implied to some that his sympathies were pro-Bolshevik. For a brief moment it even appeared that Nansen might, almost single-handedly, be keeping afloat a political wreck that would otherwise plunge to the bottom. During the years 1922–23, Nansen's agents were working feverishly in the Ukraine, establishing centers for the homeless and starving, apparently feeding between twelve million and fifteen million people. Nansen's brilliant protégé, Vidkun Quisling, practically ruled large areas of the Ukraine in 1922 and later advocated recognition of the Soviet government when he returned to Oslo the following year.[56] Quite apart from humanitarian achievements, Nansen argued that such activity would contribute importantly to postwar European reconstruction and prevent the creation of even more refugees. When Russia could at last feed itself, he argued, very much in the same vein as Herbert Hoover a few years before, trade between East and West would resume, prosperity would return to Europe, and Russia might be cured of her "long sickness" of political extremism. In the long term, Nansen wrote, the League of Nations' efforts on the refugees' behalf "helped to remove centers of disaffection and discontent; and it has helped to raise, by the distribution and settlement of industrious and highly educated refugees, the standards of civilization in various parts of the globe."[57]

Ministering to the Homeless

Nansen's new High Commission sailed immediately into uncharted waters. In addition to the Russian émigrés, other refugees began to appeal to League of Nations representatives, posing problems hitherto unknown in the international community. These were the *Heimatlosen,* or stateless persons, whose peculiar difficulties we have noted in the context of the immediate postwar settlement. In order to understand better the technical issues facing the High Commission, we will find it worthwhile to contrast the circumstances of postwar international travel and citizenship with what the world had known before.

Throughout the nineteenth century there were no serious administrative impediments to the movement of persons between states. The English author Norman Angell remembered his own youth in the 1890s, when he decided abruptly to leave the European continent for America: "I had no passport, no exit permit, no visa, no number on a quota, and none of those things was asked for on my arrival in the United States."[58] Angell simply went. Passports existed at that time, and a handful of states, including the tsarist and Ottoman empires, required them for internal travel. In Russia, the saying went, "No man can exist without a passport." But these documents had largely fallen into disuse internationally, remaining simply as diplomatic instruments to designate persons requiring or requesting special attention. Officials never asked travelers' nationality. Baedeker's tourist guides at the turn of the century advised their relatively affluent readers that no one inspected passports any more—but then suggested that the prudent voyager might carry one anyway. Up to the First World War "civilized" countries considered that no more formal arrangements were necessary to designate people moving from place to place.

All this changed with the war itself. Not only did wartime conditions make some travel hazardous, but states themselves were eager to block the departure of persons with useful skills or of military age. From the standpoint of immigration, too, there were fears of open borders, anxiety about spies, concerns about housing or food shortages. After a short while, formal restrictions on travel settled everywhere on the European continent, joining a whole host of regulations affecting economic and social life. Everywhere, as a result, passports came into use as a way of certifying nationality, regulating the flows of much-needed people, and providing checks on suspicious persons deemed security threats to countries engulfed in war. In the postwar period, passports continued to be demanded at frontiers by a new tier of government officials, who now inquired closely about travelers' identities. As with so many wartime controls, this one was never removed and became a part of everyday life.

Paul Fussell describes the shock and scandal that these little books occasioned among literate British tourists visiting the Continent in the 1920s. Affixed to this compulsory document was: a photograph, a "most egregious little modernism" that shamed or humiliated those who loathed the idea of a banal likeness of themselves being put on display; a reference to "Profession" that was an "open invitation to self-casting and self-promotion, not to mention outright fraud"; and a "Description of the Bearer" that presented unheard-of intimate details describing bodily features. The whole business fed the sense of "the nasty dehumanization of everyone" common to some literati in the postwar era. More practically, the new system of controls provoked a familiar travelers' complaint, the "passport anxiety," or rising nervous tension on crossing international frontiers. Would everything be in order? Would the traveler's credentials be acceptable?[59]

However limited or eccentric such preoccupations, there is no doubt that they marked an important change. Travel was no longer the free and easy passage that it was in Norman Angell's youth. Citizenship, once an irrelevant issue for European travelers, now assumed cardinal importance.

As soon as the wartime dust had settled, liberals attempted without success to lift the passport system. The League of Nations sponsored an international conference in Paris in 1920 on the subject of passports, and from it emerged recommendations to ease existing regulations. A meeting of the Inter-Parliamentary Union in Stockholm the following year heard denunciations of the passport system and calls for freer travel. Other meetings and conferences through the 1920s made similar appeals, but all to no avail. Although some countries, notably France, leaned toward liberalization, others wanted to limit the influx of unwanted persons. Problems of demobilization, dislocation, unemployment, fears of Communist subversion, monetary crises— all these made some governments extremely prudent. Similarly, the extension of state-provided welfare during the war and immediate postwar years increased interest in controls, seen now as serving a vital national interest.

These restrictions complicated the lives of thousands of refugees in the early 1920s and slowed the repatriation of interned civilians released at the end of the war. They posed particular difficulties for refugees from the Soviet Union, who almost invariably fell afoul of some official regulations. Carrying tattered documents issued under the tsars or equipped only with the attestations of local military commanders or municipal officials, these émigrés often found it impossible to conform to immigration procedures, to secure entry or transit visas, or even to move from the place where they first found asylum. Complicating matters further was the practice of former Russian diplomatic or consular representatives who sometimes continued to issue identity papers based on their prerevolutionary authority. Receiving countries were understandably reluctant to recognize these documents, but at the same time hesitated to acknowledge the Soviet successors to these officials. In the resulting confusion, restriction was the safest policy. Characteristically, restriction meant that groups of refugees could be manipulated at will by receiving countries. Poland and Rumania, for example, were eager to get rid of Jews who arrived from the east and freely issued documents authorizing them to obtain visas to go elsewhere. Consuls from receiving countries, however, could refuse to accept these papers, leaving the exiles stranded. Finally, all these measures rendered the lives of exiles precarious, for they were vulnerable to punitive action, even expulsion, if their papers were not in order.[60]

Such problems, moreover, did not affect the Russians alone. Armenian refugees began to appear in Europe at the beginning of the 1920s, similarly lacking necessary documentation. Having lived under the Ottoman Turks, the infant Armenian Republic, or the tsarist empire, they were often the very model of statelessness, coming from regions Westerners hardly knew existed. Often they were unable to establish of what country they were citizens—who they *were* in a juridical sense.

Nansen was desperately concerned to free the refugees from the stultifying immobility of collection points like Constantinople. To achieve this goal the High Commission sought to clear away postwar travel restrictions for refugees. In 1922, Nansen called a conference in Geneva representing sixteen nations and secured agreement on the issuing of a special travel document for Russian refugees to be known as the "Nansen certificate" or "Nansen passport." This document, written in French and the language of the issuing country, was to be administered by competent authorities in each state and used as a normal passport. No state was obliged to receive refugees bearing such certificates, but all parties agreed to recognize them as valid identity papers. Gradually, the usefulness of the Nansen passports expanded. Through an agreement of 1924, thirty-eight states extended the facilities to Armenian refugees. In 1926 twenty-three states agreed to liberalize one vital provision for refugees, that permitting the refugee to return to the issuing country if he or she chose. An arrangement of 1928 permitted representatives of the High Commission to exercise important consular functions, certifying the identity of refugees and their status as defined by international law. By that time fifty-one governments had agreed to issue and to recognize the Nansen document as it applied to Russian exiles. However limited, the Nansen passport was a significant achievement: for the first time it permitted determination of the juridical status of stateless persons through a specific international agreement; at a time when governments and bureaucracies increasingly defined the standing of their citizens, it nevertheless allowed an international agency, the High Commission, to act for those whom their countries of origin had rejected.

More than documents, however, was needed to reduce the flow of postwar refugees. Indeed, the early 1920s saw this flow maintained—to the surprise of practically everyone. Famine and civil war in Russia drove thousands westward across the frontiers of the Baltic states, Poland, and Rumania. Others took advantage of the cessation of fighting with Poland to leave Russia and seek refuge in the West. As the new regime consolidated, the Communists increasingly defined groups of people as "class enemies." Excluded from positions of status and responsibility and with their children forbidden access to state schools, many of them followed earlier waves of refugees. Mennonite communities also began to leave Russia at this time, finding the Communist system intolerable. Although the Soviets finally offered amnesty in 1921 to émigrés who had fled the Revolution, far fewer were repatriated than Nansen and others had hoped. In 1922 and 1923, inspection teams from Nansen's High Commission traveled to Russia to see how repatriates fared in the Soviet Union, hoping that a positive report might encourage the return of former exiles. Their assessments were generally optimistic, seeing whatever difficulties peasant returnees experienced, for example, as being no different from those of other peasants in the Communist state. The Soviets expressed their willingness to work for repatriation with the High Commission and seemed genuinely committed to this process. But the harsh conditions of life under the new order and the continuation of political repression dissuaded large numbers from repatriation.

By the mid-1920s, according to Sir John Hope Simpson, the Russian refugee situation stabilized and points of distribution became clear.[61] The Russians had moved from east to west in Europe, relieving the pressures on Poland and Germany and gradually emptying the refugee pools in Constantinople and the Balkans. France received the greatest proportion of refugees in Europe, about 400,000, having virtually opened her borders to all Russians willing to do menial labor.[62] After a few years, the refugees resigned themselves to not returning home. Moreover, the gradual extension of diplomatic recognition to the Soviet state helped clarify the refugees' situation and made it easier to construct new precise definitions and rights of refugees.

But the mid-1920s also saw new groups of refugees who joined the Russians in appealing to the League of Nations High Commission. Nansen negotiated with Polish authorities in 1923, when the latter threatened to expel Jewish refugees to the Soviet Union, charging that they were unwanted and that too many had already arrived. More problems arose with the first appearance of Italian anti-Fascist exiles and with the Turkish government's denaturalization in 1927 of many Armenians living abroad. In addition, Nansen eventually had to take under his wing a variety of people from Asia Minor and Transcaucasia fitting the broad definition of "Assyrian, Assyro-Chaldean and assimilated refugees"—stateless persons originally from the Ottoman or tsarist empires, to whom the Nansen passport was extended in 1928. The High Commission was, therefore, a busy bureau five years after its establishment, at the very moment when it had once been assumed that refugees would have melted away, gradually assimilating into their host countries or returning to their previous homes.

GREEKS, TURKS, AND BULGARIANS, 1919–28

For Turkey and much of the Balkan peninsula, large-scale warfare began in 1912 and continued substantially beyond the end of the First World War. For over a decade armies tramped across Balkan territories claimed in ancient national rivalries; throughout, masses of refugees swept to and fro, trying to reach safety and attempting to forecast the situation when the guns would finally fall silent. Other refugee pressures, built up during a decade of warfare, burst through the surface at the end of this period, sending hundreds of thousands more into exile, principally to Greece and Bulgaria. Once again, as with the Russian émigrés, Nansen's organization and the Great Powers tried to address these mammoth dislocations. The result was two negotiated population exchanges in the mid-1920s: between Greece and Turkey and between Greece and Bulgaria. Two million people were caught up in the process, monitored and supervised by international agencies. Troubled and incomplete though such agreements were, they nevertheless were preferable to no agreements at all; these population exchanges revealed how, when the principal parties agreed, Europeans might regulate huge flows of refugees in the common interest.

The Greco-Turkish War and the Convention of Lausanne, 1923

For a defeated country, Turkey revealed extraordinary reserves of energy and national resolve in 1919, the year following the collapse of the Ottoman armies and the flight of her discredited government. The decisive force was the nationalistic movement led by Mustafa Kemal, known as Atatürk, the military hero who organized resistance to the dismemberment of Anatolia by the victorious Allies. Turkish policy under Atatürk challenged both the postwar Allied occupation and the establishment of an Armenian state. Despite the pressure of Allied soldiers in parts of Anatolia and Constantinople and the imposition of the harsh Treaty of Sèvres signed with the Allies in 1920, Kemalist forces maintained their élan and set up a provisional government in Ankara with Mustafa Kemal as president. Aided by Soviet Russia, also eager to see the Allies driven from the area, Atatürk energized a program of modernization and nationalistic resurgence. Against him the British, French, and Italians hoped to partition the country, but clearly lacked both the necessary agreement among themselves and sufficient troops. Their war-weariness and their preoccupation with demobilization contrasted notably with the fiery enthusiasm of the Turkish nationalists.

Unlike the principal Allied powers, however, Greece was prepared to take bold action in order to pry loose for herself a part of the shattered Ottoman Empire. Near the end of the war, in 1917, the energetic and skillful Greek premier Eleutherios Venizelos brought his country into the fighting in order to win maximum territorial advantage from the defeat of the Central Powers. A champion of the "Megali Idea," Venizelos believed in the expansion of Greece to embrace all the Greeks living around the rim of the Aegean and wherever else Hellenistic culture exercised sway. Because the Americans resisted grandiose Greek claims in 1919, however, it seemed likely that only military force would resolve matters to the Greeks' satisfaction. With the blessing of the Allies, Venizelos invaded the Anatolian port of Smyrna (Izmir) in May 1919, expecting support from the heavily Greek population. The Hellenic forces did well at first, taking the city and then pushing deep into the Anatolian interior; before long their armies camped within forty miles of Ankara. At the same time Greek armies swept the Turks out of Eastern Thrace. Until the spring of 1922 they were able to hold substantial territory— to the great outrage and humiliation of the Turks.

Shortly after the first victories, however, the political basis for the Greek adventure began to erode, and three years later the entire operation collapsed in disaster. Unexpectedly, the Greek government, including Venizelos, was toppled by the war-weary Greek electorate in October 1920. The young king Alexander died suddenly following infection from a monkey bite, precipitating the return to Greece of the formerly pro-German ex-king Constantine, a bitter opponent of Venizelos and an old enemy of the Allies. Flushed with success, meanwhile, the Greeks pursued their course in Asia Minor, but henceforth without Venizelos' outstanding leadership and without support from the West. The Russians, the Italians, and the French all found reasons to back the Turkish side. The turning point came at a bloody battle on the Sakarya River in August 1921, when the Turks finally halted the Greek advance. From that moment demoralization and confusion began in the Greek ranks, and the initiative passed to Atatürk. Friendless and divided, the government in Athens was unable to reverse the situation. When the Turks took the offensive in the summer of 1922, they smashed through Greek defenses and drove the invaders into panic-stricken retreat. After a furious drive, Atatürk broke into Smyrna in September. The large Greek population of the port, swollen by masses of Orthodox and Armenian refugees, now faced the fury of the victorious Kemalist armies. Massacres occurred almost immediately. An enraged Moslem mob murdered the Greek archbishop of Smyrna and dragged his body about the Turkish quarter. Pillage and rape engulfed the Armenian district. Turkish soldiers rounded up thousands of Greeks, slaughtered them in the suburbs, and set part of the city ablaze. Behind it all, outside observers felt, was Atatürk's determination to rid his country of foreigners. "For a de-

Acquisitions.

Rumanian, from Austria-Hungary

Rumanian, from Russia (U.S.S.R.)

Yugoslav, from Austria-Hungary

Yugoslav, from Bulgaria

Greek, from Bulgaria

Greek, from Turkey, by Sèvres Treaty (restored to Turkey by Lausanne Treaty)

| 0 | Miles | 200 |
| 0 | km | 300 |

THE BALKANS AFTER WORLD WAR I

After L. S. Stavrianos, *The Balkans Since 1453* (New York, 1958).

99

liberately planned and methodically exercised atrocity," Winston Churchill wrote at the time, "Smyrna must . . . find few parallels in the history of human crime."[63]

Protected by American destroyers offshore, Greek vessels desperately embarked refugees from Smyrna, often as many as 45,000 people daily. The departure occurred amid chaos and misery, horrifying representatives of the Western press who relayed descriptions of the catastrophe. Some of the refugees went to Salonika and others to Greek islands in the Aegean; tens and eventually hundreds of thousands jammed in leaky little boats for a more hazardous voyage to Athens. Henry Morgenthau estimated the total number of evacuees at 750,000. He later witnessed the disembarkation of some of them in the Greek capital:

> The conditions of these people upon their arrival in Greece was pitiable beyond description. They had been herded upon every kind of craft that could float, crowded so densely on board that in many cases they had only room to stand on deck. There they were alternatively exposed to the blistering sun and cold rain of variable September and October. In one case, which I myself beheld, seven thousand people were packed into a vessel that would have been crowded with a load of two thousand. In this many other cases there was neither food to eat nor water to drink, and in numerous cases the ships were buffeted about for several days at sea before their wretched human cargo could be brought to land. Typhoid and smallpox swept through the ship. Lice infested everyone. Babes were born on board. Men and women went insane. Some leaped overboard to end their miseries in the sea. Those who survived were landed without shelter on the open beach, loaded with filth, without blankets or even warm clothing, without food and without money.[64]

Meanwhile, the Turkish army swept the remaining Greek forces—and even more refugees—toward Constantinople, threatening to cross the Straits near Chanak and reenter Europe. In the ensuing crisis, the British alone among the former Allies challenged Atatürk and brought him to terms. An armistice followed, yielding Eastern Thrace to Turkey and promising the evacuation of Greeks from the region.

These events prompted a repetition of the Smyrna evacuations from wherever Greeks had lived in Asia Minor and also from Turkey's newly reacquired territory across the Bosporus. Some of the Greeks followed the westward course of the Smyrna evacuees. Ships left the Black Sea ports of Trebizond and Samsun, proceeding through the Bosporus to Constantinople, where more refugees were taken on board, then through the Sea of Marmara and eventually to Greece. Others left Constantinople in precisely the opposite direction, sailing eastward along the Black Sea coast, then landing at Turkish ports and heading south, overland, to Syria.

A League of Nations committee later estimated that of at least two million Greeks living in Turkey in 1914, scarcely half survived to be evacuated in the years 1922–23. Adults had often been interned under especially hazardous conditions, and a great proportion perished. As a result, the bulk of the refugees was made up of old men, women, and children.[65] Relief workers were appalled at the dreadful conditions of the refugees en route, particularly those in the Turkish interior, where typhus was rampant and it was extremely difficult to bring to bear aid of any sort. Many died during evacuation. Near East Relief provided assistance, as it had to the Armenians a few years before. In the years 1922–23, its agents shepherded some 22,000 orphans to safety after long treks across the desert to Syria. Nansen's High Commission coordinated emergency assistance from several governments, initiated serious efforts to control epidemics, provided ships for refugees and food for those stranded on islands in the Aegean.[66]

In the autumn of 1922, when the flood was high, Nansen bombarded the League of Nations with urgent appeals, indicating that no fewer than 900,000 refugees, including 50,000 Armenians, were then in Greece. One-third of these had come from Eastern Thrace and the rest from Asia Minor.[67] Swept along in the Turkish revenge, moreover, were many who had little or no consciousness of being "Greek" and whose expulsion was prompted by some quirk of ancestry—a Greek-sounding name, for example, or a tradition of writing Turkish using Greek characters. League of Nations officials soon confronted so-called Greeks from Turkey who spoke little or no Greek and who displayed a bewildering variety of languages, dialects, customs, and loyalties. Some presumably Greek rug-makers from one resettled village refused even to incorporate Greek designs into their rugs, so outraged were they at having such a label applied to them; according to one relief worker, "they were wedded to Turkey. They sulked at the idea of attempting to do anything Greek."[68]

Witnessing the appalling crush of Greek refugees, Fridtjof Nansen was apparently the first to have proposed a formal exchange of population between Greece and Turkey to help bring order to the chaos. Asked by the powers to mediate the Greco-Turkish conflict, he began negotiations at the very moment the Lausanne Conference was assembling to achieve a final settlement between Turkey and the Allies.[69] The Turks were keen on the idea of an exchange, anxious to prevent the return of the huge numbers of Greeks already expelled and wanting also to block a Greek irredentist drive against Eastern Thrace or Asia Minor. Their goal was simply to eliminate on a permanent basis any Hellenic presence in the new, modernized Turkey. To facilitate this process, the Turks declared themselves willing to welcome the several hundred thousand Turks who still remained in Greece. Venizelos, then representing Greece, was overwhelmed by the Turkish fait accompli and reluctantly accepted an exchange as a way of removing his country's Turkish minority and assisting the settlement of the Greek refugees. Particularly noteworthy was the obligatory character of the transfer, an innovation publicly protested by the Greek side and strongly objected to by British observers, notably Lord Curzon. In reality, this was probably unavoidable. The Turks had just humiliated their enemy and seemed determined to force the remaining "Greeks" to leave. Venizelos, it has been argued, may also have privately favored the notion. Most of the refugees had already quit Turkey, and the Turks leaving Greece would abandon their land, however insufficient, for distribution to the newcomers.[70]

The resulting Lausanne Convention of January 1923 provided for "a compulsory exchange of Turkish nationals of the Greek Orthodox religion established in Turkish territory and of Greek nationals of the Muslim religion established in Greek territory." Following the traditional Greek and Turkish custom—in what was probably the only practical approach—the agreement used religion as the test of national affiliation. Excluded from the transfer were the Turks of Western Thrace and the Greeks of Constantinople, each of these being very ancient communities that were too much trouble to uproot. All others were to move within a fixed period of time, taking their property with them when possible. Compensation for abandoned property was to be paid, extending to all movement of Greeks and Turks between their respective countries since October 1912. Under the supervision of a Mixed Commission, the exchange was to take place, the abandoned property of the evacuees was to be evaluated and liquidated, and the corresponding value was to be transferred to the government of the receiving country. All property and other questions were to be adjudicated by the Mixed Commission, on which sat three neutral members as well as Greek and Turkish representatives. The migratory process, it was optimistically imagined, would take about a year and a half.

In all, over 1.5 million people from both nations were involved in the transfer, the largest such exchange in history to that point. Compared to Turkey, Greece bore a staggering burden. With just over five million people in 1920, Greece received close to a million refugees even before the Lausanne Convention began formally to operate. As we shall see, other Greek refugees were arriving at the same time from Bulgaria. Having been at war since 1912, this relatively poor country soon had to absorb this huge mass of people, perhaps one-quarter of its total population whereas only one-third that number of Turks departed.

The impact was colossal. In the first phase of resettlement, about a million new arrivals were utterly destitute. Refugees were literally everywhere—camped in boxes in the Athens Municipal Theater and in the shadows of the Acropolis and cramming railway stations, schools, public buildings, churches, sheds, warehouses, and cinemas. Municipalities were overwhelmed. In eighteen months the population of Athens more than doubled, from 300,000 to over 700,000. "Save in China," wrote one observer, "I have never been in a place where the sense of crowding is more acute."[71] Salonika, partially destroyed by fire in 1916 and with a population of 174,000, received 160,000 refugees. The many dependents among the evacuees could only be resettled and put to work with great difficulty. Malaria, typhoid fever, and dysentery took a dreadful toll. During the last months of 1923, according to a League of Nations source, mortality rates among new arrivals climbed to an astronomic 45 per cent. At that time, according to the Greeks, they had received 1.2 million refugees, 870,000 of whom needed state support.[72]

Although the Mixed Convention attempted to mitigate the effects of the transfer by permitting retention of property, this arrangement failed in practice, and once again the Greeks suffered disproportionately. The commission failed to agree on procedures and principles, tying discussions up in paper and blocking indemnities to governments and individuals. In the end, the property terms became a dead letter and simply could not be executed. Generally speaking, Greece could not be compensated for the fact that though she had to receive over 1.2 million Greeks from Turkey, only 356,000 Turks went in the opposite direction.

Moreover, although the Turks left under relatively orderly circumstances, the Greeks arrived in chaos, requiring a great deal of assistance. Imbalance, therefore, conditioned the exchange from the very beginning, subverting the hopes of a smoothly run transfer in a year and a half.

Both sides utilized the religious definition of "Greeks" and "Turks" to rid themselves of undesirable minorities and to help resolve some internal political problems. The Turks expelled groups of Armenians with this in mind, somehow assimilating them to the Greeks. Turkey, according to Macartney, also used the Greek blanket to cover unwanted Serbs, Rumanians, Russians, Gypsies, and even Arabs of the Greek Orthodox Church. The Greeks, on the other hand, seized the opportunity to pursue the Hellenization of certain regions, in one outstanding case by attempting to expel Albanian Muslims from Epirus. (The Albanians protested vigorously against this abuse of the Mixed Convention at the end of 1923 and saw the issue resolved in their favor the following year.) Similarly, Athens evicted thousands of Doenmeh from Salonika, members of a Ladino-speaking Muslim sect, descendants of seventeenth-century Jewish converts to Islam. Greek planners also employed the newcomers in Hellenizing Macedonia, to which four-fifths of the agriculturalists among the incoming refugee population were directed. In this they seem largely to have been successful, triggering not only an anticipated exodus of Muslims to Turkey but also—an unexpected bonus—a flight of Macedonians to Bulgaria.[73]

By 1923, Greece was threatening to collapse under her terrible refugee burden. Despite heroic efforts on the part of the government, disease and starvation swept refugee encampments, and mortality rates began to soar. For the Greeks, the principle of "exchange" provided little relief for the refugee crisis, and the institutions set up at Lausanne proved utterly inadequate to assist resettlement. Spokesmen for Western charities in Greece at this time were on the verge of despair. Before the Council of the League, Nansen and others demanded relief for the starving, disease-stricken refugees in Greece. The answer finally came in the form of a League-sponsored Greek Refugee Settlement Commission established at the end of 1923. Apparently at the instigation of Sir Arthur Salter, whom Morgenthau described as the League of Nations' financial wizard, this body undertook, with League of Nations patronage, to raise the funds necessary to shelter, feed, and care for the refugees. For the longer term, the

Mixed Commission drafted and implemented programs for turning the refugees into productive citizens. Morgenthau became chairman of the commission, assisted in 1926 by Sir John Hope Simpson. To set the machinery in motion, the Bank of England helped to advance one million pounds sterling to the Greek government.[74] Functioning until 1931, when the Greeks took over its work, the Mixed Commission brought substantial international loans to bear on the problem—and with considerable success. The commission directed surveying, land reclamation, drainage, sanitary improvements, and so forth, prompting considerable agricultural colonization. Within towns and cities resettlement work included massive building projects, the provision of municipal services, factories, and the stimulation of commerce. Throughout the country the commission extended medical facilities and wrestled with problems of disease control and nutrition. The Mixed Commission enlisted foreign aid and cooperation in all these projects, providing materials, advice, and markets for goods produced in Greece. In a seven-year period, according to the American journalist Dorothy Thompson, this organization displayed "a miracle of inventiveness, altruistic energy and persistence." It was, she said, "the biggest philanthropic enterprise in history."[75]

Given the calamitous nature of the expulsions from Asia Minor, the international community served Greece and her refugees remarkably well. Nansen, it has been widely agreed, spurred the League of Nations into action, and the League's involvement helped, in turn, to move governments and international banking houses to help resettle the refugees. "The problem was of colossal dimensions," Sir John Hope Simpson later wrote, "and has been dealt with drastically and on the whole with great success."[76] There was much criticism of the compulsory character of the transfer and attacks on the asymmetrical nature of the "exchange," weighted so heavily against Greece. But the League was hardly in a position to check the Kemalist minorities policy and certainly did not intend to protect the Greeks from the consequences of their Anatolian adventure. The League never proposed more than limited answers to practical problems. Of course, the solutions found were in some sense unjust. In financial terms, as Macartney notes, "the investors of Western Europe paid the piper for the tune so gaily called by the Young Turks."[77] But the solutions were real enough. By 1930, when the Resettlement Commission turned over its operation to the Greek government, its reports were optimistic about the integration of the remaining refugees into Greek society. Having begun operation in conditions of unprecedented disaster and human suffering, the commission showed that an international agency could do practical refugee work on a large scale without becoming entangled in extraneous political concerns.[78]

The Greco-Bulgarian Exchange

Within the Balkans, minorities were so intermixed and in such delicate balance that any dislocation of one group was certain to affect the others. The shattering impact of the Balkan and First World wars on the peninsula prompted the flight not only of Greek and Turkish refugees but also of substantial numbers of Bulgarians as well. Each of these movements posed problems for other minorities, beginning a chain reaction that was difficult to stop.

In 1913 refugees from once-Bulgarian territory or from land claimed by Bulgarian leaders began to stream toward the country still ruled from Sofia. This process continued with the end of the First World War, when Bulgaria was defeated along with the Central Powers. From Western Thrace and from Macedonia, refugees struggled to Bulgaria, fearing to be subject to Greece or the new state of Yugoslavia. Other refugees came from Rumania, enlarged after the war at Bulgarian expense, and from Turkey, subject to the nationalistic policies of the Young Turks. Still others came from Russia or devastated Armenian settlements. Altogether some 200,000 fugitives crowded into this small nation of 5.5 million. Leaving Bulgaria, relatively much smaller numbers of ethnic Greeks or various Slavic minorities went to neighboring countries, escaping effects of the bitter, vengeful nationalism that now pervaded Bulgarian political life.

Bulgaria's new frontiers were defined in the Treaty of Neuilly of November 1919 and in a related convention concluded a year later. Sofia was obliged to accept what was defined as a voluntary exchange of minorities—procedures largely intended to facilitate the new territorial settlement by removing Bulgarian sympathizers from the lands absorbed by Greece. An autonomous Greco-Bulgarian Mixed Commission charged with the execution of this exchange set out to compensate the émigrés for abandoned property and further determined that indemnification should be retroactive, applying to all Greek and Bulgarian refugees who had moved between the countries since 1900.

As with the Greco-Turkish exchange, however, there was a great gap between the process outlined in diplomatic negotiation and the reality defined on the ground. Greeks and Bulgarians, particularly those who were Macedonians, had been fleeing each other's countries in great numbers since the Balkan Wars. Substantial numbers of Greek and Bulgarian nationals had already migrated when the convention was signed. As a result, it seemed unlikely that much additional movement would take place. During the next three years very few volunteers applied to the Mixed Commission for transfer. Two factors, however, completely disrupted this stability. First, the flood of Greek refugees from Anatolia put extraordinary pressure on the Greek government to make room for the newly arrived refugees; the Greeks' answer was to expel "Bulgarians"—predominantly Macedonians, including many whom the Greeks wished to be rid of in order to pursue the Hellenization of the Macedonian homeland. Second, the Greeks started a refugee avalanche in 1923, when they removed a few thousand Bulgarian families from Western Thrace on the pretext of military necessity, fearing renewal of warfare with the Turks. Many of these expellees crossed the mountains to Bulgaria, where they camped in Greek villages, triggering violent incidents and prompting a southward flow of Greek-speaking refugees to Greece. By 1923, as a result, the refugee traffic moving in both directions increased significantly. Over the next two years 52,000 Bulgarians left for Greece; and 30,000 Greeks, for Bulgaria.[79] No one could pretend that the exchange was voluntary or orderly, and the machinery installed by Neuilly spun ineffectually, unable to channel the movement or settle property claims.

Overwhelmed by refugees, as we have seen, and desperate for outside assistance, the Greeks appealed to the League Council for financial help. The threat of war between the two Balkan countries gave impetus to the call for international intervention. In October 1925 a minor clash on the Greco-Bulgarian border threatened to escalate quickly into serious fighting, and with this incident the League of Nations became closely involved in relations between feuding states.[80] The League Council instructed a commission under the British diplomat Sir Horace Rumbold to examine the conflict between Greece and Bulgaria, including the passage of refugees between them. Rumbold soon reported what everyone on the spot knew: that the process of indemnification was too slow, that refugees on both sides remained helpless victims, and that future claims by former fugitives might well trouble relations between the two countries. Bulgaria's neighbors feared, in particular, that refugees were being used to build strategic railways and enlisted in the cause of Bulgarian irrendentism. In Bulgaria, on the other hand, the arrival of thousands of newcomers provoked a crisis in early 1926. As in Greece, disease, overcrowding, hunger, and the inadequacy of all services overwhelmed local authorities. And, as in the latter case, the answer came in the form of a huge loan to the government floated by the League of Nations. Massive amounts of capital began to arrive from abroad, and by May 1927 conditions for refugees began to improve. A League of Nations commissioner supervised settlement and took particular care that this be done constructively and not in such a manner as to provoke Greek or Yugoslav hostility.

Agreement on refugees between Greece and Bulgaria proved exceptionally difficult. Unlike the case of Greece and Turkey, neither country felt beaten and at the mercy of the other. The two were bitter enemies, sharply at odds since the Second Balkan War. Between them the Macedonian problem remained alive and menacing. Greece now held the lion's share of the region, and her postwar policies had propelled tens of thousands of embittered Macedonian nationalists to Bulgaria. There they joined with local Macedonians and a substantial contingent from Yugoslavia to form a powerful influence in Bulgarian life. During the interwar period, their revolutionary organization threatened constantly to destabilize the country and to embroil Bulgaria with her Yugoslav or Greek neighbors. Yet the mediation of the League of Nations achieved a good deal. By 1933, when the onset of the Great Depression in Eastern Europe prevented new loans for such purposes, most of the resettlement had been accomplished. Within a decade, nearly a quarter of a million once-destitute refugees to Bulgaria had been housed, fed, and made into productive members of society. They were part of some two million refugees settled among Greece, Turkey, and Bulgaria in this period, a not negligible accomplishment for the international community.

Only an international agency could have done so well. No neighboring state was prepared to offer assistance, and the private organizations at work in the region were clearly incapable of furnishing help on the scale required. The great powers, which effectively controlled the League of Nations through its council, saw few immediate interests at stake in the poor, hungry, and disease-infested states of southeastern Europe. Although insufficiently interested to act on their own, they were prepared to see the league sponsor the settlement commissions as charitable exercises—which would not threaten to get out of hand. The League succeeded because the requirements for aid were limited to two small countries on the European periphery for a brief duration. League officials deluded themselves, however, when they took these cases as models for subsequent refugee assistance. The refugee crises to come were not limited as in the 1920s, and the international machinery set up to deal with such disasters was to prove hopelessly inadequate.

STABILIZATION, 1924–30

With the mid-1920s, many diplomats felt that the peaceful relations so earnestly hoped for in 1919 could at last be realized. European concord was to be secured through a consolidation of the status quo, bound by diplomatic negotiation and voluntary agreements. In 1925 a clear signal of this new prospect went out from the Swiss resort town of Locarno on Lake Maggiore: representatives of Great Britain, France, Germany, and Italy agreed to resolve important differences over the Versailles settlement and thereby paved the way for Germany's entry into the League of Nations. For the European public, this "spirit of Locarno" helped condition European politics and sustained the notion that important problems were being solved through international cooperation.

Refugee questions were certainly affected by this mood. Optimism prevailed at Geneva, headquarters of the League's High Commission for Refugees. Writing in the *Encyclopaedia Britannica* in 1926, Nansen himself expressed satisfaction about the settlement of many thousands of refugees since the end of the First World War. He referred to "the generosity of a number of governments and . . . their enlightened cooperation through the machinery of the League," with results that "will be better than even the most optimistic could have ventured to expect." Nansen repeated his belief that refugee problems were finite and ultimately resolvable by international organization.[81]

International Planning in the Locarno Era

Central to the new structure of European peace was the League of Nations, whose authority and prestige grew appreciably in the decade after the First World War. Champions of this new forum attempted to achieve as wide a membership as possible while, at the same time, being sufficiently attentive to the voices of the great powers to keep those countries committed to the institution. Throughout its life, the League of Nations was obliged to strike a careful balance, striving for widespread, extensive involvement in serious issues of international concern yet ensuring that national sovereignty was not infringed. In particular, no League agency could be presented as having too much independent authority, requiring too great a financial commitment, or involving too great an interference with the existing structure of international relations. Paradoxically, the organization supposed to build a new structure of European peace had often to define its practical role as narrowly as possible.

Under these conditions, the League's High Commission for Refugees entered the world hobbled by restrictions. As we noted, the refugees eligible for protection by Nansen's agency were designated with great circumspection. At its inception, the office was only to assist the Russians. All other refugees—from Jews to the German repatriates to Magyars—fell outside the mandate of 1921. From time to time a few specific groups were added to the list, but no general definition of "refugees" ever appeared. Services were eventually extended to Greek and Armenian fugitives in Constantinople; to Bulgarian expellees from Thrace; to other Armenians; and to "Assyrian, Assyro-Chaldean, Turkish (friends of the Allies), and assimilated refugees" in 1927. In this process, we see the scrupulous concern of the League of Nations to avoid abusing the sensibilities of any member state. The Russian émigrés could be designated with impunity in 1921 because the Soviet Union was an international outcast held in diplomatic quarantine and outside the League of Nations until 1934. Armenian, Bulgarian, Greek, Middle Eastern, and other refugees could also benefit from League assistance because their fate engaged the interest of no member state to any appreciable degree.

Institutionally, the High Commission never rested on a firm foundation. There was no mention of its presence in the League Covenant. Responsible to the League Council, the High Commission was supposed to function in 1922 on a paltry 4,000 pounds sterling set aside strictly for administration.[82] Relief, as always, was to come from private charities or directly from governments. Furthermore, as we have noted, it was expected that the High Commission would gradually phase itself out of existence as the postwar refugee crises diminished and the refugees were either resettled or repatriated. Ten years, it was believed, would finish the job.

Fettered both politically and financially, the High Commission settled into a routine during the mid-1920s. It completed the evacuation of Russian émigrés from where they had been stranded after the Civil War, particularly in Constantinople, assisting their movement from various parts of Europe to France, notably in the years 1924–26. Other League of Nations agencies indirectly assisted refugees in various ways—such as when the League Health Organization combatted epidemics in Poland, Russia, Rumania, and Greece. In addition to resettling Russian refugees, Nansen's office regularly encouraged repatriation when this was acceptable to the refugees themselves. To achieve this, extensive negotiations with the Soviet authorities were necessary, resulting in a very small number of returned refugees. As we shall see, Nansen's agency also explored several possibilities for massive population resettlement in communal groups. The high commissioner arranged travel documents for refugees who were suspended in a web of international legal technicalities, and he sponsored or helped supervise the distribution of relief. Perpetually on the verge of bankrupcy, the High Commission paid its obligations by means of the "Nansen stamp" attached to identity papers after the payment of a small sum by refugees who could afford it. Finally, the League of Nations provided special grants for specific projects.

All this was achieved with a very small staff, working closely with private and voluntary agencies such as the International Committee of the Red Cross, Near East Relief, the American Relief Administration, the Joint Distribution Committee, and several others. To help coordinate the work of these agencies a Permanent International Conference of Private Organizations for Protection of Migrants assembled in Geneva in 1924. Representatives of the various agencies joined an advisory committee assisting the High Commission, and Nansen's office helped coordinate and stimulate the activity of the charitable bodies. Private or government-sponsored philanthropy, in the end, took charge of the operational tasks of assisting refugees; Nansen prodded, assisted, and cleared the political path.

In 1924, Nansen proposed to the League Council that the bulk of refugee work be transferred to the International Labor Organization (ILO). Consistent with the general view of refugee matters in the mid-1920s, the high commissioner believed that the important political issues regarding refugees had been solved. What remained were problems of employment and settlement, requiring a permanent agency. France was seen as the only country in Europe capable of absorbing substantial numbers of foreign laborers although other countries might be encouraged to receive refugees.[83] Directing the ILO was the energetic Frenchman Albert Thomas, a former collaborator of the great socialist tribune Jean Jaurès and minister of munitions in wartime France. Thomas backed Nansen's suggestion. According to the new arrangement, the High Commission continued to take charge of the legal and political questions, and the ILO devoted itself to improving the conditions of émigré workers, finding employment, and exploring settlement possibilities abroad.

Through the 1920s, refugee work continued unspectacularly. Homes and jobs were found. The ILO offered loans to migrants. At the end of the decade, just before Nansen's death in 1930, the High Commission estimated that there remained 180,000 unemployed Russian and Armenian refugees; presumably all the others were adequately settled. The high commissioner himself, now in his mid-sixties, became interested in other matters, notably the resettlement of Armenians in Erevan, by then part of the Soviet Union. By 1930, writes the historian of the League of Nations, "most of the refugees were either self-supporting, naturalized in their country of residence, repatriated to Russia, or dead."[84]

A counterpoint to the reduced activity of the High Commission in refugee work was the winding down of overseas migration—the end of what Albert Thomas referred to as "the era of the great displacements." Until the passage of the Johnson Bill by the United States Congress in 1924, setting new quotas for immigration to that country, emigration from Europe had been extremely high. In the early 1920s as many as a million Europeans migrated abroad annually. Throughout this period the League of Nations assisted both inter-European and overseas migrants by facilitating agreement on several important conventions dealing with sanitary and working conditions. At the same time,

League of Nations officials stationed in major embarkation ports offered a variety of services, and a Permanent Commission on Emigration in Geneva coordinated related activity. But by the end of the decade, there was much less work of this sort to do. Europeans still moved to South America and the British Dominions, of course, although more than half of these returned to Europe dissatisfied with their new homes. In retrospect, some have seen 1928 as the turning point after which net emigration practically vanished. In 1929 depressions struck. According to Sir John Hope Simpson, Europe showed an inward balance of migration by 1933.[85]

Closing of the gates of immigration meant that remaining refugee problems had to be solved in Europe itself. Happily, one important European country, France, was prepared to welcome foreigners. Some 1.5 million young Frenchmen had been killed during the carnage of the years 1914–18. Although both Germany and Russia suffered greater numbers of casualties, France had the highest losses in proportion to her total population—no less than 7 percent of the entire male population swept away. Two out of every ten young men had died, and another three were disabled. When the fighting ceased, there were serious concerns about the resulting diminished French productivity. In order to fill this yawning population gap, postwar French governments offered special incentive payments for large families, outlawed birth control, and assisted the integration of newcomers in French society. Above all, there was an encouragement of immigration. Georges Mauco, an authority in this field, captured this spirit in an important book he published on this subject in 1932. In the Darwinian idiom of the time, Mauco declared that the country that did not advance demographically would fall behind. Unless France were continually attentive to the population problem, she risked slipping into perpetual decline.[86]

Responding to these perceived needs, France became the most important immigration country in Europe and the second in the world, outstripped only by the United States. Throughout the five years of postwar reconstruction, French government and industry carefully recruited, organized, and regulated many thousands of foreign workers, including great numbers of refugees. In the High Commission, as a result, France could always be counted on to do far more than her share of refugee resettlement. French recruiters were on the spot in Sofia and Constantinople to induce Greek and Macedonian fugitives to move west in the early 1920s. Candidates were shipped to Marseilles, placed in camps, and put to work in various parts of the country.[87] In early 1924 the French offered to receive "all Russians who were fit and anxious to work, either as industrial or as day laborers." Answering the call, extended also to other nationalities, some 1.5 million foreign workers flocked to France before 1928, a large proportion of them refugees.[88]

When Nansen died in 1930, the recent refugee crises were just a memory, associated with the postwar convulsions of communism, demobilization, unemployment, huge deficits, and inflation. Stability appeared to have been realized. Few refugees still wandered from place to place, homeless and unwanted, without legal standing. Although America no longer received many European refugees, France had provided a widely sought asylum. Some refugees were unemployed, and Nansen still worried about their future at the end of the 1920s, when the mandate for the High Commission was nearing its close. Foreseeing the end of the institution, Nansen proposed that its work henceforth be undertaken by the League of Nations Secretariat. Several problems still remained: in particular, Jews and Armenians were only precariously settled in new homes. Colonization schemes appeared to some the answer, presumably the finishing touches to the refugee achievements of the Locarno era.

Resettlement Schemes in the 1920s: Jews and Armenians

One of the most durable assumptions about refugee questions in the interwar period was that solutions might be found by establishing rural communities, preferably in thinly populated parts of the globe. "There seems to be a widespread conviction," Sir John Hope Simpson observed in 1939, "that permanent refugee resettlement on a large scale must necessarily take an agricultural form."[89] To many Europeans, farming still projected the image of a healthy life, of rootedness, of freedom from political turmoil—everything, in short, that refugees presumably sought. Furthermore, from the standpoint of potential receiving countries, particularly in the revolutionary atmosphere of postwar Europe, the massing of refugees in large cities or in densely populated areas threatened to be unsettling politically. Thus, their dispersal in the countryside might better serve domestic tranquillity. Finally, from the geopolitical perspective, many felt that moving refugees to remote, uninhabited parts of the globe would be conducive to European peace. Albert Thomas, for example, shared the widely held view that population pressures lead to war. Certain countries, he argued, notably Poland, Italy, and Turkey, were clearly overpopulated. In the last years of his life, at the end of the 1920s, he developed a plan to send surplus populations to League-sponsored agricultural settlements far from the centers of inter-European conflict.

Throughout the interwar era, the outstanding example of this kind of colonization was Palestine, where since the 1880s European Jews had been at work building what was referred to as a "National Home." The Jews themselves did not necessarily see their settlement as a means to rescue the downtrodden, let alone as a recourse for Jewish refugees. On the contrary, the Zionist movement, which followed the lead of the Austrian publicist Theodor Herzl in the early 1900s, declared its aim in classic nationalist terms: the goal was a renaissance of the Jewish people—a political, cultural, social, and economic transformation of the Jews to be achieved through settling the ancient Jewish homeland. Occasionally, in moments of crisis such as at the time of the Kishinev pogrom in 1903, Zionists talked of a *Nachtasyl*— a temporary haven for refugees. For those who supported this concept, the haven could be anywhere, and the Zionist movement was even briefly committed to exploring settlement prospects in East Africa. But the long-term goal remained Palestine and the Jewish homeland, which was to arise there through national regeneration.

By the end of the First World War, Palestinian Jewry—known as the *Yishuv*—numbered only about 55,000, of whom merely a fraction were fully committed to Zionist goals. Set among an Arab population of about ten times that number and having only the most fragile of institutional foundations, the Jewish enterprise seemed extremely precarious. The immediate postwar period, however, was a turning point. Pursuing the opportunity presented both by the Paris Peace Conference and the Balfour Declaration of 1917, in which the British had indicated support for the Jewish cause in Palestine, Zionists strenuously argued on behalf of the National Home. To a remarkable degree they succeeded. When the League of Nations formally defined the British mandate for Palestine in 1922, it did so on the basis of an obligation to the Jews. The League Council referred to "the historical connection of the Jewish people with Palestine," in-

structed Britain to facilitate the establishment of a "Jewish National Home" by encouraging Jewish immigration and settlement.
For the Jews, presumably, there was now a place for their refugees to go.

In practice, the *Yishuv* provided only limited options for Jewish settlement and did not satisfactorily address the postwar plight
of Jewish refugees. For one thing, life was extremely hard in the
fledgling National Home during the early 1920s, when modern
institutions had barely been established and when there were
frequently shortages of work, housing, and the amenities of Europe. Immigrants camped on the beaches in Tel Aviv, living only
in tents, and new settlers had generally to adapt to a rough, pio-
neer existence. At first there was a sudden and very substantial
wave of new arrivals whose passage was made possible once the
First World War had ended in Europe. Between 1919 and 1923
some 37,000 Jews joined the *Yishuv*, many of them fleeing civil
war, pogroms, revolution, and related upheavals in Eastern Europe. However, the newcomers often returned to Europe after a
short stay: about a quarter left within a few years, and by 1923,
when the postwar wave was noticeably diminishing, reemigration reached 43 percent.

In addition, the British became fearful of massive Jewish colonization, particularly following violent outbreaks of Arab discontent in 1921. From the beginning, the British set a ceiling on Jewish immigration. Codifying this policy, the White Paper of 1922,
prepared under the auspices of the colonial secretary, Winston
Churchill, declared that such immigration should not exceed the
country's "absorptive capacity"—thereby setting a precedent both
for greater future limitation and also for stricter control over the
kinds of people permitted to come. In practice, the mandatory
power transferred the selection process to the Zionist Executive,
which then determined the criteria for immigration to Palestine.
Before long this process hardened into a distinctly political mold.
The Zionist Organization had become intensely politicized by this
time, and the predominantly left-wing groups in charge, notably
the Labor Zionists, did their utmost to use this process to
strengthen their own following in the *Yishuv*.

Thus, Palestine was limited as a refuge for Jews in the 1920s.
It is perhaps remarkable under these circumstances that the *Yishuv*
grew to the extent it did and that it still managed to provide some
outlet for persecuted Jews during that period. Following the
postwar wave of immigration, the great menace to Jews ap-

peared to be in Poland. In the turbulent political environment of the new Polish republic, Jews came into repeated conflict with a middle class uncertain of its recently acquired standing in economic life and with an assertive Polish nationalism insecure about the one-third of the nation that was not Polish in language or culture. For the National Democrats, or Endeks, the prominent political voice until 1926, Jews were seen as hostile and unassimilable, incapable of ever being part of the Polish nation. Pursuing this line in the mid-1920s, the government of Wladyslaw Grabski launched a broadly based assault on the Jews' economic positions, encouraging economic boycotts, nationalizing branches of industry in which Jews predominated, and dismissing Jewish employees. The results for the Jewish community were disastrous, impoverishing tens of thousands at the very moment when emigration to America became impossible. To escape such conditions, some 32,500 Jews, often middle-class victims of government harassment, came to Palestine in the period 1924–26. More followed, although at a lesser rate, owing to economic troubles in Palestine, bringing the *Yishuv* from 84,000 in 1922 to 154,000 in 1929.[90]

Within the Zionist camp this wave of immigration was viewed with mixed emotions. Zionists strongly favored Jewish settlement in general terms, of course. But many were apprehensive about the lack of pioneering spirit among the "economic fugitives," and many doubted that such bourgeois outcasts could seriously contribute to the Jewish national resurgence. Even Chaim Weizmann, the moderate president of the World Zionist Organization, fretted at the time about "the old Diaspora habit of creating towns to receive an urbanized immigration."[91] Never was it more evident that, under existing conditions, Palestine was not seen by the Zionist movement as a haven for Jewish refugees.

Given the massive dislocation of Eastern European Jewry after the First World War, together with the limited possibilities of emigration, Jewish relief workers came to see that realistic settlement projects would have to be directed to Eastern Europe. "The Jewish masses in Russia," said one Jewish official associated with the American Relief Administration, "whether they wish it or not must remain in their country."[92] Having already indicated its receptivity to aid provided through Hoover's American Relief Administration (ARA), Russia became the target of the one important mass resettlement effort—significantly, by building agricultural colonies.

The great concern of Jewish relief workers was the fate of some 2.5 million Jews who remained in war-torn Russia when the fighting ended. Though freed from tsarist legal and political disabilities, many Jews suffered acutely from the undercutting of their economic existence. Jews frequently found themselves classified among the social groups discredited by the new revolutionary regime. Petty traders, middlemen, or small manufacturers, these Jews were among the class enemies the new order set out to replace. Many managed to leave Russia, as we have seen. But many thousands still wandered about the Soviet state and survived thanks only to Jewish communal services. In practical terms, the period of War Communism hurt Jews particularly badly. The Joint Distribution Committee funneled 4 million dollars to Jewish relief up to 1922, working through the ARA, trying to restore a basic level of subsistence to a desperately impoverished and sometimes bewildered Jewish world.

In 1924, attempting to provide a more permanent solution to this economic and social disaster than soup kitchens, Joseph Rosen, an agronomist working for the Joint, began an agricultural settlement project known as the American Jewish Joint Agricultural Corporation (Agro-Joint). Working with the Soviet Government, Rosen had the idea of rescuing large numbers of Jews from threatened starvation, homelessness, and the lack of a productive employment by installing them on the land in designated regions of the Crimea and the Ukraine. This effort proceeded through the 1920s with a fair degree of success. By 1928, Agro-Joint had built 112 colonies in the Ukraine and 105 in the Crimea, involving many thousands of people. Agro-Joint apparently settled about 60,000 Jews in its settlements, and the total population of Jewish colonies may have been as high as 100,000.[93] Other Jewish charities in the West contributed similarly to these Jewish agricultural colonies, a strategy reinforced by the winding down of overseas migration and the increasing difficulty of leaving the new Communist state. Coordinating fund raising for these ventures was an American Society for Jewish Farm Settlements in Russia, which worked closely with the Soviets. The effort flagged in the early 1930s, however. Money from America was no longer forthcoming during the Great Depression, and in Russia political problems associated with the drive to collectivized agriculture interfered with further development. Another colonization scheme for Jews—settlement in the Birobidzhan region, across the border from Manchuria—soon preoccupied officials in

Moscow. Soviet leaders of the Agro-Joint were eventually swallowed up in the great purges of the mid-1930s. The program's urgency, moreover, declined as the Soviet Union recovered economic strength and Soviet industry absorbed an ever larger proportion of the Jewish population.

One other resettlement scheme, addressing the problem of Armenian refugees, was floated in the mid-1920s. Behind this project lay more than four years of dispersion of large numbers of Armenian émigrés scattered after the catastrophe that beset the Armenian Republic just after the First World War. In September 1923, when the League of Nations took responsibility for the protection of these refugees, there were said to be 320,000 of them living often miserably in Europe and the Middle East.[94] Clearly, the international community had not performed brilliantly on behalf of Armenia or the Armenians. Although the League of Nations Assembly repeatedly supported an Armenian national home at the beginning of the 1920s, none of the great powers was prepared to exert itself to provide the necessary protection. When the Russians and Turks came to final agreement on the subject in the summer of 1920, there was no force ready to interpose itself and prevent the extinction of Armenian independence. Within months the Soviets had absorbed the greater part of traditional Armenia, leaving only some western portions to the Turks. So ended the sovereign Armenian state, prompting Lord Curzon, a witness to these events during the negotiations for the Treaty of Sèvres with Turkey, to term the Armenian question "one of the greatest scandals of the world."[95]

Armenian spokesmen continued to appeal to the League of Nations, often seen as the last hope for the Armenian refugees. Finally, in 1924, the Assembly voted to look into agricultural settlement as a possible remedy. Nansen apparently first raised the idea with Ankara, proposing that a portion of Asiatic Turkey be set aside for Armenians and meeting with a blank refusal. Responding to what was then felt to be the most realistic prospect, League of Nations representatives considered installing Armenian refugees in the Caucasus under the auspices of the Armenian Soviet Republic. Nansen reluctantly took charge of the effort, heading a committee of experts organized by the ILO. The high commissioner for refugees certainly had sympathy with the Armenians, but he seems to have felt that League involvement would permit states to escape their responsibilities toward the refugees—particularly given their earlier abandonment of Armenian independence.[96] Nevertheless, the first soundings seemed positive. The Soviets were willing to use outside assistance, agreed to receive the Armenians who wished to return to Armenia, and promised to provide land and tax relief to assist the new immigrants.

Nansen's commission set off for Soviet Armenia in the summer of 1925, assisted by experts in several fields and the high commissioner's trusted Norwegian assistant, Vidkun Quisling. The investigators concluded that Soviet Armenia should be considered "the sole existing and possible national home for the dispersed remnant of the Armenian people." Coming to grips with the significant technical difficulties of resettlement, the report nevertheless envisioned the repatriation of at least 25,000 refugees.[97] The main problem, it emerged, was money. The Soviet Armenian Republic was willing to guarantee loans sought abroad, but ultimately significant sums would have to be raised to finance the effort in the Caucasus.

On several occasions the high commissioner appealed to League members for support—all in vain. No amount of financial juggling could make up for the fact that by the late 1920s none of the member states was prepared to contribute significantly to the exercise. Quite likely, this was because the Soviet government would ultimately be a beneficiary of subsidized migration to the U.S.S.R., a prospect hardly appreciated in the major European capitals. At the same time, aid from private Armenian sources was plainly insufficient. Nansen himself was bitterly disappointed and offered to step down as high commissioner in 1927. In the end, he settled for a very small version of the original project, bringing a few thousand Armenians to the Caucasus in 1928. The following year the resettlement scheme was finally abandoned.

Nansen's failure in this specific area illustrated the limits of his achievement with refugees. His High Commission did much to assist those who had been made homeless in the wake of the First World War. His office hammered out procedures for certifying large numbers of people as refugees, entitled to some basic consideration in the jungle of international relations. Under Nansen's inspired leadership, the League of Nations did manage to assist resettlement of Russian refugees, Armenians, and many hundreds of thousands of fugitives in southeastern Europe. Behind all this lay the suggestion that more was involved than charity; Nansen helped to cultivate a political will, however fragile, to do something about refugees. Of course, states continued to pursue their own interests in this respect as in others. Against the massive power of the state to designate masses of people as undesirables, to expel the unwanted, and to block entry to refugees, he had no mandate to proceed. Finally, this political will drew on the expectation that refugee problems were relatively simple and easily solved. The next decade presented different kinds of refugee crises, in which this political will was cruelly tested.

3

In Flight from Fascism

Fascism burst on the European scene in the wake of the First World War—bellicose, energetic, and determined to challenge existing/political structures. Vague and even contradictory in ideology, fascism was a revolutionary movement promising radical change, to be achieved by its taking control of the nation-state and imposing its authority everywhere. Writing on fascism in the *Enciclopedia Italiana* after ten years of power, Mussolini emphasized the centrality of the state for Italian fascism: "The foundation of Fascism is the conception of the State, its character, its duty, and its aim. Fascism conceives of the State as an absolute, in comparison with which all individuals or groups are relative, to be conceived of in their relation to the State."[1] True to its promise, fascism in Italy turned brutally on individuals, attempting to dragoon an entire society into confrontation with those deemed enemies of the nation. In Italy, Portugal, Germany, and also in Spain, where fascist variants were established during the interwar period, fascism meant the persecution of designated groups and personalities held to be outside the national community.

One result was the creation of refugees. Individuals fleeing fascism became the characteristic refugee type in the interwar period. For more than ten years after the First World War, such fugitives were mostly Italians, fleeing the regime installed by Mussolini beginning in 1922. In the next decade they were mainly German-speaking, driven into exile after Hitler became chancellor of Germany in January 1933. Immediately after the fascist successes of 1922 and 1933, the exiles resembled those of the

122

nineteenth century—political personalities taking up residence abroad and waiting for the moment of return. Then masses began to follow the first wave, including families and people uninterested in politics—victims of the portentous fascist claims on individuals. Mainly of the political left in the early period, the victims included all of those detested by the fascists in power. And as Nazism took hold in Germany in the mid-1930s, these victims were overwhelmingly Jews, coming from every part of Central Europe where Nazism held sway. Finally, a few months before the outbreak of the Second World War, hundreds of thousands of Spanish Republicans fled across the Pyrenees after the triumph of Franco.

Fewer in number than the refugees following the First World War, these refugees posed significantly more difficult problems. The greatest proportion tried to make their way to societies sunk in economic depression—quite unlike those coming to Europe in 1919, when postwar reconstruction and expansion were just beginning there. Behind the fugitives from fascism, moreover, there appeared hundreds of thousands of other petitioners, mainly Eastern European Jews, whom no one wanted to encourage by a widespread extending of asylum. These refugees, in addition, entered into the complex political equations of the age of appeasement: some onlookers accepted the fascists' denunciation of these outcasts as Communist troublemakers; others saw them as likely to precipitate war in Europe or overwhelm a fragile national identity. Given these predispositions, national and international machinery set up to deal with an earlier refugee experience was bound to break down. The League of Nations as well as individual states failed adequately to address refugee problems—even to the extent that this had been done in the immediate post-World War I years. Refugee problems continued to worsen, finally to be engulfed in the even greater challenge of the Second World War.

FASCISM AND ITS ENEMIES BEFORE 1938

Although in retrospect fascism appears to overshadow the entire interwar era, in practice it advanced only slowly, with Italy for a decade being the only case of a fascist movement in power. Economic depression, however, gave fascism new vigor. In 1933 it triumphed in Germany and seemed headed for success in France,

Austria, and parts of Southeastern Europe. Even where unsuccessful, fascism by this point was admired by millions and appeared utterly different from the seedy, disreputable, sometimes comic movement of its earliest phases. Refugees from fascism, therefore, increasingly came on the European stage as losers—not the proud and distinguished outcasts who went abroad in the first moments after Mussolini or Hitler took power, but rather beaten, harassed, sometimes bewildered outcasts.

Refugees from Italy

In the early 1920s, the Fascist Party *squadristi* created a climate of terror for those who would not succumb to the new order in Italy. The first anti-Fascist émigrés generally came from the political elite, a very small group of committed opponents to Mussolini and his movement. They included eminent personalities of the previous regime—the former Liberal prime minister Francesco Nitti; the head of Italy's Catholic party, Don Luigi Sturzo; and the Liberal leader Giovanni Amendola. With them came many Socialists and anarchists, prominent left-wing and liberal journalists, and even a group of dissident Fascists—people who had lost out in the furious scramble for power among provincial Fascist chieftains in the months following Mussolini's March on Rome. These exiles normally assumed that their stay abroad would be short. Like many conservative backers of Fascism in the autumn of 1922, they reasoned that political warfare would soon diminish in intensity, that the spasmodic violence of the Blackshirts would come to an end, and that the Duce himself would likely be brushed away by more experienced political hands. For many, the decision to live in exile was taken only after anguished reflection.

Mussolini may have been pleased to see the earliest anti-Fascists go abroad although his movement does not seem to have systematically forced large numbers to do so. It took time for Fascism to fasten its hold on Italy. Until 1924, an active parliamentary opposition remained in place, and the Duce ruled through a coalition government containing non-Fascists. Gradually, however, the screws of totalitarianism tightened. Parliamentary opponents were beaten and occasionally killed; the *squadristi*, now turned into an officially designated "national militia" under the command of Fascist henchmen, roamed the countryside intimidating dissidents; critical newspapers were silenced, and their editors imprisoned. A particularly important step was the murder of the Socialist opposition leader Giacomo Matteotti by Fascist thugs in June 1924: thereafter, violence was increasingly directed at the elimination of non-Fascists from parliament and public life. Following the Matteotti murder, some of the most important political opponents of Fascism went abroad, mainly to Paris, which became a major center of anti-Mussolini agitation.

By 1926, the anti-Fascist emigration became politically significant, involving more than the presence of a small number of dissidents outside Italy. Occasionally infiltrating Italy to communicate with the opposition, they appeared well organized, and capable of doing important mischief to the Duce's regime. Mussolini referred disparagingly to the exiles as *fuorusciti*, implying that they were a band of self-interested outlaws; at the same time, he intensified the persecution of non-Fascists at home, ensuring that even greater numbers would leave.

Anti-Fascist exiles abroad seemed numerous and important because they merged into the vast sea of Italian emigration. During the first five years of Fascism, more than 1.5 million Italians emigrated, mainly for economic reasons; by 1927 about 9.2 million Italians were living outside Italy.[2] Within the huge émigré communities there were always pockets of warm support for the former luminaries of the previous regime. Such support strengthened over time as the exile community received fresh infusions from Italy and as the list of illustrious refugees lengthened. Concerned about opposition centers created by emigration, Mussolini took steps to curb their importance. Fascist agents abroad engaged in running battles with the émigrés. Mussolini maintained special punitive squads for terrorizing vocal émigré opponents of the regime, and the *fuorusciti* in turn struck out at the Fascists. In the mid-1920s these conflicts seriously threatened public order in France, the European home both of most refugees and also most of those who migrated to earn a better livelihood. In 1924 an anarchist refugee murdered Nicola Bonservizi, the chief Fascist agent in France, while he was seated on the terrace of a Parisian café; in response, Mussolini dispatched to Paris his gunman Amerigo Dumini, leader of the assassins who killed Matteotti.[3] More public shootings followed. Two years later France was racked by a scandal concerning the Italian refugees: police investigations disclosed that an anti-Fascist army was being recruited and organized in the Nice region by General Giuseppe

("Peppino") Garibaldi and his brother Ricciotti—grandsons of Garibaldi, the nineteenth-century nationalist hero. This effort attracted several hundred adventurers and appeared important enough to damage Franco-Italian relations at the time. Soon it emerged that the Garibaldi brothers, in turn, were being subsidized by a Fascist agent in France, giving the entire affair the allure of a provocation—creating even further tension between the two countries.[4]

The Fascist regime sent a new wave of refugees out of the country after November 1926, when, as part of his consolidation of power, Mussolini issued a Decree on Public Safety to reinforce the Italian police state. By its provisions, opposition to Fascism became an offense, and Fascist terror was legalized. Furthermore, Fascism established a new crime, "abusive emigration." Leaving the country was now a matter for strict governmental control. All passports were annulled, and no one was permitted to emigrate without permission. Frontier guards were instructed to shoot anyone attempting to do so. The law also permitted the government to denationalize opponents abroad even if they committed no crime; all that was needed was that, in the view of the government, they had injured Italian interests or brought Italy into disrepute. In the wake of these moves, hundreds of new fugitives turned up in France, Spain, Switzerland, Belgium, or England—without papers and under immediate danger if sent back to Italy. About the same time wives and children began to accompany the émigrés, a sign that many now believed in the durability of the Fascist regime. From having been limited largely to the political elite, therefore, political emigration had become a mass movement.

Eventually, some ten thousand anti-Fascist exiles left Italy, the overwhelming majority heading for France.[5] France, of course, appealed to the refugees because of its proximity to Italy and because of the affinity of its culture. Refugees to France also benefited from the presence of a huge Italian population that had recently settled there. There were almost 900,000 Italians living in France in the mid-1920s, about one-third of the entire immigrant population in that country. Italian cafés, libraries, newspapers, and political parties abounded, creating a natural home for uprooted dissidents. One Italian refugee noted that Italian exiles never felt isolated there: "former Italian members of parliament found in France . . . tens or hundreds of their electors, known journalists many of their readers, mothers of families fine Italian shops and food, and everyone is finding friends or even relatives." Networks of villagers or relatives existed to find jobs and provide accommodation or moral support.[6] The French, moreover, though anxious lest provocations and assassins run riot, proved remarkably receptive to the *fuorusciti*. Although Mussolini repeatedly pressured the French government to silence the refugees' anti-Fascist activity, Paris refused to comply. French governments certainly were not looking for trouble with the Duce and did not relish receiving large numbers of Italian leftists, but they both kept the gates of immigration open and refused to outlaw the anti-Fascists' political activity. Paris became an important center of opposition to Mussolini, home to a series of anti-Fascist organizations.

The presence of the *fuorusciti*, as a result, remained a sore point between France and Italy throughout the interwar period. Mussolini may well have manipulated the issue of anti-Fascist exiles as an artificial problem, useful in picking a quarrel with his detested neighbor.[7] Provocations, however, were not limited to one side. In the summer of 1930, the French-based anti-Fascist group *Giustizia e Libertà*, founded a year earlier, sponsored a spectacular coup against Mussolini's regime by sending an airplane to drop leaflets over Milan in broad daylight—the first of several such ventures. And in the early phase of the Spanish Civil War, Italian exiles organized the Garibaldi Battalion to fight in Spain, clashing victoriously with Mussolini's soldiers at Guadalajara the following year. The Fascists, for their part, heightened their activity against the *fuorusciti* when France was ruled by the Socialist-led Popular Front, notably sponsoring the murder of the anti-Fascist Carlo Rosselli and his brother by the extreme right-wing Cagoule movement in 1937.

Unlike other anti-Fascist refugees, the Italian exiles generally remained a buoyant lot in the interwar period, quixotically optimistic about their prospects for return to a post-Fascist Italy. "No other political migration has such an advantage in exile," one of them cheerfully observed in 1938. Italian émigrés were not widely disliked and lived surrounded by their compatriots. "In such conditions humanitarian and social problems are for the Italian refugees considerably attenuated and the political work is not being carried on in a desperate vacuum."[8] Until the climate of international opinion turned definitively against refugees in 1938, the most serious danger they faced came from Fascist agents abroad. Given the economic value of Italian emigration in post-war France, their security was never really in doubt.

Refugees from Germany

In the earliest phase, refugees from Germany appeared not unlike their Italian counterparts of the 1920s. Here, too, the first émigrés included prominent political personalities, distinguished scholars, journalists, and the like. Norman Bentwich, then working for the League of Nations' High Commission on Refugees, was struck by the distinguished roster of over 1,200 German professors who went abroad shortly after Hitler took power: "In the academic world there had been nothing comparable to it since the emigration of the Greek scholars or the capture of Constantinople by the Turks in the fifteenth century."[9] As in the case of Italy, hardly anyone foresaw a permanent resettlement, and few considered that the first exodus would lead to a mass migration. Emigrants could take a great deal of their property with them, and some retained capital in Germany, believing it would be safe.

The exodus began in the spring of 1933 in a hailstorm of street violence and legislative assaults on those presumed to be anti-Nazis. In the first weeks, Hitler's Storm Troopers unleashed their own campaign against the enemies of Nazism, randomly beating and harassing Socialists, trade unionists, Communists, and Jews. The Nazis considered their program a revolutionary enterprise and saw emigration as a logical consequence of revolutionary politics. In the flush of victory in 1933, however, they seemed to have devoted little concerted effort to driving people out of Germany. Terror, indeed, was decentralized. State governments and local administrations went their own way, with some maintaining civilized standards of law and order and others crumpling before the Hitlerian onslaught or zealously pursuing victims on their own. During the first weeks and months, refugees left Germany and returned, members of families went back and forth, and individuals made their plans without the assistance of organizations. For Thomas Mann, harassed by Nazism in Berlin, exile began in Munich, once the headquarters of Hitler's movement. Although terror had swept successfully through Prussia and other parts of the Reich, Mann reasoned in early 1933 that he and his family would be safe in Bavaria, considered for a fleeting moment able to hold out against the Gestapo.[10]

Seen in retrospect, Nazi terror fitted into the broad strategy of Gleichschaltung, the mobilization of every branch of German government, administration, and society in the service of the Hitlerian movement. Gleichschaltung meant the bludgeoning of all independent spheres of interest and authority in an effort to solidify the new regime and extend its power. Following the burning of the Reichstag building in Berlin at the end of February 1933, Hitler secured passage of emergency decrees that suspended individual liberties and permitted unlimited repression of the enemies of Nazism. In March he bullied the German legislature into passing the Enabling Act, giving the Führer dictatorial power unfettered by opposition or the inhibitions of law, government, or bureaucracy. Turning now to the task of consolidation, Nazism created a vast police network eager to stamp Germany with the Hitlerian imprint. Thousands became victims of a reign of terror because they refused to be "coordinated," or because they simply got in the way of the S.A. or Nazi party activists.

Consistent with the ideological foundations of the movement, the Nazis focused particularly on Jews. Thirsting for violence and pillage, Hitler's brownshirted Storm Troopers threw themselves into a violent antisemitic campaign, which stunned onlookers and horrified international opinion. Officially, the net of persecution embraced all "non-Aryans," snagging many who never considered themselves Jews at all. The Reestablishment of the Professional Civil Service Law of 7 April 1933, for example, stipulated that non-Aryans included everyone descended from one "non-Aryan" parent or grandparent. Thousands were purged from the bureaucracy, teaching, and the professions. Thereafter a series of laws secured removal of Jews from the schools, excluded them from German cultural life, and denaturalized those who had obtained citizenship under the Weimar republic. And more anti-Jewish laws were promised.

Everywhere in German society terror and repression found victims—filling the concentration camps, removing people from positions of influence and prestige, robbing them of their livelihood. But Jews were the foremost victims. According to Bentwich, about four-fifths of the 65,000 refugees who left Germany in 1933 were Jews.[11] In 1933 there were approximately 525,000 people in Germany identifying in some way with the Jewish religion and

about 600,000 "non-Aryans," about 1 percent of the entire population. Between 1933 and early 1938, when Jewish emigration
accelerated drastically, 150,000 of these people left the country—
roughly one-fourth of the non-Aryans in Germany. Accurate statistics are extremely difficult to establish because Jews emigrated
under a variety of auspices, because some shunned the German-
Jewish organizations that helped keep track of the Jewish emigrants, and because many exiles returned to Germany after a short
period. It seems likely that the proportion of non-Jews among
the refugees declined significantly after 1933, following the stabilization of the Nazi regime and with the increased persecution
of Jews.[12]

The first waves of refugees fled to neighbouring countries, often
unaware that most would never see Germany again. Most ended
up in France although substantial numbers came to Belgium,
Holland, Switzerland, Czechoslovakia, and Austria. Others traveled to towns and cities in the Saar Valley, then under French
administration, to Yugoslavia, or even Hungary. A few thousand German Communists made their way to Moscow although
as we shall see, the Russians not only blocked entry to most of
them but also persecuted the few who did find refuge in the Soviet Union. Finally, from 10,000 to 20,000 foreign Jews from Eastern Europe, mainly Poland, were expelled by the Germans in 1933
and forced to return home.[13]

Given the nature of Nazi terror, increasingly directed at the
entire Jewish population, the refugee exodus soon became a mass
migration. The first wave was predominantly of political opponents and distinguished notables, and it included many non-Jews;
succeeding waves were overwhelmingly Jewish and came to include a much greater proportion of the young, the very old, and
entire families. Although the pace of emigration slowed in 1934
and most of 1935, it increased substantially after the Nuremberg
Laws of September 1935 deprived Jews of their German citizenship and led to extensive new exclusions from economic and
public life. To slow the reflux of émigrés back to Germany, the
Gestapo issued new orders to apprehend returnees and put them
in concentration camps. To a greater degree than ever before, Jews
responded by emigrating, now with the sense that resettlement
abroad might indeed be permanent.

Despite the gravity of the situation, however, the exodus from
Germany remained limited before 1938. Until then, German Jewry
was relentlessly worn down by economic desperation rather than

driven abroad at the point of a gun. Terror singled out the Jews, but before 1938 there were always grounds for believing that the situation might stabilize or even improve. Anti-Jewish rioting flared in April 1935, climaxing in mass attacks in Berlin during July. Thereafter order was reestablished, and some took the Nuremberg Laws, announced that autumn, as a sign that the murderous brutalities would cease, that law and order would prevail. During 1936 violence in the streets diminished, with the German effort to present a favorable international image for the Olympic Games of that year. Four and a half years of Nazism punished the Jews economically and stripped them of civil rights, but the most important drive to ruin them entirely awaited the removal of economics minister Hjalmar Schacht in September 1937. In the face of creeping persecution and harassment, therefore, Jews retained a foothold in Germany.

Nazi policies ensured that the bulk of the refugees would leave with practically all their capital drained away. In the end this proved disastrous for German Jews, who were seeking to enter countries caught in the grips of the most serious economic depression of the century. The Nazis made full use of the *Reichsfluchtsteuer*, an emigration tax first introduced by the Weimar Republic in 1931 in order to help conserve foreign currency reserves in Germany and also to deter people from leaving the country. The tax soon became a means of wringing from the Jews their last remaining resources. German income from this source rose spectacularly before the outbreak of war—from one million marks in 1932–33 to 342 million marks in 1938–39. In all, the tax may have yielded as much as 900 million marks to the German Treasury. Beyond this, Nazi regulations restricting remittances of income from Germany hit hard at émigrés living abroad. Material losses through emigration climbed from 30 percent to 50 percent of the capital of the refugees for the years 1933–37, and for the period 1937–39 they reached from 60 to 100 percent. As a result, German Jewry was practically ruined, and those who left increasingly migrated as paupers unable to support themselves.[14]

Refugee Movements in Europe, 1933–38

Our own picture of the refugee situation in Europe under Hitler is heavily colored by developments in the two years before the outbreak of war in 1939, when the exodus became a panic-stricken flight. During most of the interwar period, however, the number of refugees unsettled in Europe was relatively small—certainly under 50,000. Refugees continued to sail overseas in numbers approximately equal to those who remained on the Continent. In December 1935, when he resigned his position as high commissioner for refugees coming from Germany, James G. McDonald indicated that some 80,000 refugees had already left the Reich, approximately three-quarters of whom had found new homes or been repatriated to their countries of origin.[15] At the end of 1937, the Council for German Jewry estimated that out of a total of 135,000 Jewish refugees, 100,000 had gone overseas, of whom 43,000 went to Palestine.[16] Refugees often remained in Europe, expecting to return before long to Germany. This was particularly the case with those who emigrated for political reasons. Some remained unsettled while awaiting the opportunity to enter specific countries overseas or because they sought a better situation for themselves than in their first country of asylum. And some, tragically misreading the political oracles, returned to Germany.

A small stream of refugees from other regimes of the authoritarian right joined the great course of those in flight from Nazism. We have already noted the important refugee movements associated with the establishment of new countries of Eastern Europe following the Paris Peace Conference. Occasionally, Hungarian or Rumanian dissidents turned up in Western European capitals, living the life of political exiles. Driven out of his country by a successful insurrection in 1924, Albanian ruler Ahmed Zogu found refuge and support in Yugoslavia. Authorities in Belgrade matched him with antitsarist exiles, enabling Zogu to restore his rule in Tiranë by the end of the year. Portuguese opponents of Salazar's dictatorship emigrated in the late 1920s after the failure to overthrow his government in 1927, and more came following the establishment of the Portuguese Estado Novo, with its antidemocratic constitution of March 1933. The next year saw a political emigration from Austria. Under pressure from Austrian Nazis as well as the new regime in Berlin, the government of Chancellor Engelbert Dollfuss moved sharply to the right in 1934, dispensing with a parliament and tying his country close to Fascist Italy for protection. Partly at the urging of the Duce, Dollfuss struck brutally at the Austrian Social Democrats, ultimately banning the party entirely. This policy provoked a near civil war in February, resolved by the government with the help of artillery and organized Viennese rightists grouped into the Heimwehr. As a result, thousands of Austrian leftists decamped, most heading for Paris.

About 7,000 refugees left the Saar Valley after January 1935, when a plebiscite held in the region overwhelmingly favored its return to Germany. Previously, under League of Nations administration, the Saar had been an important political refuge for dissidents fleeing Hitlerian Germany. The 1935 vote boosted Nazi morale and signaled to the world the exuberance of the Hitlerian regime. Saarlanders who feared to live under Hitler promptly abandoned the region, together with prominent Social Democrats, Communists, and Jews who had found sanctuary in the region during the previous two years. Reasoning that they could afford to be conciliatory now that they had won a major foreign policy coup, the Germans allowed the refugees to retain or sell immovable property and to remove movable property. Between 38 and 40 percent of the émigrés were Jews, and the great majority crossed the frontier to France.[17] Having received the bulk of the Saar refugees, the French then championed their case before the League of Nations, securing for them the use of the Nansen certificates previously reserved for Armenians and Russians. France also insisted that Germany not punish opponents of reunion with Germany who chose not to become refugees in 1935. The French delegation to the League urged fellow members to help assume responsibility for the exiles and secured a special grant by the League Assembly to resettle Saar refugees in Paraguay. The numbers involved in the latter project, however, were derisory—under 200, in the end.[18]

Across Europe, the Free City of Danzig provided an unusual case—both a refuge from fascism and, later, a place from which refugees tried desperately to escape as war approached in 1939. A semi-independent Baltic seaport, with a population of 357,000 in 1919, Danzig was separated from Germany by the Versailles Treaty and tied to Poland by a customs union and by Polish control of its foreign policy. Placed under League of Nations administration, Danzig remained continually in the public eye—a nagging source of dispute between Poland and Germany and an embarrassing obligation assumed by the international community in 1919. Particularly awkward for the League of Nations was the strong support for Hitler among the huge German population of the city—sufficient to turn the local government toward

Nazism as early as 1933. Despite this delicate situation, some 5,000 Polish Jews migrated to the Baltic seaport, preferring its relatively free economic climate to boycott and persecution in Poland or Germany. Interestingly, Hitler urged local Nazis not to radicalize local politics or to turn too sharply against the Jews. In the early years of the Third Reich, the Führer was eager to conciliate Poland rather than force a confrontation. The Nazis signed a nonaggression pact with Warsaw in January 1934, and this relationship restrained German moves to absorb Danzig until 1938. During this period, therefore, Danzig continued to receive refugees.

Local Nazis managed, nevertheless, to maintain pressure on Danzig's Jewish community, which numbered over 10,000 in 1929. In 1935 the Permanent Court of International Justice in The Hague declared unconstitutional a series of antisemitic laws passed by the Danzig Senate—all to no avail. Thereafter the Nazis in the Free City began to move against their opposition. The next year local authorities banned the Social Democratic Party and imprisoned most of the opposition leaders or forced them into exile. By 1937 about 3,000 Jews had left.[19] In October of that year there were serious anti-Jewish riots, with the sacking of stores and the arrest of Jewish merchants. Danzig's Nazi party threatened to introduce the Nuremberg Laws into the city. Although the League of Nations' high commissioner, the Swiss historian Carl Burckhardt, managed to win a reprieve, he failed to block the Nazis' assault. Therefore, 1938 saw a refugee crisis in Danzig as well as in the rest of Europe.

Given the rising crescendo of anti-Jewish attacks before 1938, it may seem odd that Jewish refugees returned to Germany. We know, however, that German return migration of other groups was often quite important and that Jews came back to the Reich in substantial numbers at least until 1935, when the Gestapo moved to punish this process severely. In part, return was a sign of the failure of integration elsewhere. Norman Bentwich noted that until 1935 philanthropic agencies outside Germany encouraged repatriation of those refugees who could not properly adjust to countries where they found refuge. The Joint Distribution Committee had inadequate resources to assist the emigrants abroad and in 1934 measured its failure in the reflux of between 1,200 and 1,500 refugees from Holland and almost twice that number from France.[20] Unemployment provides part of the explanation. In the early and mid-1930s, joblessness climbed to over

30 percent in the Netherlands, for example, when neighboring Germany, in the flush of rearmament, appeared full of opportunities. From across Europe and the United States people of German background flocked to the Reich, seeing a dynamic, vigorous country that appeared to have solved the problems of economic depression.[21] At the same time, Jews were affected by the legendary homesickness that beset German émigrés abroad. Karl Jaspers once told a story about a German-Jewish refugee in New York who kept Hitler's picture on the wall of his room; only that daily reminder of the horror of the Hitlerian regime could keep him from pining for his native land. There were refugees, Jews and non-Jews, who failed to overcome such sentiments and returned to the Reich.

It is impossible to know how many voluntarily abandoned their exile; it was certainly not wise to declare publicly one's intention to do so. Historian Yehuda Bauer states that of 53,000 Jews who fled in panic in 1933, 16,000 returned. He further notes that between 1933 and 1938 some 6,000 Jews were reported to have abandoned a harsh asylum in Poland to live again in Germany.[22] According to a German-Jewish source, which had no reason to inflate the size of the reemigration, 10,000 Jews who had fled abroad were beginning to return by early 1935.[23] We must remember that Jews had no way of knowing, at this point, of the genocide that awaited them. Four years after the Nazis took power, Jews maintained an economic role as we have seen, and they hoped that legislation would relieve their plight by ending sporadic violence and what seemed like gratuitous harassment. This was an illusion, of course, but in their wishful thinking returning Jews did not differ from many non-Jewish counterparts or from European onlookers. Hitler and Nazism, they felt, would not last forever. Unhappy or disappointed abroad, some were prepared to await the demise at home.

REFUGE IN THE DEPRESSION ERA

International Climate

Refugees, one might argue, always arrive at the wrong time. Seldom has this been more true than with those fleeing Nazism in the 1930s. Throughout Western Europe, where most anti-Nazi

refugees were headed, economic depression conditioned attitudes toward immigration. Depression undermined every European economy west of Russia following the financial crisis of 1931,
triggered by the collapse of the Credit-Anstalt Austrian banking
house in May of that year. In the ensuing panic, difficulties fed
one another: international credit evaporated, financial institutions suffered huge losses, and the repercussions for domestic
economies were catastrophic. Production fell, and unemployment climbed to spectacular levels. The ensuing hardship put
extraordinary pressures on democratic societies and institutions.
Nervousness about economic conditions, moreover, was com-
pounded by weakened confidence in liberal society and a sense
of dread about the future encouraged by postwar literary and
cultural trends. Weimar Germany, bearing the heaviest weight
of depression with close to six million unemployed in 1932, succumbed to Nazism by the beginning of 1933. In Southeastern
Europe the allure of fascism grew appreciably. Even in Western
Europe pools of angry demonstrators grew larger; and left- and
right-wing extremists, more strident, threatening to topple liberal parliamentarians.

To deal with the crisis, these governments looked within their
own societies and saw the remedy in restrictions—preventing the
growth of the labor supply, reducing government expenditures,
stabilizing currencies. In a Europe as yet uninfluenced by the
British economist John Maynard Keynes, advisers to every government recommended prudence and restraint—prescriptions
hardly attuned to the appeals of increasingly desperate refugees.
Accordingly, governments equipped themselves with emergency powers and buckled on the armor of protectionism. Tariffs
went up—the first sign that a narrowly defined national self-
interest would define policies in the international arena. Doubters
were told to be patient; patience would see markets readjusting
themselves, eventually restoring economic health.

Impatient by definition, refugees ran into a wall of restrictions
hastily erected in Western European countries in the early 1930s.
For those contemplating flight, the problem was not so much the
inability to enter countries of refuge as it was securing permanent residence rights. Governments were reluctant to consider
refugees as anything but visitors, soon to pass through their
countries on their way to somewhere else. Jobs were the crucial
factor. On this score, rather than on the issue of temporary asylum, gloom enveloped all evaluations of the Jewish future in

Germany: "The prospect of increased emigration in the near future is small," wrote Arthur Ruppin in 1934. "At the very best years will pass before the countries which in the past have been open to immigrants will have absorbed their own unemployed and before a shortage of labor will result in a revision of the official immigration policy."[24] As early as June 1933, French, Dutch, and Belgian representatives to an international labor conference in Geneva deplored the deleterious effects of refugees on swollen labor markets. States like these had previously welcomed thousands of Jewish emigrants from Eastern Europe to work in the boom years of the 1920s. Now they sought ways to discourage new entrants. In time, these countries obliged immigrant refugees to satisfy local authorities that they would be no burden on domestic economies, would offer no competition for jobs or livelihood, would pose no problems for law enforcement.

Similar policies were applied in the United States. Since 1929 the Americans carefully scrutinized applicants for visas, requiring consular authorities to assess whether the immigrants were "likely to become a public charge"—a policy hardened by administrative usage into what was known as the LPC clause. As a result, the Americans continually failed to fill the annual quota of immigrants from Germany, set in the 1920s at 26,000. In 1935, responding to pressure from Jewish leaders, Roosevelt eased procedures somewhat and allowed greater numbers of refugees to immigrate. But the fundamental policies of restriction remained in force, and the numbers never did exceed the original annual quotas: 4,392 came in the years 1933–34; 5,201, in 1934 and 1935; and 6,346, in the years 1935–36. Of these, between 80 percent and 85 percent were Jews.

Temporary havens certainly existed in Europe in the early 1930s. Between April and September 1933, for example, 10,000 German refugees crossed into Switzerland at one frontier station alone. Three-quarters of the German Jewish refugees remained in Europe in 1933, and about a fifth went to Palestine. Gradually, the European component diminished as restrictions on the Continent came into force and as Roosevelt liberalized American policies somewhat. Jewish relief agencies worked strenuously to facilitate overseas migration of refugees. In time, the refugees abandoned hopes of a prompt return to Germany and cared less about remaining close to relatives in Europe. British restrictions on entry into Palestine had an important effect in 1937, and Jewish immigration there fell off drastically. Just before 1938, when the crush of refugees fleeing in haste or attempting to leave the Reich became overwhelming, the greatest proportion of Jewish refugees, about 38 percent, were bound for the United States. The Americans received 10,895 Germans in 1936 and 1937, 70 percent more than in 1935.[25]

For a Europe still in the throes of depression, the refugee situation appeared serious, but certainly under control. In his widely read report prepared for the Royal Institute of International Affairs in 1937, Sir John Hope Simpson estimated that there were 165,000 refugees from Germany at the end of that year. Most of these had managed to find places overseas. Current refugee problems, he noted, "could be solved by concerted efforts and the best use of existing institutions." For this to be accomplished, however, the German exodus would have to be reduced, and the Nazis would have to abandon their practice of stripping the emigrants of everything they owned before they quit Germany, thus ensuring that they would be paupers the moment they crossed the frontier. Hope Simpson also worried about Spanish Civil War refugees, the financing of the resettlement of Russians in the Far East, and threats to Middle Eastern minorities. As the crisis year 1938 approached, he looked in several directions for solutions and drew a reasonable conclusion: "the difficulties were not insuperable."[26]

However resourceful in his search for solutions, Hope Simpson did not bother to look east to the Soviet Union. It is well for us to keep in mind that whatever obstacles countries placed in the paths of refugees, these shrink into insignificance when compared to the unbending exclusion imposed by the Soviets. The Russians, indeed, were generating their own refugees during this period. This, in addition to the persecution of the few left-wing exiles who did gain access to Soviet sanctuary, forms an ironic commentary on the internationalist rhetoric of the Communist movement in the 1930s.

Still isolated from the West in economic as well as political terms, Russia remained unaffected by the financial collapse of the early 1930s and the economic repercussions we have noted. Nevertheless, this was a particularly troubled period in Soviet history, associated with the forced collectivization of agriculture built into Stalin's Five Year Plan of 1929. In practical terms, this plan meant a remorseless assault on a section of the Soviet peasantry, virtual civil war in parts of the U.S.S.R., and a terrible famine provoked by Stalin's own policies. The Ukraine endured the worst of this catastrophe, with deaths from starvation and repression numbering several millions. To achieve their goals, Soviet officials deported millions of well-to-do peasants or kulaks to work camps in remote regions of the Soviet state. At the same time, in an effort to reorder the Soviet economy, administrators shifted vast numbers of peasants and industrial workers to new settlements and new employment. The bureaucratic nets dragged millions from one corner of this vast country to another while the machinery of repression swallowed up countless others. And finally, just when the Soviets showed signs of recovering from these horrors, Stalin began a massive new wave of political persecution known as the Great Purges, which extended from 1936 to 1938.

Escaping this devastation, refugees occasionally emerged from the thick fog of isolation that hid Soviet reality from the West. Trotsky was the most famous émigré, expelled to Turkey in 1929. That same year thousands of religious dissidents, German-speaking and mostly Mennonites, appealed from Moscow to be allowed to leave Russia. The German government and public opinion intervened on their behalf, pressuring the Kremlin to permit an exodus. Eager to maintain good relations with the Weimar Republic, Moscow complied, allowing some to emigrate.[27] Meanwhile, kulaks and others in flight from the rigor of collectivization or Stalin's OGPU police network, continued to trickle into Poland, the Baltic states, or Rumania. To stem the tide and also to clamp down on any hint of opposition, the Soviets imposed strict controls on internal movement and attempted to seal their frontiers. Internal passports were introduced in 1932—reminiscent of the tsarist empire. A law of 1935 set the death penalty for flight abroad and put families members of illegal émigrés severely at risk.[28]

Although little was known in the West about these events, the Soviet Union presented a sufficiently grim exterior to render her unappealing for most refugees. Nevertheless, thousands of persecuted Communists looked hopefully to the U.S.S.R. when Hitlian persecution became unbearable in 1933. Most were disappointed. Even after Hitler took power, the Russians were eager to maintain good relations with Germany; Moscow assumed that giving sanctuary to hounded Marxists would be considered an unfriendly act by the new Nazi regime. As a result, Soviet diplomats signaled that Stalin was willing to sacrifice the German Communists to help preserve the Rapallo Treaty of 1922 with

Germany. Although the Soviet constitution offered asylum to foreigners persecuted for political reasons, the Russians limited very strictly the number they would receive. Jewish Communists were particularly unwelcome in Russia, especially during the purges of the mid-1930s, and found themselves sweepingly accused of Trotskyite sympathies.[29] Some Communists did receive asylum in the first two years of Nazism, but the purges brought this to a halt.

By the late 1930s the Soviet Union officially rejected the notion of receiving refugees from Hitler. The Soviet foreign minister at that time, Maxim Litvinov, indicated that "sympathies" for the sufferings of German comrades could hardly prompt him to interfere in German internal affairs—and presumably also to defend German Marxists, who were then bearing the brunt of Nazi persecution. Orthodox Communists accused leftist opponents of showing "humanistic fatuousness"—allowing policy to be dictated by mere sentiment rather than geopolitical wisdom. Russian spokesmen referred to the unfavorable occupational structures of the mainly middle-class German refugees and to their capitalist conditioning—factors that would complicate their absorption into a socialist state. Because the refugee problem really derived from the turmoil of capitalist society, they charged, it was incumbent on the latter to resolve the problem. One German Communist, Wieland Herzfelde, argued that the refugees represented a thorn in the eye of fascism in so-called democratic countries—a thorn that should not be removed by transferring them to the East.[30] Before the League of Nations Assembly in 1937, Soviet delegates implacably opposed strengthened refugee assistance and blocked proposed League of Nations action to pursue that objective.[31] During 1938, the crisis year for refugees, the British government wanted to explore the possibility of having the Russians receive Jewish refugees. This was the least propitious of times for such a humanitarian proposal, however, with an officially sponsored terror campaign swallowing up hundreds of thousands of lives in the U.S.S.R. From the British embassy in Moscow came a disappointing reply, noting that though there was no antisemitic policy in the Kremlin, "an infinitely larger number of Jews have in all probability been executed in the Soviet Union during the last two years than in Germany under the National Socialist Regime."[32]

Within the Soviet Union the small communities of refugees who had managed to enter the country suffered brutal repression during the Great Purges. Moved by Stalin, the Soviet police apparatus translated the dictator's intense paranoia and xenophobia into a policy of state. In the wake of the 1934 assassination of Sergei Kirov, a high-ranking Leningrad Party official, arrests, denunciations, deportations to work camps, and show trials became institutionalized in Soviet life. Huddled together in a handful of cities, exiles were among the most vulnerable targets of official terror. The NKVD swooped down on refugee groups in 1936, branding German writers as Trotskyites, Zinovievists, and fascist spies. Refugees were regularly summoned for interrogation. Emigré publications were silenced. Bewildered German Communists, some of them outstanding personalities, disappeared into the Gargantuan Lubianka prison in Moscow and were never seen again. Terrified exiles left at large were encouraged to spy on their fellows, to concoct charges against them, and to confess their own wrongdoing. Repression broke all bounds of rationality. Erich Mühsam, for example, was a famous anarchist murdered by the Nazis in Oranienburg in 1934. His wife left Germany that same year to lecture on the horrors of Nazism. Invited to Moscow, she was arrested as a Trotskyite, released, then rearrested and dispatched to a labor camp.[33] Life for the exiles who remained untouched was soaked in fear: they dared not speak German on the streets, could not approach a foreign embassy, panicked when they received mail from abroad. As Thomas Mann observed at the time, the socialist sanctuary was becoming like a prison cell. According to one scholar, over 70 percent of the writers, artists, and intellectuals who fled to the Soviet Union were ultimately arrested in the purges, and few of these survived.[34] Finally, the Russians sealed even the small chinks that had permitted a handful of exiles to travel abroad. NKVD agents tracked down Communist exiles living elsewhere in Europe and murdered them. Alternatively, the Soviets denounced them to local authorities and had them sent to Nazi Germany.[35]

New Threats Against the Jews in Eastern Europe

Although Nazi Germany appeared the immediate threat to Jewish existence in Europe during the 1930s, observers sometimes feared that a far greater menace existed in Eastern Europe, where governments in Poland, Hungary, and Rumania were increasingly hostile to their Jewish populations. Many more Jews were involved than under Hitler: Poland had over 3 million Jews; Rumania, almost 757,000; and Hungary, about 445,000—a total of close to 4.3 million, as opposed to the 525,000 in the Reich. Moreover, unlike the relatively prosperous, well-integrated German Jews, those of Eastern Europe were mostly poor and traditionally minded and suffered from the deep-seated anti-Jewish prejudice still widely diffused in Eastern European societies. There were fewer checks on violence and persecution than in the West, and some felt that in the long run the threat of catastrophe there was greater. As the storm clouds for Jews gathered in the second half of the 1930s, the Zionist Revisionist leader Vladimir (Ze'ev) Jabotinsky was only one of those who anticipated a much more extensive disaster for the Jews than was occurring under Hitler: "I am sure that elemental floods will soon break out all over Eastern European Jewry," he wrote, "so terribly powerful that the German catastrophe will soon be eclipsed."[36]

Those concerned about Jewish refugees frequently worried that German Jews were merely the first installment of a massive Jewish exodus or attempted exodus. "What if Poland, Hungary, Rumania also expel their Jewish citizens?" the London *Daily Express* asked rhetorically in March 1938, echoing these concerns.[37] The specter haunting refugee policy during the 1930s was that encouraging Jewish fugitives from Nazism might well provoke the flight of hundreds of thousands more from Eastern Europe. As the economic and political situation of Jews weakened during the depression years, a Jewish crisis of vast proportions loomed. Reinforcing these fears, Hungarian, Rumanian, and Polish representatives made repeated interventions abroad during the decade about their "Jewish problems" and proposed massive evacuations of these undesirables to other continents. At the same time, Jewish personalities urged vastly increased emigration for East European Jews. At the existing rate, noted Arthur Ruppin in 1934, a mere 40,000 to 45,000 migrated annually—only one-third of the Jewish natural increase. His conclusion was grim: "It is an open question whether under these circumstances the economic position of Jews in Eastern Europe, especially Poland . . . can be preserved from a catastrophic collapse."[38] From Poland, where the situation appeared most acute, the foreign minister Józef Beck calmly referred at international meetings to his country's "surplus" Jewish population and looked forward to an annual emigration of between 80,000 and 100,000—all of whom would have to leave their assets behind. The Rumanians similarly told the world that their own Jews had to go. In 1938 they approached the British, for example, calling for "a radical and early solution of the Jewish question in Rumania by means of internationally assisted migration." King Carol spoke directly to the British representative in Bucharest, envisioning the departure of some 200,000 Jews.[39]

Within the East European countries themselves, evidence suggested that the pessimists might be right and that the Jews faced a grave crisis. Concentrated in cities and market town, Jews grouped conspicuously in local commercial life. Often the only middle-class elements in local society, they drew sharp hostility during periods of economic modernization and emerging national self-consciousness. Their situation worsened critically in the depression years, when the economic nationalism of the underdeveloped countries of Eastern Europe blossomed into boycotts of Jews and intense discrimination. The added political antisemitism in several countries rendered the position of the Jews extremely insecure.[40]

In Poland the Jews' material circumstances declined steadily throughout the interwar period, accompanied by a Jewish emigration of about 400,000 from 1921 to 1931—a rate roughly five times higher than among the general population. Then came the depression, which hit Poland severely and at the same time vastly reduced opportunities for settlement elsewhere. The total number of emigrants from Poland fell steeply—from 243,000 in 1929 to 21,000 in 1932. Only 8,632 Jews left Poland in 1931.[41] Polish antisemitism began a powerful revival in the early 1930s, likely linked to economic difficulties and to the close relationship with Nazi Germany from 1934. Following the death of Marshal Józef Pilsudski the next year, political leadership drifted to the right, and anti-Jewish voices received official sanction. From 1935 to 1939 a kind of "war against the Jews" existed in Poland, involving economic boycotts, the segregation of Jewish university students, and exclusion of Jews from certain professions. Pogroms flickered across Poland in 1935 and 1936, recalling the anti-Jewish riots of the tsarist era. Although officially opposed to anti-Jewish violence, the Polish government became increasingly committed to the idea of removing large numbers of Jews from the country. Meanwhile, boycott and depression ruined a good part of Polish Jewry economically. By the late 1930s Poland was a disaster area for Jews. An English observer offered the following appraisal: "There seems little doubt that overwhelming opinion in Poland today favors the elimination of the Jew from economic life and from the 'polonization' of commerce. The first step in this direction is the economic boycott, the second emigration. The third step of out-and-out Nazi legislation depends upon whether Poland becomes totalitarian."[42] In 1939 this appeared to hang in the balance.

Things were much the same in Hungary and Rumania although crises there were perhaps slower to develop. Both countries saw right-wing regimes emerge in the course of the Great Depression, and both swung into the orbit of Nazi Germany, which extended its influence into Southeastern Europe in the second half of the 1930s. In Hungary the advent of the radical rightist Gyula Gömbös as head of government in 1932 highlighted the Jewish issue. By this time traditional Magyar conservatism, once protective of Jews up to a point, had weakened considerably. Hungary was prey to more adventuristic nationalists, eager to revise the Treaty of Trianon and sympathetic to fascism. As prime minister, Gömbös exuded contempt for liberal democracy, moved openly to an alliance with Hitler, and prominently displayed his dislike of Jews. Although in practice Gömbös proved more restrained than his rhetoric, his government signaled a growing acceptability of anti-Jewish policy. Thereafter succeeding Hungarian leaders actively pressed their case against the Jews. Hungary passed an anti-Jewish law in the spring of 1938, intended to remove Jews from industrial, commercial, and professional spheres of activity. More such statutes followed, cutting down even further on the Jewish role in Hungarian life. Nazi propaganda against Jews flooded the country, and the local fascist equivalent, the Arrow Cross, achieved an important political breakthrough on the eve of the Second World War.

Interwar Rumania followed a similar pattern. The principal focus of attack was the Jews' purported disloyalty—the continuing allegiance to Hungary of those in Transylvania, the "Bolshevism" of others accused of leaning toward Moscow, or the supposed cultural betrayal of those who refused to be authentic Rumanians. Mainly limited to student groups and ideologues in the 1920s, radical political antisemitism reached the center of power by 1937, with the extreme right-wing government of Octavian Goga. Violent antisemitism and pogroms were the result. King Carol dismissed Goga early in 1938, but the succeeding two years of royal dictatorship provided only limited relief. The king, as we have seen, believed that Jews should be eliminated—although he was silent on details. In an effort to ease the situation, the prominent Jewish leader Wilhelm Filderman negotiated an exodus of 50,000 Jews annually. Similarly, the Jews won government support for illegal Jewish emigration to Palestine.

The Jewish refugee question of the late 1930s, therefore, hardly was limited to Germany. It is true that many observers understood antisemitic campaigns elsewhere as being the result of "infection" with German ideas. But those concerned with the political impact of antisemitism could not help being preoccupied with what Sir John Hope Simpson called "the potential problem." Even in Czechoslovakia and the Baltic countries there were serious signs of trouble during the 1930s. "The whole of Eastern Jewry is in an insecure position," Hope Simpson wrote in 1938, "and both in Poland and in Rumania at different times government spokesmen have suggested that measures will be adopted to induce emigration. . . ." To the east of Germany, Hope Simpson observed, there were nearly five million threatened Jews—excluding those of Russia.[43] To the more gentle imaginations of the interwar era, five million was a refugee figure impossibly large to contemplate. Briefed on the subject in early 1939, Franklin Roosevelt had precisely this reaction. Flight on such a scale was inconceivable, the American president and most other political leaders felt, and there was no point encouraging people to believe otherwise.[44]

Closing the Doors: Some Examples

FRANCE

Throughout the interwar period, France was the most important European sanctuary from fascism, receiving more refugees than any other country—even before over 450,000 Spanish Republican exiles tramped across the frontier in the spring of 1939. Geography and political tradition dictated that France would play this role. Not only was France a neighbor both to Germany and Italy, but she also had a long tradition of granting asylum to the persecuted fugitives of Central and Eastern Europe as well as of the Iberian Peninsula. For generations France had received the political outcasts of repressive Continental regimes, and Paris had been for émigrés a political Mecca where their revolutionary faith could be kept alight. In 1933, with a new wave of fugitives from fascism, many assumed that France was the obvious place to go, to organize, and to await the collapse of Hitler.

During the 1920s, French immigration policy reinforced French liberality toward refugees. Following the First World War, as we have seen, the French were eager to make up their catastrophic manpower losses by encouraging the immigration of foreign workers. From Poland, Belgium, Italy, and other countries, these came in great profusion, raising the immigrant population to almost three million in 1931. Among the immigrants was the bulk of the *fuorusciti*, the anti-Fascist enemies of Mussolini's regime. With the Great Depression, however, both businesspeople and politicians had serious second thoughts about the previously open immigration door; economic disaster argued strongly for restrictions. In 1932 industrial production plunged to a third of what it had been in 1929. Official statistics listed well over a quarter of a million unemployed in 1934—an understated total that was still extremely high by French standards. Theoreticians from the French right began to be heard, arguing that the presence of so many outsiders weakened rather than strengthened French society. Some who had scoured the Continent for laborers during the previous decade now clamored for foreigners to be sent home; professional groups demanded quotas to limit competition from recently arrived outsiders. Even before Hitler, therefore, the question of foreigners in France became a major national issue.

Economic concerns had no effect on the very first wave of refugees from Nazism, some 40 percent of whom came to France. In the spring of 1933, these poured across the frontier unimpeded, even if without proper passports. In April, Interior Minister Camille Chautemps encouraged prefects of border departments "to give to German refugees the same hospitality formerly offered in analogous circumstances to Italian, Spanish and Russian citizens." For a brief moment, sympathy for the victims of Nazism was widespread.[45] After six months of beneficence, however, restrictions began to pour down from Paris, where politicians worried about the accumulating mass of fugitives and rising opposition to them. Near the end of 1933, the Ministry of the Interior ended the exceptional measures to receive refugees and tightened the conditions for entry of new exiles from Germany. Despite widespread support for anti-Hitler personalities and the first victims of Nazism, French officials began to turn some of the refugees back. Other bureaucrats exercised their legendary French capacity for delay and obstruction. Significant numbers of applicants were refused entry, and others were expelled when their illegal status was discovered. Once in France the refugees found it increasingly difficult to obtain work permits or the necessary residency documents. Despite the welcome for the Saar refugees in 1935, this general approach continued up to the time of the Popular Front government in the spring of 1936. Nominally, France maintained an official policy of asylum; in practice, she received more refugees than any other European country; but the immigration machinery, once well oiled to receive refugees, was now working to keep their numbers down. Norman Bentwich, representing the League of Nations' High Commission for Refugees, went from one ministerial office to another in the early 1930s and sensed an unfriendly shift: France, he was told, would not become a dumping ground for refugees.[46]

Behind these restrictions lay a serious crisis in French society. Particularly after 1932, when a government of the center-left took power, many French people were persuaded that the Republic was on the verge of disintegration. Sharp-tongued critics on the far right delighted in exposing the worthlessness of France's liberal heritage and the ruin of society by humanitarian ideals. Meanwhile, unemployment invigorated protest from the semifascist leagues and helped poison the public atmosphere. Under these conditions, those who championed refugees were put on the defensive. Opposition to the French policy of asylum merged with a wider current of xenophobia, antisemitism, and a deep distaste for liberal democracy. Helpless refugees were variously accused of stealing the jobs of Frenchmen, undermining France's cultural purity, or trying to embroil France with Hitler. The country had entered what the historian Claude Fohlen would later call "the era of hatred."

With the advent of the Popular Front government of Léon Blum in the summer of 1936, the situation eased somewhat for refugees, particularly for individual émigrés attempting to cross the bureaucratic obstacle course of immigration. Solemnly, a government spokesman promised that France would "remain faithful to all humanitarian principles and pay special tribute to traditions of hospitality." Expulsions, which reached a peak in the years 1934–35, fell off appreciably. Blum, who had been the victim of an antisemitic campaign of unprecedented violence, stoutly protested his support for the refugees, and the Socialist Party seriously endeavored to improve conditions for them.[47] Interior Minister Roger Salengro ordered prefects to speed the reception of refugees at the frontier, and the Ministry of Labor defended the issuing of work permits to exiles already in France. Nevertheless, the Socialist-led government remained cautious, attuned to the intense popular feeling over the issue and widespread fears that France could become swamped with outsiders. Much of the French working class, Simone Weil noted in 1937, remained hostile to the arrival of outcasts from abroad, strangers to trade union activity and willing to work for practically nothing.[48] Antisemitism, moreover, remained deeply embedded in much of the French popular consciousness and surfaced often in sweeping attacks on Jewish refugees.

The Popular Front made no structural change in French immigration policy. In Geneva in July 1936, while signing an international agreement legalizing the situation of refugees from the Reich, Salengro insisted that this engagement did not apply to new arrivals. Proposals for distributing refugees throughout the country reached the drawing boards and might conceivably have permitted the reception of significantly more exiles. But there was no time to develop these ideas. Support for the government drained away rapidly during 1937, and the economy failed to revive. The Spanish Civil War and the continued success of Nazism provided ammunition for the opposition. In 1938 the Socialist experiment collapsed, and a new refugee crisis began.

Under Edouard Daladier in 1938, the French government moved as sharply against the Popular Front liberalism on refugee issues as it moved against the forty-hour week or used emergency powers against trade unionists. Concerned mainly with defense and foreign affairs, Daladier had for many years judged the refugees a serious security problem, a kind of Trojan horse that would someday spill spies and subversives into French society. As the flood of refugees mounted, notably after the German Anschluss with Austria in March 1938, the French government tightened procedures for entering France. A decree-law of May 1938 made it more difficult for refugees to obtain temporary residence permits and allowed low-level frontier officials to turn them back summarily. "Repatriation, resettlement and interment were what the prime minister had in mind," concludes one historian, "not assimilation and economic assistance."[49] In addition to the tighter immigration rules and the new round of expulsions, the government armed itself with the means to denationalize immigrants who had obtained French citizenship and to intern undesirables in special camps. Officially, France proclaimed to the world that she was "saturated" with refugees—the word used by Henry Bérenger, France's representative on the Inter-Governmental Committee set up to assist German and Austrian refugees in 1938.

In the nervous tension preceding the outbreak of war in 1939, there is no doubt that both the public and officials exaggerated the importance of Central European refugees in France. The end of 1938 saw a new wave of exiles coming from Czechoslovakia, even more desperate German-Jewish fugitives who were virtually expelled from the Reich in the wake of Kristallnacht, along with Spanish Republicans fleeing the advancing armies of Francisco Franco. No one knew how many refugees had come, but it was certain that many simply passed through the country, as the French government encouraged them to do. Official statistics suggest that no more than 30,000 refugees from the Reich may at any one time have been in France in the years after Hitler took power. On the other hand, a government report on Jewish refugees prepared in late 1938 pointed to another 30,000 German and Austrian Jews illegally in France—*les clandestins*, as they were called. Sir John Hope Simpson probably came close to the mark estimating 40,000 refugees from Greater Germany on the eve of the war, although there is no way of knowing for sure.[50] Clearly these refugees were a tiny handful compared to the great mass of foreigners—still about 2.5 million in 1936. Refugee policy was caught up in the intense political conflict of the mid-1930s in France: the severe restrictions that unfolded during 1938 similarly fitted the broader pattern of reaction to the Popular Front.

GREAT BRITAIN

For the British, the really dangerous strangers were outside, rather than inside, the country. During the 1930s the public mood was far more suspicious of foreign governments than of an enemy within, and this was reflected in a growing anxiety about the prospects of war and a guilt-fed urge to respond favorably to German demands. At home, refugees never became the focus of attention that they were in France. Despite her reputation for liberalism, few refugees had come to Great Britain after the First World War. Among the exiled Russians, probably about 15,000 at one time, the largest group were those who escaped from Murmansk on British ships after the collapse of the popular government in Archangel. About a thousand Armenians lived in

England in the 1930s, only a fifth of whom were technically considered refugees. Most anti-Fascist Italians preferred Paris to London. Because of her insular position, moreover, rendering control extremely easy, practically all outsiders entered Britain legally; therefore, few dark references to *clandestins* obscured public discussion of the refugee question. The numbers also made it difficult to claim that Britain was being swamped by outsiders: whereas France, for example, had an alien population of close to three million when the depression began, there were only about 250,000 aliens then living in Britain.[51]

In contrast to the French, British officials had no pretense about a long-standing policy of asylum. Such liberality as had indeed existed in times past sprang far more from official indifference to foreigners in general and from a sensitivity to individual liberties than it did from official hospitality for the politically oppressed. For the Home Office, hearing the appeals of refugees in 1933, policy flowed along neat utilitarian channels. "We do not . . . admit that there is a 'right of asylum,' " explained one official dryly, "but when we have to decide whether a particular political refugee is to be given admission to this country, we have to base our decision . . . on whether it is in the public interest that he be admitted."[52]

Few among the first wave of refugees from Nazism wanted to go to Britain, and few ended up receiving asylum there. For these exiles, as we have seen, France was a more logical place to await the expected early collapse of Nazism: it was close at hand, more familiar to Continentals than England, requiring less psychological uprooting. For those who did cross the Channel, the reception was reserved. Officially, the government declared that Britain was not a country of immigration. In view of the densely settled population and high unemployment (over 2.5 million, or one-fifth of the labor force in 1933), officials carefully screened applicants for admission, permitting only a few thousand "desirable, industrious and acceptable persons" to enter the country. In practice, the Home Office declared its readiness to exclude any alien who did not have a visa to another country affixed to his or her travel documents; Britain was not to be a place where a fugitive waited for some immigration opportunity to come along.

This cautious posture toward refugees was formulated before the refugee crisis and in a climate of general sympathy toward those persecuted by Hitler. According to A. J. P. Taylor, "Englishmen of all classes and of all parties were offended by the

Nazi treatment of the Jews," and "this did more than anything else to turn moral feeling against Germany."[53] Moral feeling, however, did not imply an open door to masses of fugitives, and until 1938 there was little pressure to change this policy. To encourage admission of German-Jewish refugees in 1933, the Anglo-Jewish community promised the government that "all expense, whether in respect of temporary or permanent accommodation or maintenance, would be borne by the Jewish community without ultimate charge to the state." British Jews assumed a maximum of only 3,000 to 4,000 refugees, vastly underestimating the eventual numbers, as did practically everyone at the time.[54] No one predicted a great outpouring of victims, and other places were presumed more appropriate and more able to receive fugitives of the Nazis.

Within Britain, policies and procedures set in this early period remained in force until the refugee crisis of 1938. The government expressed sympathy for the victims of Nazism; administratively, immigration machinery was often clogged by muddle and an insensitivity to the desperate straits of some of the fugitives. Limiting admission, the government picked its way between competing pressures—prorefugee arguments on the one hand, but also fears lest increased liberality generate more expulsions from Germany; concerns about unemployment; and occasionally antialien and antisemitic feelings in England. By the summer of 1938, Sir John Hope Simpson believed that only about 8,000 had arrived, "apart from a floating element at any one time of about 2,000 persons either undergoing training or awaiting immigration." About 80 percent of the refugees were Jews.[55] The great majority were also of the middle class, having escaped from Germany sufficiently early to have taken some of their capital with them. Although the refugees had a few champions, such as Viscount Cecil of Chelwood, the issue was largely ignored in political circles. "Polite antisemitism" of the upper classes may well have saturated English society at the time, but the question of Jewish immigrants scarcely arose in the House of Commons and seems to have troubled social relations very little. Opposition to Jews did not appear to be a potent political issue. Oswald Mosley's British Union of Fascists eagerly adopted antisemitism in 1934, as did other right-wing extremist groups at the time. There was certainly a constituency for such appeals, but they do not seem to have contributed much to the movement's popularity.[56]

The real focus of debate concerning Jews and refugees alike was

Palestine, where important decisions were soon taken, partly under the impact of a growing refugee problem. After an economic slump in the late 1920s, Palestine became a booming, bustling place, with a Jewish population of about 400,000 in the mid-1930s—about one-third of the total population of the country. Jewish immigration began to soar following Hitler's taking power in Germany. From 9,500 immigrants in 1932, the number of arrivals climbed steeply: over 30,000 in 1933, 42,000 in 1934, nearly 62,000 in 1935.[57] Most of these came from Poland, but an increasingly large proportion, as many as one-fifth between 1935 and 1936, were fleeing Germany. Reacting to this important wave of settlement and development, Palestinian Arabs launched a violent revolt against British rule in 1936. In response, the mandatory authority began a serious reconsideration of its original commitment to favor a Jewish national home. Jewish immigration, chief among the Arab grievances, became a critical issue. Previously, the British exercised loose controls over numbers—largely determined by an assessment of the economic capacity of the country to absorb new inhabitants. And up to this point the real problem for Zionists had not been British restrictions, but rather the scarcity of Jews willing to go to Palestine. Now everything changed. Desperate Jews began to clamor for entry certificates. Just as the Jews were coming to depend heavily on Palestine as a land of refuge, just as Jews were beginning to press seriously on official immigration ceilings, the British began to restrict Jewish entrants.

From 1936 the British reduced Jewish immigration to Palestine substantially. In 1937 Jewish arrivals fell to 10,600, and there was considerable talk about closing the Palestinian gates entirely to Jews. Most colonial administrators favored a significantly lowered immigration ceiling. Another option was proposed by the Peel Commission in 1937—partitioning most of Palestine into a Jewish and an Arab state. (The plan included a proposal for a population exchange between Jews and Arabs, modeled on the Greco-Turkish transfer of fourteen years before.) Failure to agree on this suggestion, however, prompted the British to seek other solutions, less favorable to Jewish refugees. In an international atmosphere dominated by the fear of war and the need to prepare for it, they sought increasingly to appease the Arabs in order to ensure their support for the coming contest with Germany. After several proposals and counterproposals, the British finally fixed on their program in the White Paper of the spring of 1939: there was to be an annual quota of 10,000 Jews for the next five years, plus another 25,000 refugees. Following this five-year period (and presumably the arrival of 75,000 immigrants), no more Jews would be admitted except with the agreement of the Arabs. With this policy, as we shall see, conflict between the Jews and the British broke into the open. Immigration became the major point at issue between Jewish representatives and the mandatory power, and illegal Jewish immigration became a regular feature of Palestinian life.

In Palestine the British were particularly hard; at home, as if to compensate, there was some relaxation on the eve of war. Stoutly opposed to Zionist demands, officials responsible for foreign and colonial affairs resisted the appeals on behalf of Jews in flight, hoping thereby to reduce the exodus from Germany and to help defuse the refugee crisis. Faced with the rising tide of refugees after the Anschluss of 1938, the government first set new restrictions on those seeking entry to Britain. Remarkably, however, these moves drew extensive criticism, both in Parliament and from the public at large. Sensitized by the Palestine question, segments of British opinion campaigned vigorously for the admission of more refugees. In 1939 the government grudgingly responded. During that year, as other countries imposed ever tighter immigration restrictions, the British sent four million pounds to Prague to assist refugees from areas taken by Germany; Social Democrats from the Sudetenland were admitted en bloc to Great Britain. Ten thousand children, mostly Jewish, reached Britain from Greater Germany between December 1938 and the outbreak of war. Extremely sensitive to international criticism of its Palestine policy, the government instituted a special cabinet committee on refugees to streamline procedures for receiving Jews and to offer assistance. Since the beginning of 1939, Sir John Hope Simpson reported later that year, the entry of refugees was being accelerated.[58]

Between 1933 and the end of 1939, about 50,000 refugees from the Reich and 6,000 from Czechoslovakia received asylum in Britain, most arriving on the very eve of the war. In Palestine, meanwhile, the British were as firm as ever, complicating the drawing of a balance sheet. Through this period over 215,000 Jewish immigrants came to Palestine, doubling the size of the Jewish community there. The great majority, as we have seen, came from Eastern Europe. The gates were swinging closed in 1939, with practical restrictions even tighter than those required by the White Paper of that year. As fighting began in Poland, special patrol vessels of the Royal Navy were busy turning Jewish refugees away from Palestine, often their last refuge. In comparison with other countries, writes Joshua Sherman, British refugee policy may well have been "comparatively compassionate, even generous."[59] But this is more a reflection on the international rejection of refugees than a comment on British benevolence.

SWITZERLAND

No European country had a longer tradition of receiving persecuted refugees than Switzerland. Virtually every Swiss commentary on the subject began with a reference to the Reformation and Counter-Reformation, when Protestants hounded out of Catholic countries flooded those Swiss cantons committed to Reform. The anti-Protestant spasm in France associated with the revocation of the Edict of Nantes sent some 140,000 refugees into the Swiss Confederation between 1685 and 1700, many of them continuing on to German states, the Netherlands, or England. This tradition of receiving outcasts continued through the nineteenth century, when Zurich, Geneva, Basel, and other Swiss cities usually harbored hundreds of colorful exiles from abroad. Russian revolutionary intellectuals were particularly drawn to Swiss universities in the decades before 1914. During the First World War many thousands of deserters and escaped prisoners of war reached Switzerland, more than 110,000 French civilians were evacuated through Swiss territory, and the total number of interned military and civilians reached about 75,000. At the end of the war many representatives of the defeated regimes in Eastern and Central Europe sought asylum there, crossing paths with formerly exiled Russians who were rushing home to assist the Soviet revolutionary enterprise.[60]

Although tradition suggested an open door to refugees, practical considerations pushed in the opposite direction by the end of the nineteenth century. As the Swiss discovered after 1848, harboring prominent troublemakers from neighboring countries could prove a supreme embarrassment, particularly when governments made strenuous efforts to recover those accused of violent revolutionary acts. Preserving Swiss neutrality in an increasingly volatile European continent was a delicate undertaking and one that could be compromised by the presence of large numbers of exiles committed to spreading disorder in their home countries. The First World War only reinforced Swiss concerns to stay out of the European turmoil. In addition, the liberal Swiss policy on residents encouraged a long-standing concern about *Uberfremdung*, what was perceived to be an excessive proportion of foreigners among the population. Over 15 percent of Swiss inhabitants were aliens in 1914, and although this dropped to 8.7 percent in 1930, Switzerland had the highest proportion of outsiders of all European countries except Luxembourg at that time. Foreigners in major cities, such as Geneva or Basel, numbered as high as 40 percent of their populations.[61] Given the sometimes precarious balance of linguistic groups within the country, as well as occasional gusts of xenophobia, the issue nagged continually at politicians. Swiss territory was seen as too small, her food supply as too vulnerable, and her national culture as too tightly defined to permit such a high proportion of outsiders.

Despite these concerns, however, the Swiss Confederation admitted large numbers of refugees from Germany in 1933, allowing them to reside temporarily in Switzerland pending reimmigration elsewhere. For about five years, this remained the essence of Swiss policy: the Confederation was seen as a place of transit through which refugees might pass on their way to more permanent sanctuaries. Responsibility for this policy was divided. Federal authorities based in Bern approved the admission of foreigners and directed police surveillance of them; cantons, however, had the right to accord temporary residence permits for a period of up to two years. Local authorities had wide latitude to refuse entry to the country, and Bern could not force them to accept political refugees. Once admitted, the refugees were forbidden to exercise any lucrative activity. Deeply preoccupied with unemployment in the first half of the 1930s, Swiss officials scrutinized new arrivals rigorously to ensure they did not compete in commerce or the labor market. Only in a few, very exceptional cases were refugees allowed to support themselves.

Under these circumstances the number of refugees seeking entry to Switzerland declined sharply after the first wave of 1933. In the succeeding period, the great majority of those crossing the frontier were Jews, as was the case everywhere. Swiss Jews were not in a strong position to lobby on behalf of their coreligionists. Only about .04 percent of the population, they numbered about 18,000 in 1930, and almost half were foreigners. In an effort to encourage new admissions, the local Jewish organizations offered to support and maintain Jewish refugees and raised 150,000 Swiss francs for this purpose after the first wave broke across the border. At the same time, Swiss Jewry managed to assist the substantial reemigration. By the end of December 1937, only 103 refugees received assistance from the Jewish community, down from almost 6,000 who had originally been helped. Refugees continued to trickle into the country, but others left—some returning to the Reich and others migrating elsewhere in Europe or abroad. Notably, the total number of Jews in Switzerland scarcely rose at all during the entire decade.[62]

Swiss policy began to shift under the impact of the furious Nazi assault on Jews following the German absorption of the Austrian state in March 1938, accompanied by widespread fears about an inundation of refugees. Immediately, Swiss officials ordered frontier guards to reject any foreigner whose papers ware not in order. Two weeks later the Federal Council required Former Austrians to present an entry visa obtained from Swiss consular authorities. Regulations concerning transit through the Swiss Confederation were tightened; orders went out to expel persons discovered to have gained asylum improperly. Nevertheless, the Swiss felt acute pressure by the summer of 1938. Anti-Jewish violence was rampant in Austria, and the Nazis demanded that the Jews leave the newly absorbed territory. Thousands of Jews who had been arrested during the Nazi takeover were suddenly released from concentration camps and abruptly ordered out of the country. Storm Troopers sometimes rounded them up and simply thrust them across the frontiers. Moreover, unlike refugees of a few years before, the Jews now fled penniless and often without identity documents to the nearest refuge. Hungary and Yugoslavia kept their frontiers closed, and France and Italy offered only doubtful prospects for rescue. Switzerland suddenly felt the brunt of a wave of largely Jewish refugees. Swiss consular offices in Germany and Italy reported that they were besieged by desperate crowds of Jews, panicking under the Nazi pressure. According to the federal chief of police, 10,000 refugees were in Switzerland by mid-1938, and 3,000 to 4,000 had come from Austria since Anschluss. Border guards were told to be vigilant. Refugees arriving by train were almost all turned back; more began to come on foot, over remote mountain paths, evading frontier posts. Swiss Jews felt overwhelmed by events; they desperately tried to reassure their government, to extend assistance to new arrivals, and to hasten their reemigration.

In mid-August 1938, the federal government decided to close Swiss frontiers to all those not strictly entitled to enter the country and to expel those who had done so illegally. German moves, however, threatened to upset Swiss restrictions. According to existing procedures, Swiss and Germans could visit each other's countries without visas; Austrians, however, required these documents to visit Switzerland, thus enabling Swiss immigration officers to filter out Austrian Jewish refugees. Following Anschluss, the Germans began to replace Austrians with German passports, practically eliminating this useful mechanism for Swiss restriction. How would Bern respond? Unwilling to cancel the reciprocal Swiss-German arrangement, Swiss negotiators approached the Nazis with a remarkable proposal: Heinrich Rothmund, the top Swiss police officer, suggested that the Germans affix a special stamp on the passports of all "non-Aryans" and "Aryans" whom Germany wished to expel permanently. This would enable Swiss consular authorities to ascertain which immigrants or visitors from Germany were in fact refugees and thus to reject them at the frontier. Rothmund wanted to prevent Switzerland from becoming "saturated with Jews," as he put it in a report on 15 September. Pleased at the offer, the Nazis characteristically insisted on an even higher price for agreement: they demanded that the Swiss reciprocally stamp the passports of *Swiss* Jews traveling to Germany. At this point, Rothmund balked. Reluctant to alienate Swiss Jews and fearing hostile international reaction even more, the Swiss police chief refused. Backing down, the Germans henceforth marked the passports of all those deemed to be Jewish according to the Nuremburg Laws with a red "J," three centimeters high, on the left-hand side of the first page. The Germans promised not to allow Jews to leave for Switzerland unless they had visas for that purpose and thus to desist in imposing their Jews on their neighbor. The final link in the chain was hammered into place on 4 October, with an announcement from the Swiss police: Germans bearing passports indicating that they were not Aryans now had to receive special authorization to enter Switzerland.[63]

As a result of this Swiss agreement with Germany and the stamping of Jewish passports, other countries now had the technical means to discriminate against Jews seeking to leave the Reich—even when the Jews attempted to hide their status as refugees. As early as 20 October, for example, the Paraguayan consulate in Berlin refused a visa to a Jew whose documents were stamped with the letter "J." In Switzerland it was now possible to arrest the flow of refugees even while keeping normal relations with Germany and a free flow of tourists, businessmen and other visitors. The Swiss imposed even more stringent police regulations in February 1939 and reduced even further the number of refugees allowed in. Swiss police tracked down fugitives within the country and occasionally deposited them forcibly on the German side of the frontier. Everyone continued to urge re-emigration for those who had already arrived, but opportunities were drying up quickly. At the end of 1938 it was estimated that there were between 10,000 and 12,000 refugees in Switzerland; at the outbreak of war this fell to between 7,000 and 8,000, 5,000 of whom were Jews. A handful of exceptional cases could still pass the scrutiny of Swiss police and immigration officials, but the path of sanctuary had been narrowed drastically.[64]

THE FAILURE OF INTERNATIONAL ORGANIZATION

The League of Nations

Given its important work on behalf of refugees in the early 1920s, some expected the League of Nations to move boldly after 1933, when a new refugee crisis loomed. The two situations, however, were strikingly different. Nansen's League of Nations High Commission dealt first with Russian refugees at a time when the Soviet Union was not a League of Nations member; the League also assisted Balkan refugees who came from weak and highly vulnerable countries that swung no weight in the international arena. During this period there were many countries, both in Europe and America, willing to receive new immigrants. In circumstances like these, the League was relatively successful, and the great bulk of the refugees were settled thanks to agreements worked out by the new international body. A decade later, the exodus provoked by Germany affected all major European countries, and everyone proved reluctant to accept more than a few thousand refugees. No member state wished to see the League granted authority that would circumscribe sovereign rights to exclude immigrants. Although Germany left the League in the fall of 1933, she was hardly isolated from the European diplomatic arena as had once been the case with Russia. Most Western diplomats hoped that the Germans could be induced to return to the international body and were reluctant to have the League appear anti-German. Thus, while it was embarrassing for the League to do nothing, any action threatened to collide with fundamental international interests.

Some of the first international responses to Nazi persecutions and expulsions occurred outside the League framework. Academics in the United States and Western Europe organized spontaneously to assist German scholars and artists who were removed from their posts and propelled into exile. Socialist and labor leaders responded similarly and used their network of political and trade union organization to coordinate assistance in various countries. By 1935, European leftists established an International Solidarity Appeal to assist all political refugees. American Jews, acting at first against the advice of the American Jewish Congress, began an international boycott of German goods, hoping to reduce German exports and to persuade the Hitler regime to treat Jews humanely. This movement won substantial European Jewish support, organized through the headquarters in Geneva that was established in 1934.

One League of Nations agency might well have acted on behalf of the new wave of fugitives. This was the International Office for Refugees, known as the Nansen Office, established in 1931. Following Nansen's death in 1930, the League Assembly consigned legal and political work on behalf of refugees to the League Secretariat, an indication that such activity was thereafter at a low priority for the international body. At the same time, the League Assembly established an autonomous new institution, the Nansen Office, to take charge of "humanitarian" work until the end of 1938—following which it was assumed that it would no longer be needed. This office was to serve as the agent for the distribution of relief, to encourage the absorption of some refugees, and to hasten the repatriation of others. Its first president was Max Huber, head of the International Red Cross Committee, and it was assisted by an International Advisory Commission replete with representatives from the International Labor Organization, governments, and other private charities. Bound by previous decisions, however, the Nansen Office was able to deal only with "Nansen" refugees—Russians, Armenians, and so on—to whom were added the émigrés from the Saar in 1935. However, the organization slipped inexorably into decline once Nansen's inspired leadership was gone. The Nansen Office remained in place even after it was supposed to close down in 1938, but this was hardly the address to which growing numbers of desperate refugees in the 1930s could turn.

Another possibility was League of Nations action in protection of internationally guaranteed minority rights. For five states with substantial "national minorities"—Poland, Czechoslovakia, Rumania, Greece and Yugoslavia—there existed specific treaty obligations that theoretically prevented the persecution and expulsion of minority groups. Unfortunately, this system did not work particularly well in the best of circumstances, often breeding deep resentment on the part of governments that had assumed the obligation. Unfortunately, no such clauses were imposed on Germany in the 1919 Versailles Treaty. One treaty, however, the Polish-German Agreement of 1922 over Upper Silesia, did contain clauses protecting minority rights. These provisions enabled the League of Nations to speak for residents of this territory who were transferred to German rule after a 1921 plebiscite. In 1933 a group of German Jews from Upper Silesia petitioned on behalf of a Jewish refugee named Franz Bernheim. Dismissed by the Nazis from his job in the Silesian town of Gleiwitz and forced to flee to Prague, Bernheim claimed that Germany had violated rights guaranteed by the 1922 accord. Following a detailed examination, the League of Nations backed Bernheim, and the Germans were forced to retreat: for another four years the Nazis' reign of terror passed by the region, and the Nazis compensated Bernheim and others for having been wrongly persecuted. This unusual incident showed how, under very particular circumstances, the young Nazi regime felt obliged to heed international opinion in highly circumscribed areas. Success was clearly limited, and the League of Nations hardly felt in a position to tackle the wider question of Jewish persecution in the rest of Germany. In the summer of 1933, indeed, distinguished League of Nations supporters such as Viscount Cecil, one of the founders of the organization, and Sean Lester, soon to become high commissioner to Danzig, advised Jewish leaders that nothing more would be gained by moves like the Bernheim petition. As they told Norman Bentwich, "every endeavor should be made to obtain the cooperation of the German government in a scheme of ordered emigration of Jews and the adjustment of Jewish economic life in Germany."[65]

Shortly after the Bernheim decision, the Dutch delegate to the League of Nations raised the question of the worsening refugee situation before the League Council, inviting its members to seek an international solution. Confrontation loomed. Anti-Nazi feeling in liberal countries was riding high in 1933, particularly under the impact of the widely publicized first wave of persecutions. Meanwhile, the German representative, Konstantin von Neurath, declared that the League had no competence in the matter of German refugees. Here was a classic League of Nations dilemma: on the one hand, it seemed obvious that some international response was necessary as refugees streamed from Germany into neighboring countries; on the other hand League delegations were extremely concerned not to interfere in the internal affairs of a member state or to criticize German policies too violently. League negotiators worked out a neat solution that did not appear to engage the organization too prominently against Germany and to which the Germans agreed: the Council would establish a High Commission to assit the German refugees, but this body would not act in the name of the League of Nations and would not receive directives or regular funds from that institution. The High Commission would be responsible to an administrative council made up of delegates of fifteen governments designated by the League. Its headquarters would be Lausanne, emphasizing its distinction from the Geneva-based international body. According to Bentwich, one of its high-ranking officers, "the High Commission was treated somewhat as a cast-off child of the League."[66] The Assembly finalized these arrangements on 11 October 1933 with the German representative abstaining. In a fit of displeasure, Germany abruptly quit the League of Nations three days later, leaving the impression that even a weak High Commission was an intolerable affront.

From the smoke-filled meeting rooms in Geneva emerged a somewhat cumbersome designation—the High Commission for Refugees (Jewish and Other) Coming from Germany. Intended to do for refugees from Germany what the Nansen Office had done for "Nansen refugees," this body was supposed to intervene with governments on behalf of individuals, to make general recommendations to the League of Nations on refugee matters, and to assist in resettling those in flight. To head the new body the, League Council elected James G. McDonald, a prominent American scholar in the field of international relations, chairman of the American Foreign Policy Association, a man widely respected by Jews and Gentiles in his own country.

McDonald had high hopes for the enterprise. Physically impressive and a plain but effective speaker, McDonald was a devout Christian and humanist, an energetic optimist eager to get to the root of the refugee problem. He had highly placed German acquaintances in Berlin, had visited the Dachau concentration camp, and believed himself under no illusions about Nazism. More important, perhaps, McDonald was extremely sympathetic to the Jews in flight. He was a friend of the banker Felix Warburg, who was the founder and honorary chairman of the Joint Distribution Committee; ever more favorable to Zionism during his tenure, he eventually became the first American ambassador to the State of Israel. One of his first gestures as high commissioner was to invite Leo Baeck, a revered Reform rabbi in Germany, to join the High Commission's administrative council. (Baeck felt obliged to refuse after the German Foreign Office opposed the idea.) McDonald also canvassed widely to win financial backing for the organization, approached Western governments seeking to expand settlement possibilities, and sought to contact the German government to bring order to the chaotic conditions for refugees.

In every quarter, McDonald met disappointment, and two years later he left his post in frustration. On 27 December 1935, McDonald sent a sensational and widely commented on letter of resignation, drafted with the aid of influential Jewish refugee workers. As he explained, conditions in Germany responsible for creating refugees had worsened "catastrophically" over the preceding two years. McDonald complained that his organization had been weakened from the beginning by its dissociation from the League. But his central emphasis was on the need to apply moral pressure on the Nazis. From the vantage point of 1935, this appeared as a strong suggestion, a significant gesture on behalf of refugees. The high commissioner called on the League to bombard the Hitlerian regime with powerful exhortations, mentioning "a determined appeal to the German government in the name of humanity and the principles of the public law of Europe." He urged this "friendly but firm intercession with the German government" in order to get to the bottom of the problem. Significantly, McDonald issued no ringing call to open the gates of immigration. "In the present economic conditions of the world," he declared, "the European states, and even those overseas, have only a limited power of absorption of refugees. The problem must be tackled at its source if disaster is to be avoided." Few people at that time considered restrictive immigration policies a universal fait accompli, and few thought this the central issue in the refugee drama.

Of course, McDonald could not say everything that was on his mind. The high commissioner may well have envisioned much more forceful action against the Nazis than a moral appeal from the League of Nations. But few were prepared to go further in 1936. Certainly, there was no sense of imminent catastrophe. Finding a home for the refugees had been difficult, but the scale of homelessness was still small. Palestine, by absorbing some 62,000 Jews in 1935, relieved most of the pressure. In the depths of the Depression, Palestine was singularly prosperous, able to absorb substantial numbers of immigrants. And British restrictions were not imposed before 1936. As McDonald indicated at the Zionist Congress the previous year: "The daily grace in the High Commissioner's office was 'Thank God for Palestine.' "[67] The real apprehension, the high commissioner made clear, was the *prospective* crush of German refugees who would soon be propelled outward by antisemitic persecutions. To achieve solutions, McDonald felt he needed a far stronger agency with international backing. McDonald's assistant, Bentwich, referred to the governing body of the High Commission as composed of "nondescript diplomatists, who knew little, cared little, and wanted to do as little as possible about the cause." Viscount Cecil had much the same reaction, calling the supposed Advisory Council "useless."[68] Furthermore, because the High Commission was cut loose from the League of Nations, it barely had enough funds for administrative needs and constantly depended on private charities, especially Jewish organizations, for handouts. Consequently, its prestige and political authority were considerably reduced. Much the same was to be the fate of the Inter-Governmental Committee on Refugees established in 1938. Finally, the high commissioner himself may well uave been too cautious, too indecisive, unwilling to press Western governments, particularly the United States, to do more for the refugees. McDonald never attacked American immigration restrictions, which were being felt by 1935. He never seriously confronted Western representatives and may simply have lacked the capacity for tough diplomatic bargaining.

It seems likely that the weak and low-keyed High Commission was very much to the taste of its League of Nations sponsors. In 1935, responding to criticism of the agency, the Norwegian government recommended to the secretary-general that the League establish a central refugee organization at Geneva, properly funded and supported by the international body. But the Norwegians found little enthusiasm for the idea. The British representative Sir John Simon shared the hopes of many that Germany would rejoin the League and was reluctant to antagonize Hitler by moving so boldly on the refugee front. At the same time, there ware fears of serious financial burdens from such a move.[69] The British Foreign Office even opposed McDonald's letter as being too strong. His statement, said one memorandum, was "an unwise document, which did a disservice to the real interests of the Jews in Germany"—presumably because of its extremely pessimistic conclusions. In McDonald's letter, the report went on, "the guiding hand of Zionism was apparent."[70]

To replace McDonald, the League Council selected the elderly Sir Neill Malcolm, a retired general who had once commanded the British garrison in Malaya. At the same time, the League Council altered the terms of the high commissioner's office. There is little doubt that League members preferred to avoid the contentiousness of the previous incumbent. To them, the refugee issue remained awkward, but certainly manageable: if properly handled, it was hoped, the refugee problem could be liquidated by the end of 1938. Despite strong appeals from Jewish refugee agencies, the World Jewish Congress, and the Jewish Agency in Palestine, the League Council carefully circumscribed the new High Commission's sphere of activity. The League did bring the refugee body under its own wing, ending its autonomous status and subjecting it to the close scrutiny of the League's secretary-general. The High Commission was to continue to perform the tasks of consular bureaucracies, to review attendant legal problems, and to effect liaison with private organizations pursuing settlement and relief. Malcolm was kept on a tight budgetary leash and was forbidden to receive funds from private organizations. Internal German policy was declared out of bounds. Neither the Nazis nor the Western governments had much to fear from the High Commission at that point.

Working within these limits, Malcolm attempted to improve the status of refugees who had escaped from the Reich in the preceding period. At an intergovernmental conference at Geneva in July 1936, he secured agreement on a provisional arrangement providing a certificate of identity and affording certain legal guarantees to fugitives abroad. However, this accord treated *existing* refugees and was silent on new arrivals; it did nothing to

facilitate mass settlement and scrupulously avoided any reference to German policy. The refugees continued to flow. In 1937,
with a year in this post still before him, Malcolm reported pes-
simistically to the League Assembly. Everywhere, as more refugees appeared, immigration restrictions were being imposed.
These developments now affected emigration prospects. Some
potential refugees were now trapped in Germany with no place
to go; others were wandering from place to place in Europe, unable to alight. The High Commission's most important achievement was the Geneva Convention on Refugees Coming from
Germany, concluded in February 1938, which solidified the legal
status of those refugees who had managed to leave Germany to
that point.[71] This agreement repeated previous measures of protection, improved conditions for refugees in countries of asylum,
and limited the recourse to expulsion. Once again, the document
mainly protected refugees who had already escaped Germany.
Signed by the major West European countries, it probably
emerged at the last possible moment—just before a new flood of
refugees occasioned by Anschluss. Henceforth, as we shall see,
governments resisted any move that might possibly suggest new
obligations to refugees.

Although the crisis year 1938 and the Evian Conference generated a new refugee organization outside the League of Nations, its own refugee machinery continued to function, however inadequately. Working within the international body, the
Norwegian representative urged that the League consolidate its
refugee work and pump energy into the High Commission. The
Soviet Union, which entered the League in 1934 and had a permanent seat on the League Council, stoutly opposed such moves,
objecting to the protection afforded anti-Bolshevik émigrés. Given
the need for unanimity, this sidetracked the proposal until the
Soviet representative was persuaded to abstain from voting on
the issue in May 1938.[72] Finally, after diplomatic maneuvering, a
scheme for a new refugee agency won support at the end of that
year. Yet another League High Commission for Refugees emerged
in early 1939, bound and shackled much as its predecessors had
been. France and other European governments agreed to see another Englishman placed in charge, a convenience given the increasing importance of London in refugee negotiations then underway. The British Foreign Office, still smarting from McDonald's
letter of three years before, scrutinized a field of candidates carefully. Once again, the unspoken international consensus was to

avoid disturbing existing arrangements. The Foreign Office, therefore, worried about Sir John Hope Simpson, who circulated his important refugee report at the time of Evian and was an obvious candidate to head the body. A highly visible critic of restrictive policies toward refugees, Hope Simpson was judged "unnecessarily critical" of British policy and likely to cause trouble. Upon request, the India Office came up with an assumed safer choice. This was Sir Herbert Emerson, a veteran civil servant about to retire from the governorship of the Punjab. Now more than ever, it was assumed, the High Commission was not to forge policy on its own. Joining the former Nansen Office and the High Commission for Refugees Coming from Germany, the new League of Nations organization was supposed to process the considerable paperwork associated with refugee conventions, coordinate humanitarian assistance, and promote resettlement. But with a small budget, no political direction, and caution being urged from every quarter, there was no hope it would achieve much in the emerging refugee crisis.

The Crisis Year, 1938

The crisis year was 1938, when refugee pressures building since the Nazi takeover in Germany burst the dikes, spilling tens of thousands of refugees into a Europe unwilling to receive them. It was also a year when illusions gradually eroded. Before 1938, sincere and knowledgeable persons believed that the refugee situation was manageable, that problems posed, though serious, were nevertheless on their way to resolution. During 1938 the Nazis made their boldest moves on the European chessboard thus far: in March, with Anschluss, the Germans absorbed Austria; early in September, Hitler provoked a month of crisis over Czechoslovakia, when Europeans believed they teetered on the brink of war; then came the Munich Diktat, the Nazis' incorporation of the Sudetenland, followed by their march into Prague the following year. With these events went intensified persecution of Jews on a scale previously unimagined. Throughout Eastern Europe, meanwhile, Jews experienced a series of new disasters. Each of these events generated many thousands of new refugees and ruined the hopes of those who had foreseen early solutions.

Writing his famous refugee report near the beginning of 1938,

Sir John Hope Simpson struggled to keep his subject in perspective. Some 150,000 German refugees had left their homes since 1933, he noted, but this represented only .23 percent of the German population. This was less than half of the Armenian exodus, even excluding the flow of Armenians east of the Black Sea. It was vastly smaller than either the Greek flight from Asia Minor or the Russian emigration following the Bolshevik Revolution. No intolerable concentration points of human misery had appeared, as with the Russians in the Levant just after the First World War or as with Armenian refugees in Transcaucasia. Although "uncompromising to be rid of the Jewish element," the drive of the Third Reich had nevertheless "been gradual and the means used [were] not those of massacre and large-scale deportation but of steadily increasing economic, social and legal pressure." Settling the 150,000, Hope Simpson went on, "would not present a problem of extraordinary difficulty." Most of the refugees had already left Europe, and "the number requiring migration from the first countries of refuge had been fairly constant about the level of 30,000 to 35,000 persons." In a carefully drawn conclusion, Hope Simpson maintained a degree of optimism. Solutions could be found, he stressed, "by concerted effort and the best use of existing institutions." To do so, of course, the German exodus would have to be kept at a manageable level; some method was needed to compensate refugees for confiscated properties; and other refugee issues such as those in Spain and the Near East would have to be addressed.[73]

Visiting Berlin shortly after Anschluss, the veteran demographer Arthur Ruppin, a director of settlement for the Jewish Agency in Palestine, noted in his diary that "many Jews in Germany do not yet fully and seriously comprehend what has befallen them." Yet even Ruppin, who knew more than anyone about the sociology of world Jewry, envisioned a relatively modest emigration: he looked to the departure of 20,000 Jews per year from the newly expanded Reich (which he estimated then to have 530,000 Jews), thereby solving the Jewish question in about two decades. Ruppin gave no thought to emergency evacuation. His assessment illustrates how even experienced, well-informed observers of the refugee scene had difficulty looking into the future and how it took time to grasp the magnitude of the catastrophe in store.[74]

Anschluss set the crisis in motion. About 180,000 Jews lived in Austria when the Nazis swept into the country on 12 March 1938, and of these, 165,000 were concentrated helplessly in Vienna. During the weeks that followed, the Nazis began a furious assault on their exposed and stunned victims, concentrating into a few months the persecution and exclusion from the economy and society that had taken five years in Germany. According to outside observers, anti-Jewish excesses exceeded anything that had taken place before. Nazi Storm Troopers roamed the streets and pounced on Jews at will—beating, robbing, humiliating, and packing them off to prison, work teams, or concentration camps. Some of the victims were simply massacred. Tens of thousands were taken away to camps. Overnight, Jewish institutions were plundered and destroyed, businesses and property were liquidated, and Jews were driven from their homes. Dozens of Austrian Jews committed suicide—about ten times the rate that would correspond to their proportion of the total population.[75] Focusing on their objective, the new Nazi rulers pressed for Jewish emigration. To head this effort, Berlin dispatched to Vienna a recently promoted SS Untersturmführer Adolf Eichmann, an expert in such matters. Headquartered in a former Rothschild residence in Vienna, Eichmann organized a Central Office for Jewish Emigration in April to speed Jews on their way. His goal, quite simply, was to rid Austria of its Jews. More violence flared in August. Before long, Eichmann's men simply took people to the frontiers and dumped them across the border. On departure, the Jews were allowed to take with them mere pocket money in Austrian currency. Ten thousand Jews left in the month following Eichmann's arrival; between April and November 1938, 50,000 Jews had fled—over 30,000 more than had left Germany in the same period. Up to the end of November 1939, when emigration became especially difficult owing to the outbreak of war, about 126,500 Jews managed to flee elsewhere.[76]

Escape from Austria could be extraordinarily difficult. For a handful of distinguished artists, scholars, and famous personalities, the path was cleared. President Franklin Roosevelt expressed a personal interest in the aged and frail Sigmund Freud, who was reluctant to leave Vienna; in the end the founder of psychoanalysis went to London, thanks to the effort of his English disciple Ernest Jones. But ordinary Viennese Jews desperately sought the necessary documents with which to enter another country. Foreigners were particularly exposed. About 20,000 Polish Jews lived in Austria at the time, plus thousands more from Czechoslovakia, Hungary, and Rumania. The British Jewish writer

Israel Cohen came to Vienna not long after the Nazis arrived and saw people lining up at various diplomatic offices before midnight, seeking an appointment the next day. At the British consulate he found the courtyard, staircase, and waiting room packed with frightened people; thousands were seeking to enter England, Palestine, and the British Dominions. Local Jews were stunned and unsure where to turn. "The feeling of terror could be sensed most keenly in cafes, where the light-hearted hubbub of conversation of former days had sunk to a timid whisper, for everyone suspected his neighbor, and nobody spoke without first glancing over his shoulder."[77]

Each of the European countries that had previously received refugees took countermeasures to reduce earlier rates of acceptance or to shut out the refugees altogether. Neighboring Hungary and Yugoslavia immediately closed their frontiers as we have seen. Fascist Italy, which had received about 5,000 German refugees in the preceding period, swung sharply against the Jews in mid-1938, closing off one more avenue of emigration. A decree of September of that year ordered foreign Jews out of the country in six months, provoking an emergency for some 20,000 who had come since 1919. The Dutch, Belgians, and Swiss admitted only small numbers of refugees and reinforced frontier guards. Several countries, including Britain, imposed a special visa requirements to screen out refugees. Meanwhile, Jews who had contemplated escape before now rushed to complete formalities. Panic swept diplomatic missions and chancelleries in Central Europe as Jews pleaded for sanctuary. The persecution in Austria had occurred in the full glare of publicity and undoubtedly horrified Western opinion. Almost everyone was prepared to sympathize with the plight of the victims. What terrified officials, however, was the thought that they might lose control: the stream of refugees was now becoming a great rushing torrent; no one could say when it would end or how a line should be drawn. In the British House of Commons, the home secretary, Sir Samuel Hoare, maintained that the general desire to offer asylum still existed. It was essential, however, to prevent "indiscriminate admission,"

> to avoid creating the impression that the door is open to immigrants of all kinds. If such an impression were created, would-be immigrants would present themselves at ports in such large numbers that it would be impossible to admit them all, great difficulties would be experienced by the immigration officers in deciding who

could properly be admitted, and unnecessary hardship would be
inflicted on those who had made a fruitless journey across the continent.[78]

Authorities struggled to maintain business as usual, to protect
the processes of selectivity worked out in more peaceful circumstances. "Shall All Come In?" asked the *Daily Express* in a leading article on 24 March 1938. In one European country after another, authorities worked out how to say no.

Under pressure from prorefugee opinion in the United States,
President Roosevelt took the initiative at the end of March to call
an international conference on refugees to meet in the summer
of 1938. Like his European counterparts, Roosevelt wanted to
stand firm against admitting more refugees and to deflect sentiment favoring a liberalization of immigration laws and procedures. Carefully defining the terms of discussion, the original invitation proposed that twenty-nine governments consider
"facilitating the immigration from Germany and presumably
Austria of political refugees," but at the same time emphasized
that "no state would be expected to receive greater numbers of
emigrants than is permitted by its existing legislation." Moreover, there would be no increased financial burden: private organizations would bear all new costs involved. However neatly
tailored this proposal may have been for the American political
scene, it did not fit well in other countries. The Swiss refused to
host the meeting, apparently wanting to avoid embarrassment
over their own increasingly restrictive policies. The British lacked
enthusiasm, sensitive that their Palestine policy was evolving in
an anti-Jewish direction and suspecting that electoral concerns
guided American policy favorably toward Jews. The Foreign Office also expressed the fear that generosity might prompt even
more persecution and lead to expulsions from Poland and Rumania.

Roosevelt's meeting assembled in the agreeable French resort
of Evian-les-Bains early in July. The chief American delegate,
Myron C. Taylor, former head of United States Steel, presided
over the first session and was elected permanent president. The
Americans clearly guided the assembly. Contrary to Roosevelt's
intentions, however, the proceedings proved singularly unim-
pressive. William L. Shirer, a young journalist attending the conference for a U.S. radio network, learned that his New York office was not interested in covering it. "I doubt if much will be
done," he confided in his diary as the speeches began.[79] Shirer's

journalistic instincts were right. One delegate after another read statements into the record, justifying existing restrictive policies and congratulating themselves on how much had already been accomplished for refugees. Many were distracted by the attractions of the resort—its golf course, the gambling casino, summer skiing in Chamonix, or the celebrated stables of the locality. Most of the thirty-two delegations present seemed to be addressing their home constituencies rather than the pathetic representatives of refugee organizations and persecuted Jews who, uninvited, hung about the conference. All governments concerned worked together to ensure that delegates would not explore existing immigration rules and would not pressure any delegation to do more. Notably, the British managed to keep Palestine off the agenda—to the great annoyance of the Zionists who were present.

The conference recommended the establishment in London of an Inter-Governmental Committee, an American suggestion that proved to be the sole tangible result of the deliberations. Its job was to negotiate with the Reich so as to end the current chaos of expulsions and permit the refugees to take some of their property with them. Beyond this, the committee was empowered to approach the governments of receiving countries "with a view to developing opportunities for permanent settlement." The door was thus kept open for expanded immigration possibilities, but only in the vaguest terms. What probably sold the delegations on the American scheme was the prospect of negotiations with Germany. To increasingly alarmed Western countries, genuinely apprehensive about the prospect of war, the path to European peace was to be sought through talks with Hitler. Refugee problems also, it was assumed, could be solved rationally, through discussions with the Nazis. The League of Nations' own refugee organization was, of course, not empowered to undertake such talks, and other League agencies were ill-equipped for this purpose once the Germans decided to quit the organization in 1934. An independent body, it was thought, might achieve something. To direct the committee, the British and Americans agreed on George Rublee, a close friend of Roosevelt and an experienced negotiator. He was to work with a British chairman, Earl Winterton, a staunch anti-Zionist and member of the House of Lords. France and the Netherlands also received representation with two vice-chairmen.

To the consternation of Jewish representatives in mid-1938,

Evian simply underscored the unwillingness of the Western countries to receive Jewish refugees. The entire course of the discussions focused on German policy rather than on that of potential receiving countries. Most delegates probably agreed with the meanspirited Canadian deputy minister of immigration, Frederick Blair, who wanted Evian to hold the line on refugees so as to force the Nazis to solve their Jewish question internally.[80] The French host of the conference took the opportunity to announce that France was "saturated" with foreigners and was no longer a haven for the oppressed. To the Germans the conference was probably also a disappointment. A Wilhelmstrasse spokesman exultingly announced that Western countries wanted the Jews no more than did the Germans themselves; privately, however, the meeting signaled to State Secretary Ernst von Weizsäcker that emigration might not be facilitated as Nazi tacticians desired. German authorities had permitted Jewish representatives from the Reich to travel to Evian in the hope that they could clear the way for mass expulsions; these Jews returned empty-handed, and Berlin now faced the prospect of protracted bargaining with George Rublee in London.[81]

German and Austrian Jews were not the only refugees on the minds of Evian delegates. Poland and Rumania sent observers to the conference, a constant reminder that the Jewish issue in those countries might soon explode. Indeed, for over 4.5 million Jews in East Central Europe, the situation was becoming extremely serious. The British foreign secretary, Lord Halifax, predicted in a private letter Evian "might well be a prelude to an international negotiation of great magnitude concerning the future of the Jewish population in Europe."[82] Events in the course of 1938 underscored a gathering crisis affecting Jews throughout the region, with obvious implications for refugee policy and future settlement.

Besides Germany, Poland remained the greatest source of concern. Coinciding with Anschluss, a Polish law of 31 March 1938 authorized automatic cancellation of Polish citizenship for Poles residing more than five years outside the country. This was Warsaw's answer to the threatened repatriation of about 20,000 Polish Jews then living in Austria. At a stroke, this move threatened to render Jews stateless and to forbid their return to Poland. Thereby these Jews, plus thousands of Poles living elsewhere in Europe, became the most troublesome sort of refugee for Western countries—the kind who could never be sent home. About the same time, Polish antisemites intensified their campaign against Jews in Poland. Anti-Jewish riots flared that spring in major Polish cities, leaving some Jews killed, hundreds beaten, and much property destroyed. In the autumn the sudden worsening of Polish-German relations had particularly ominous implications for Jews. Early in October the Polish government recalled all Polish passports for inspection in a move to enforce the earlier denationalization law. As Hitler pressured the Poles on the Danzig issue at the end of the month, the Wilhelmstrasse suddenly initiated the expulsion of Polish Jews living in the Reich. Mass arrests of Polish Jews occurred throughout Germany, and thousands were forced east across the Polish frontier. After receiving these refugees for a few days, the Polish authorities suddenly refused, claiming that thousands were now stripped of their nationalities and that Poland was not obliged to accept them. Ejected by the Germans and refused entry by the Poles, some 5,000 Jews found themselves stranded in the border village of Zbaszyn. Supported by charitable organizations from abroad, these refugees camped for months in crudely fashioned shelters before finally entering Poland in July 1939. Poland thus seemed a harbinger of even more terrible refugee problems. Would the over three million Polish Jews soon be pounding on the gates of the West as were those of Germany and Austria? Many observers in 1938 feared precisely that.

Similarly, dangerous indications came from Rumania and Hungary during 1938. At the beginning of the year, King Carol took firm control of the Rumanian government by abolishing political parties and issuing a corporate constitution that extended royal power and enshrined right-wing nationalism. To outside observers, the monarchy was drifting inexorably into the arms of the Third Reich. Throughout the decade the ultranationalist Iron Guard fed a long-standing popular antisemitism. Although King Carol faced the Guardists in bitter rivalry, he continued to enforce a law that they sponsored in January 1938, effectively denationalizing tens of thousands of Jews. Under royal dictatorship some 225,000 Jews, representing 36 percent of the Jewish population, were deprived of their citizenship. Many thousands were removed from their positions by a numerus clausus and lost their livelihood. Democratic newspapers owned by Jews were suppressed. Jews were removed from government service, forbidden to obtain foreign currency, and denied permission to compete commercially with non-Jews in small villages. On several occasions Rumanian representatives broached with Western representatives the possible emigration of hundreds of thousands of Rumanian Jews. Thus, fears about a mass expulsion were not based on fantasy. As the Rumanian foreign minister announced in London the following year, the Jewish issue was a "capital question" for his government, requiring an international framework for its resolution.[83]

In Hungary, with about 445,000 Jews, we have already noted the signal that went out with the advent of the radical right-wing nationalist and antisemite Gyula Gömbös as prime minister in 1932, accompanied by the growing strength of fascist sentiment. Anti-Jewish feeling intensified after Anschluss, stimulated in part by Nazi propaganda imported from Germany. In May 1938, Hungary passed an anti-Jewish law cutting down the proportion of Jews in businesses and professions to 20 percent, and more such moves followed under Prime Minister Béla Imrédy later that year. In the wake of the Munich Accord, Hungary annexed substantial parts of Czechoslovakian territory, adding 150,000 Slovakian and Galician Jews. The campaign to Magyarize the newly acquired land of Subcarpathian Rus gave new pretexts to Hungarian antisemitism. A second anti-Jewish law carried persecution further, drastically reducing the Jewish involvement in economic and cultural life. As in Rumania, there were broad moves to denationalize Jews and deep fears about eventual expulsion. Jews in both countries were frequently demoralized, pauperized, rendered helpless before an onslaught that appeared Europewide.

Previously marginal to political and diplomatic events, refugees moved to the center of attention during the crisis over Czechoslovakia. In his demands that Germany acquire the disputed Sudetenland, Hitler fulminated about supposed Sudeten-German refugees: over 200,000, he told a crowd in the Berlin Sportspalast, had to flee the "barbaric persecution" of the government in Prague. Real refugees flowed in the other direction, however. Immediately after the Munich decision, ceding the area to the Reich, thousands of Jews and other Czechs left their homes so as not to live under Nazi rule. Most headed for the capital. In Prague, the mood was distinctly unfavorable toward over 130,000 now homeless fugitives from the region. Humiliated and abandoned, the Czech government was in deep disarray, desperately trying to salvage scraps of support. Fearing renewed German pressure over anti-Nazi dissidents on Czech territory, the government also worried about creating a substantial new German minority inside the reduced frontiers. Adding to these troubles, moreover, was the displacement of some 10,000 Czechs when the Poles absorbed Teschen early in October. Some 25,000 refugees camped in Prague schools and in the Masaryk Stadium, some of them in desperate need. According to Sir Neill Malcolm, the crush of refugees placed grave pressures on food supplies and accommodation in the beleaguered state. An extraordinarily diverse group of refugees attempted to leave the country during the next few months. Some of these were refugees for the second time, having fled to Czechoslovakia from Germany or Austria. Together with the former residents of the ceded territories, moreover, tens of thousands of Jews, Czechs, and Slovaks sought to escape the predicted onslaught of the Germans.

Western governments made token offers to help, but here, too, the refugees were often disliked. In Britain and France there had never been much sympathy for the Czech case in the Sudetenland, and there was a widespread sense that such regions ought not to be ruled from Prague. Flight from these territories, it was assumed, was part of the price for peace—a price the Czechs should mainly pay themselves. Sometimes the fugitives, especially the Jews among them, were assumed to be part of a war party, eager to provoke conflict between Germany and the democracies. Partly as a result, efforts to support Czech refugees remained halfhearted in many cases. The British did offer 4 million pounds sterling to help resettle the refugees. The French similarly extended financial aid, which, as in the case of British assistance, implied that the Czechs would cut down on their export of fugitives. To underscore the point, the British pressured the Czechs directly over the issue, chastising Prague for its rejection of Sudeten Jewish refugees expelled by the Nazis.

For Jews in Czechoslovakia, the prospects for escape evaporated quickly. Anti-Jewish disturbances spread throughout the country, and on the eve of its elimination the Prague government was under strong German pressure to heighten antisemitic activity. The Czechs managed to conclude an agreement with the Jewish Agency for the emigration of 2,500 Jews, mostly to Palestine. As everywhere, the major problem was the difficulty of finding visas to enter countries abroad and the slow processing of applications. The final blow came in March 1939, when the Germans seized Bohemia and Moravia. Some 117,000 Jews lived in these areas. Thousands had managed to escape by that point, but, according to one estimate, 32,600 refugees remained, some of whom had already obtained entry visas for other countries but had not yet managed to leave. The Nazis permitted continued emigration and even encouraged it in some cases. Fugitives frequently feared arrest, however, sometimes hopelessly complicating their situation; their exits, in many cases, had to be undertaken secretly. Finally, disaster at home had its impact abroad. The extinction of a sovereign Czech state automatically made refugees of tens of thousands of Czechs living outside the country, mainly in France. These soon joined the accumulating mass of stateless persons whose former national identity no longer corresponded to geopolitical realities.[84]

Kristallnacht, the furious nationwide pogrom against German Jews that raged on 9 and 10 November 1938, provided the most graphic illustration of worsening conditions for Jews in the Third Reich and plunged the refugee world into crisis. The context was the fresh round of persecution that had been underway for most of that year. In retrospect, 1938 marks a significant turning point in the dynamics of Nazism, as the regime radicalized the technique of control, dropped the conservative guise that had obscured many of its goals, and proceeded more adventurously both at home and abroad. Certainly in the assault on Jews, the Nazi apparatus moved more boldly than in previous years. Jewish policy became increasingly centralized in the offices of the SS rather than being pursued independently by a variety of agencies. Economic repression accelerated, rendering the Jews practically helpless. Anti-Jewish laws ravaged the supposedly autonomous structures of Jewish community life, once nominally encouraged by the Hitlerian state. Beginning in the summer, decrees prevented Jews from practicing medicine and law and from engaging in retail trade. New moves simply confiscated Jewish property wholesale. In October the expulsion of Polish Jews provided the pretext for the Kristallnacht riots. Among those who were brutally pushed across Germany's eastern frontiers was the family of the German-born Herschl Grynszpan, a young Jew who had been living as a refugee in Paris. Grynszpan's protest against the expulsion induced him to assassinate a German diplomatic official in the French capital. This prompted coordinated attacks by Nazi Storm Troopers throughout the Reich, with massive destruction of property, looting of Jewish shops, beatings, arrests, and murders. All this savagery was displayed the following day in the Western press, which gave the story considerably more prominence than it often did with mass murder a few years later. To most observers, these Kristallnacht riots appeared as the culminating barbarity, the definitive expression of Nazi implacability toward the Jews.

Kristallnacht sent a tremor through refugee organizations, which mobilized for what now seemed to them an emergency emigration. Everywhere under Nazi rule the story was the same. In Austria and the incorporated Sudetenland, the German masters pressed the Jews to emigrate. Kristallnacht had its equivalent in Danzig, the supposedly Free City under the League of Nations where the local Nazi administration introduced the Nuremburg Laws on 23 November 1938. Here, too, there was a crisis—in this case affecting about 15,000 Jews and other "non-Aryans" under direct Nazi threat.[85] Jewish officials throughout Europe began in desperation to cut corners, often under direct pressure of the Gestapo. They sent refugees abroad with incomplete documentation, obtained other papers through dubious channels, and connived to dispatch emigrants illegally into Palestine. Chartered vessels left German ports with uncertain destinations; sometimes these went from port to port in Central and South America trying to find a chink in the bureaucratic wall that prevented the passengers from landing. One famous ship, the *St. Louis*, left Hamburg with 930 passengers in May 1939, heading for Cuba. At Havana, Cuban authorities questioned the validity of the entry documents and refused admission. Eventually, after appeals everywhere fell on deaf ears, the ship finally returned to Europe, and its passengers were dispersed in several countries, most to be swallowed up by Hitler. All previous concerns for orderly emigration and the export of capital now appeared unrealistic: in the desperate rush after November 1938 the overwhelming priority was escape—one way or another.

The Crisis Deepens, 1939

In the last year before the war, it was impossible to maintain the tone of cautious optimism that suffused Sir John Hope Simpson's first refugee report. Hope Simpson reviewed the refugee situation in June 1939, noting how dramatically worse things had become following Anschluss, the Czechoslovak crisis, and also the collapse of Republican forces in Spain. Previously, refugees had been leaving the Third Reich at a rate of about 25,000 to 30,000 per year—managed efficiently, though with difficulty, mainly by Jewish and other private organizations. In 1939 the flow became much greater; the crush of people was now comparable to movements during and after the First World War and the Russian Revolution. Moreover, unlike the earlier period, restrictions had a serious effect: everyone realized that without stringent immigration controls that were everywhere being imposed, the refugee totals would have been much higher than they were. The core of the crisis was the Nazi persecution of Jews, and by 1938, indeed, the general tendency was to see the refugee problem as a "Jewish problem." By the spring of 1939, according to Hope Simpson, Jews in Greater Germany were almost completely impoverished and in need of evacuation. Over 400,000 still remained, including 120,000 over sixty years of age and 51,000 under twenty. Outside Germany, the number of Jewish refugees doubled from the beginning of 1938 to mid-1939. Like many observers, Hope Simpson accredited the claim of countries adjoining the Reich that they were "saturated" and that new admissions were only possible if overseas countries accepted more refugees for permanent settlement. As many as 100,000 exiles, he estimated, were unsettled in Europe as he completed his survey.[86]

Symptomatic of the new refugee situation was the phenomenon of statelessness, which became a much-discussed problem in the late 1930s. As we have noted in Chapter Two, redrafting the European map after the First World War left significant groups of people without formally certified national identities—persons for whom the French designation *apatride* became increasingly used.[87] So matters might have stood if several states had not begun to use denationalization as a political weapon against exiles. The Soviet Union stripped close to a million Russians of their citizenship in the early 1920s. Turkey did the same to Armenian refugees, proceeding informally from 1923 and by sweeping statute in 1927. Fascist Italy assumed the same power to denationalize in 1926, moving against the *fuorusciti* Mussolini found so annoying. Significantly, this wave of denationalization did not focus on people who had become nationalized citizens; these actions were meant to punish political foes living abroad whatever their personal history or national allegiance.

Nansen passports assisted the Russians and Armenians who became stateless, and relatively few persons suffered from vindictive denationalization policies in the 1920s. The Italian anti-

Fascists were probably the main victims, and their numbers were not large. The real difficulty arose in the Depression era, when permanent refuge became so difficult to find and when European states took particular care not to assume obligations to refugees in flight. Germany, at this point, became the main producer of stateless persons. By a law of 14 July 1933, the new Nazi government assumed the authority to denaturalize citizens whose naturalization was now deemed to have been "undesirable" and to denationalize Germans living abroad who "prejudiced German interests by an attitude contrary to the duty of loyalty" toward the Reich.[88] A new generation of stateless persons now began to appear in Europe, for whom the loss of citizenship posed the most serious technical difficulties. Jews were hit particularly hard. From 1933, the Nazis made it clear that racist principles were to be fundamental in determining who was to be denationalized. Eastern European Jews received special mention in the ordinance implementing the 1933 statute, and in the following years many became *apatrides*. In addition, many thousands of Czechs and Austrians found themselves made stateless following absorption of their countries into the Reich. By the end of the decade, therefore, the number of stateless persons increased astronomically.

Stateless people became entrapped in a bureaucratic web far more dense and restrictive than that set for other immigrants or refugees. Victims of Nazism suffered so acutely from the loss of citizenship because it made settlement abroad even more difficult than otherwise. Often such people had fled without passports or documentary evidence as to their citizenship. Sometimes German consulates simply refused to renew passports or other documents that had expired. Without a passport or with a passport that had expired, the *apatride* found it impossible to meet the requirements of states for even temporary residence. Constantly questioned about who they were, what their status was, and what was their destination, these people could not cross international frontiers, could not remain where they were, and were often not supposed technically to be at liberty at all. People without papers now suffered particularly from the ill humor of governments grown hostile and indifferent to the plight of refugees. No one felt obliged to receive them. The nub of the problem, of course, was that they could not easily be expelled. Having no recognized nationality, there was no state that would ultimately take responsibility for them. Reflecting on this odd predicament, the political philosopher Hannah Arendt was particularly sensitive to the plight of the *apatrides* and their peculiar significance in the European state system. Stateless persons, she wrote, were "the most symptomatic group in contemporary politics," pointing directly toward totalitarianism. They were outlaws by definition, completely at the mercy of the police, who could expel them at any moment. Cut off from civilized society, they were a signal of how principles of equality and of free existence could be trampled by the machinery of the modern state.[89]

The League of Nations proved singularly helpless before this problem. In retrospect it seems plain that no progress was likely so long as states refused to budge on receiving more refugees. The Assembly of the League recognized the difficulty and from 1933 appealed repeatedly against the summary expulsion and maltreatment of persons without papers. An international convention of 1933 reinforced the arrangements for the Nansen passports, but did not extend their applicability to other than "Nansen" refugees. The February 1938 convention did move in that direction, providing a travel document for refugees coming from Germany. States were slow to accede to this agreement, however. In practice, stateless persons continued to live outside the law, and no state permitted international obligations to override the fundamental unwillingness to protect the *apatrides*.

By the end of 1938, those attempting to escape Nazi Germany were desperate, prepared to go to extraordinary lengths to overcome immigration restrictions. A diplomatic dispatch from Vienna reported that every kind of document could be obtained for a price—passports, visas, tickets, whatever. In November a British diplomat signaled from Berlin that hundreds of Jews were smuggling themselves across frontiers and that suicide rates had climbed alarmingly.[90] About this time, thousands of Jews discovered an avenue for escape in China, and during 1939, oddly enough, the Far East became a focus of attention for European Jews.

The goal for thousands of fugitives was Shanghai, a teeming metropolis of over four million Chinese and 100,000 foreigners in 1938. Shanghai was the fifth largest port in the world—crowded, dirty, cosmopolitan, with a reputation for crime, violence, and intrigue. But Shanghai had one extraordinary advantage for refugees in the late 1930s: it was an open city, practically the only place in the world requiring no visas or other documentation for entry.[91] During the early 1930s, Shanghai Jewry numbered about

4,000 mainly Russian and Polish refugees who came following the Bolshevik Revolution. In 1937 the Japanese arrived, occupying most of the city in the course of their war against China. The Japanese presence had little effect on the Jews at first, however. Most lived in the area known as the International Settlement, nominally ruled by a municipal council elected by non-Chinese residents and practically controlled by representatives of the major powers with influence in China.

Jewish refugees began to enter Shanghai in large numbers following Anschluss, usually coming by sea, via the Suez Canal. The pace quickened at the end of 1938 with the events in Czechoslovakia and the *Kristallnacht* riots. By August 1939, it was reported, some 14,000 German and Austrian refugees had arrived, at which point the Japanese began to apply restrictions. Apparently, these regulations were not enforced, and the refugee community continued to grow, reaching 17,000 at the end of 1939. Wartime travel and shipping problems reduced the numbers of arrivals at that point; German refugees now had to proceed overland by train, through Siberia and Japanese-held Manchuria; others, especially Poles, went first to Kobe, Japan, and then to Shanghai. Soon even the Siberian route was blocked. From the outbreak of war, reemigration to more congenial places of asylum became almost impossible, and the refugee community stabilized at about 17,000. Within the city, visitors reported a distinct flavor of *Mittel-Europa*, especially among more affluent refugees—with German newspapers, coffeehouses, concerts, and so on. At the same time there was abject poverty despite the fact that refugees could live more cheaply in Shanghai than anywhere else—on about five cents a day. The Joint Distribution Committee provided 25,000 dollars a month, supporting about half the Jewish refugees during most of the war.[92]

Far from encouraging Jewish immigration to Shanghai, the various governments in charge of the international settlement—Britain, France, and the United States—worried that a flood of refugees might upset the delicate balance of interests that permitted their odd relationship with the Japanese to continue. As a result, they did their best to stem the tide at the very moment when the refugee pressure was greatest. British and American representatives even asked the Wilhelmstrasse in February and March 1939 to help stop Jews from going to Shanghai. The Nazis, who were well informed about the problems of the city, refused to act: for them the priority was to rid the Reich of its Jews.[93]

During 1939, as the case of Shanghai illustrates, governments in Western countries did more to hinder than to help the headlong flight of refugees from the Reich. The Inter-Governmental Committee, as we shall see in the next chapter, had nothing concrete to offer refugees, spending most of the year in fruitless negotiations with Nazi leaders, pursuing the "orderly exodus" so much hoped for at the time. It was left to private organizations to assist the escapees, and without their aid there is no doubt that many thousands more would have been swallowed up in the Nazi Holocaust.

Always anxious to prevent refugees from becoming a Nazi-imposed economic liability, Western governments regularly insisted that private charities, especially Jewish organizations, finance emigration from the Reich. Surveying what had been accomplished by the end of 1937, Sir John Hope Simpson noted that the bulk of refugee aid had been handled by Jewish institutions. Non-Jewish agencies were also involved, of course, although on a vastly smaller scale. Norman Bentwich mentioned the Quakers, in particular, who he felt were outstanding in assisting refugees. Generally speaking, the associations working for political refugees, trade unionists, and Christian victims of Nazi persecution provided much less comprehensive assistance than did the Jewish bodies, limiting themselves to small subsistence allowances. Jewish agencies, as in the period after the First World War, were preoccupied with a vast range of problems of emigration, relief, and resettlement.[94] Observers were stuck by the professionalism and smooth efficiency of these institutions. The Joint, the largest body at work, disbursed some 3 million dollars in the period 1933–36, half of which was spent on relief in Germany and on the assistance of emigration and half in contributions to emigration work in non-German countries. The Joint worked closely with the Jewish Colonization Association (JCA), HICEM (an amalgam of several Jewish emigration and immigration organizations established in 1927), the Jewish Agency in Palestine, and other groups.[95]

HICEM soon became the principal agent for Jews leaving Europe. Growing rapidly after 1933, HICEM extended its previous consultative activities into full-fledged operational assistance of all kinds. Drawing heavily on the Joint for financial assistance, it established a chain of refugee aid committees in countries adjacent to the Reich and began to finance the dispatch of Jews overseas. In 1936 it reported that it had helped over 14,000 refugees; by the end of 1940 the total had grown to almost 40,000.

Relief agencies required huge sums during the crisis period from late 1938 up to the end of 1941, when the Nazis blocked the Jewish exodus from occupied Europe. The Joint, for example, which raised 11 million dollars for refugee work between 1914 and the end of 1939, spent five million in 1939 alone. Jewish organizations together disbursed between 30 and 40 million dollars on refugee aid from 1939 to the end of the 1941.[96] By 1939, European Jewish communities had nearly exhausted their own resources, and American aid became vital to the continued support of refugees. But even in the United States funds became more scarce in 1939: The Joint had a deficit of 1.8 million dollars when the year closed, and income actually decreased in the years 1940–41.[97] For the major Jewish agencies, cruel dilemmas weighed on decision-makers: Polish Jewry was in dire straits, greater emigration from Germany without permanent resettlement was extremely costly, and competing Jewish organizations advanced rival claims for limited resources. Furthermore, the outbreak of war raised the serious possibility that assisting emigrating Jews could help assist the belligerents—likely to infringe American neutrality laws. However great their achievements in the preceding period, therefore, Jewish agencies showed signs of severe strain as the refugee crisis deepened.

Resettlement Schemes for Jews

During the late 1930s, the search for solutions to the Jewish refugee crisis produced a series of schemes for mass resettlement, ranging from well-intentioned efforts to outlandish concoctions. Sometimes these were seriously advanced and sometimes not; only rarely did they emerge from the drawing boards, and only in the case of Palestine was there a reasonable prospect that these proposals would actually assist Jewish refugees. In retrospect, one is impressed by the extraordinary variety of these projects and the curious combination of cynicism and credulity with which people responded to them.

Behind most of these plans was the hope that the refugees could be sent far away—to remote, unsettled regions where they would not pose serious difficulties. This was what the American columnist Walter Lippmann had in mind when he reckoned in 1939 that Europe's "surplus" population would best go to some "unsettled territory where an organized community life in the modern sense does not yet exist." Even champions of the refugees could argue in the same vein. The journalist Dorothy Thompson, for example, one of the strongest refugee advocates in America, called in 1938 for the establishment of an International Settlement Company to establish colonies of exiles in "this still empty world."[98] Remarkably, such appeals also accommodated official German emissaries who periodically contended that their brutal policies were guided by broad, European interests. The Jewish problem, Hitler told the South African defense minister Oswald Pirow in November 1938, was not a German problem alone; the Führer stressed the need for a Europewide solution and pointed toward the establishment of some sort of reservation to which the Jews could ultimately be sent.[99]

Of the many settlement ideas that were discussed as the refugee crisis deepened, Palestine was the most substantial. By the mid-1930s, as we have seen, there was a thriving Jewish establishment there of about 400,000 people and also a Jewish political structure capable of organizing large-scale immigration. James McDonald quickly saw the importance of Jewish settlement in the Holy Land; so also did a variety of pro-Zionist elements in England. One Conservative member of Parliament, Oliver Locker-Lampson, proposed that Jewish fugitives simply be granted Palestinian citizenship, resolving at a stroke their threatened statelessness. In 1935 several important Jewish organizations in Britain elaborated a scheme for the annual migration from Germany of 16,000 young people and 7,000 children, with about half of the total going to Palestine. This project floundered because the Germans refused to allow significant transfers of capital from the Reich.[100]

Until the mid-1930s, however, Zionist leaders were extremely cautious about colonization schemes, remaining committed to the slow, patient construction of the National Home. They were particularly concerned that Zionist pioneers bring to the colonization enterprise the requisite political commitments and professional skills to achieve a vision of Zionism that was heavily marked by European socialist ideals. "In order that the immigration not flood the existing settlement in Palestine like lava," Arthur Ruppin declared at a Zionist congress in the summer of 1933, "it must be proportionate to a certain percentage of that settlement." Two years later Chaim Weizmann, head of the World Zionist Organization, told the Zionist Executive that the movement had to choose between immediate rescue of Jews and the broader project of Jewish national redemption: Weizmann favored the latter view.[101]

So long as there were few Jewish candidates for emigration to Palestine, these calculations made some sense. But conditions changed following the rise to power of Nazism, the Arab revolt in 1936, and the British moves to reduce even the existing number of Jewish immigrants. One of the first to sense the new reality was an outstanding opponent of the Zionist mainstream and the socialist establishment in Palestine, Ze'ev Jabotinsky. Jabotinsky was a brilliant orator and leader of the Revisionist dissidents within the Zionist movement; his followers broke away from Weizmann's movement and set up their own New Zionist Organization in 1936. That year Jabotinsky launched a controversial program for a mass "evacuation" of Jews to Palestine: according to this plan, formalized by the Revisionists in early 1938, a million and a half Jews would be transferred to Palestine in ten consecutive yearly instalments.[102] Few had spoken before of emigration on such an enormous scale. Particularly in Poland, Jabotinsky was bitterly attacked by Jews who thought his scheme overly fatalistic with respect to antisemitism; others judged his ideas grandiose and self-serving, likely to fuel anti-Jewish forces in Poland who constantly accused the Jews of disloyalty. But with the refugee situation worsening massively and with the imposition of new restrictions, even mainstream Zionists began calling for mass settlement in Palestine. As Weizmann told the Peel Commission in 1937, "the world is divided into places where [the Jews] cannot live and places where they may not enter." Zionists everywhere demanded that the British open the gates of Palestine.

The struggle over immigration certificates for Palestine was no longer a theoretical exercise; it reflected the desperate reality of fugitives with no other place to go. The British recorded almost 30,000 Jewish immigrants in 1936, but only 10,500 in 1937. About 40 percent of the latter were from Central Europe. During the next two years, legal immigration rose slightly to 12,900 and 16,400, and the proportion of refugees from Central Europe reached 75 percent just before the outbreak of war. But this was woefully insufficient to meet the new emergency. As pressure mounted, the Jewish leaders began to circumvent the process of legal immigration, seeking ways to smuggle refugees into Palestine. This movement, known by the Zionists as "Aliyah Bet," had previously been frowned on by Zionist leaders but now won official support within the movement. Coordinated mainly by the underground Jewish military force, the Haganah, and the Irgun Zvai Leumi, followers of Jabotinsky, an estimated 18,100 illegal immigrants came to Palestine via this route, mostly in the period 1938–41.[103] About half came from Central Europe. Defying the new British restrictions, Jewish agents hired ships in the Balkans and attempted to sneak through British maritime surveillance. In response, colonial authorities reinforced offshore patrols and scrutinized immigration more carefully. The ensuing clashes between terrorized refugees and resolute British officials, often on the high seas, frequently shocked outside observers and poisoned relations between the Jews and the mandatory power on the eve of the war.

Besides Palestine, the most persistently discussed territory for Jewish colonization was Madagascar, a French island possession in the Indian Ocean off the east coast of Africa. Ruled from Paris since 1896, Madagascar had an indigenous population of 3.8 million in 1936, with 36,000 Europeans. Few locations could have been more remotely situated or less congenial to Europeans than this distant outpost, and perhaps for this reason Madagascar had been bruited about since the late nineteenth century as a possible home for unwanted European Jews.[104] Madasgascar came forward once again in the mid-1930s, when, at the instigation of Foreign Minister Józef Beck, the Polish government began to investigate Jewish migration possibilities. The following year the Polish delegate to the League of Nations urged recognition of the Jewish right to go to Palestine, but also stressed the need to find other territories for Jewish settlement. Among the options, the Poles judged Madagascar a likely location. They obtained permission from the Popular Front government in Paris to explore the prospects in detail, apparently with the cooperation of Polish Jewish representatives and some Zionist leaders.[105] Simultaneously, interest widened. French and German newspapers discussed the idea favorably, the British looked into the scheme, American Under Secretary of State Sumner Welles became interested, and Hitler himself seems to have favored a Madagascar project. When the German foreign minister Joachim von Ribbentrop met with his French counterpart, Georges Bonnet, in December 1938, the latter opined that France might be able to send

10,000 Jews abroad, possibly to Madagascar.[106] Given the inhospitable tropical environment, the lack of development, and the strong opposition of local settlers, the idea was stillborn. Nevertheless, the Madagascar scheme had a remarkable life. It reemerged in mid-1940, following the German defeat of France, when the Nazis speculated on a "Final Solution" of the Jewish question. Remarkably it was still alive in 1946, when the British Foreign Office sounded out their consul general on the island about possible Jewish settlement.[107]

Competing with Madagascar, dozens of ingenious suggestions were advanced at the end of the decade as the refugee crisis worsened. In the wake of Anschluss, Sir Neill Malcolm proposed that a tiny state be created somewhere (he favored North Borneo, where he had business interests), of which stateless persons could automatically become citizens. Equipped with documents provided by this new entity, the refugees would escape the special handicaps of statelessness. The British Foreign Office scuttled the plan, however, arguing that it could not practically improve the prospects for emigration or resettlement.[108]

The ebullient Franklin Delano Roosevelt entertained a series of frequently exotic resettlement ideas, some of which were discussed with European leaders. Friends of the refugees and some who were clearly not friends at all bombarded the American administration with geographic sites where Jewish refugees could be sent: the Dominican Republic, British Guiana, Cyprus, the Philippines, the Belgian Congo, Ecuador, Mexico, Haiti, and Surinam were all considered, to mention only a few. At the end of 1938, Roosevelt signaled to Mussolini that refugees might be dispatched to southwestern Ethiopia, then effectively subdued by the Italian army. Mussolini, however, flatly refused: not a square inch of Ethiopian territory, he told American Ambassador William Phillips, could be made available for Jewish settlement.[109] Shortly after, Roosevelt approached the British about the possibility of large-scale Jewish implantation in the Portuguese colony of Angola. Through Whitehall, the American president hoped to induce the Portuguese to contribute "to the creation of a supplemental Jewish homeland."[110] Embarrassed by their own restrictive policies in Palestine and unwilling to disturb their colonial order in Africa, the British poured cold water on the idea. Meanwhile, the British entertained a chain of resettlement proposals, often in response to American prompting. In London these suggestions sometimes awakened interest because of the possibility that such colonies might remove some of the pressure from Palestine. Unfortunately, there was also substantial opposition to the refugees in places like Kenya or Northern Rhodesia that had been favored by the Colonial Office. By 1938 the British realized that these schemes were long-term suggestions, unable to address the existing refugee emergency. Armed with this appreciation, officials considered such plans dubiously. Suggestions went from one office to the next. "The world map was scoured repeatedly for resttlement possibilities," according to one writer, "the correspondence columns of the British press resounded with polite controversies of ex-Colonial administrators and others debating the climatic and other merits of obscure islands and jungle-covered valleys whose salubrious delights had otherwise gone unremarked."[111]

Politicians, statesmen and occasionally Jewish leaders grasped at resettlement schemes like straws, unable to contemplate any more adequate response to the refugee emergency. Ill-suited to the task, the various projects suffered continually from a lack of serious planning and governmental commitment. The American administration persistently ignored skeptical advice from leading experts in the field of population resettlement, seizing instead on proposals that gave the allure of energetic, imaginative action.[112] No state could possibly have advanced the massive amounts of capital necessary to start such settlement enterprises, and the Nazis remained obdurate before requests for a transfer of Jewish property. Few of the officials who entertained these ideas bothered enough about the viability of the projects—the suitability of the various regions for a large influx of Europeans, prospects for long-term financing, or the ability of these colonies to function economically. Meeting the preconceptions of the time, government leaders and civil servants warmed to the image of young Jews tilling the soil of sunny, distant lands, improving their own lives immeasurably in the process. In reality, three-quarters of the German Jews were over forty, most lived in large cities, only 2 percent knew anything about agriculture, and over half were involved in trade and commerce. One can only conclude that the projects were seldom designed with Jews clearly in mind at all. At best, they were a product of wishful thinking, generating the kind of widely shared illusion of the 1930s in which people imagined the world as they wished it to be. At worst, they were a cynical manipulation, used by proponents of restrictive immigration policies to keep the desperate petitioners at bay.

LAST CHANCE, 1939–41

Feared and predicted since the advent of Hitler, war in Europe began in September 1939 with the Nazi onslaught on Poland. Few anticipated a great European disaster on the order of the First World War, and leaders on both sides of the conflict remained confident of success. In alliance with the Soviet Union, Nazi Germany smashed the Polish army in less than three weeks while all remained quiet in the West. Although helpless to save Poland, both Britain and French believed that ultimately Germany would be brought to her knees. The Western allies felt protected against Blitzkrieg by the English Channel and the Maginot Line fortifications and judged themselves better able to sustain a long conflict than the Hitlerian state. In British strategic thinking, economic pressure and blockade were likely to provoke economic collapse and disintegration in the Reich. In France, General Gamelin opposed launching any serious offensive until 1941 at the earliest; by then, he felt, the Allies would be significantly superior to Germany in men and equipment. For both countries this strategy dictated a high priority for "economic warfare" and a careful attention to the extensive rearmament effort that was already under way. Relatively little changed in the Western democracies, therefore, after the first, awkward steps of mobilization and civil defense. Once the Wehrmacht ground to a halt in Poland, formal hostilities seemed practically finished. Bombs did not fall in the west, troops did not cross frontiers, and few shots were fired along the fortified front.

This inactivity stood in remarkable contrast to the frantic flight of refugees set in motion by the opening of hostilities. The Wehrmacht scattered hundreds of thousands of Poles, sending many of them into the arms of the Red Army, which advanced from the east. In the west, however, the fighting did not mark a sharp break in refugee history. War reinforced the general reluctance to receive refugees from Hitler; countries at war with Germany were particularly loathe now to lighten the economic burden of the Reich by receiving her outcast indigents. Jews continued to leave Germany although departure was made vastly more difficult by wartime immigration controls and the scarcity of transport. France was awash in refugees during most of 1939, following the agony of the Spanish Republic. The great upheaval there came when Hitler's armies turned westward in the spring of 1940. During the next two years millions of Europeans were forced to move, dispersed by a war that upset all the confident assumptions of 1939.

The Collapse of Republican Spain

To many Europeans the Civil War in Spain was a dress rehearsal for the wider European war that broke out in 1939. Ideologically, the insurgence led by Francisco Franco presented an image of fascism: superficially and in the eyes of the left, the Nationalist leader appeared as the Spanish version of Hitler and Mussolini. In political terms the alliance struck in 1936 between the anti-Republican forces and the Axis powers confirmed this judgment. The Soviet Union entered the contest a few months after it began, and, throughout Europe, Communists popularized the notion that Franco represented yet another dimension of European fascism. As the war went on, piling up staggering numbers of casualties, its significance for the democracies shifted: the Western powers proved helpless to organize nonintervention in Spain and unable to prevent Germany and Italy from exploiting the fighting to their advantage. France, in particular, was rocked by the issue of Spain, which paralyzed the government of Léon Blum and deeply divided public opinion.

The first refugees from the conflict moved within Spain, fleeing from one side or the other, usually spurred on by reports of atrocities committed in the first weeks of struggle. The Popular Front government assisted the departure of thousands of middle-class Spaniards, terrified by the flurry of anticlerical attacks and revolutionary measures on the Republican side. British and French warships removed large numbers from Barcelona, where a separatist Catalan government only barely maintained order amid rampaging anarchists, militant unionists, and revolutionaries. Fearing repression in the Nationalist regions, on the other hand, thousands of peasants from Estremadura and Andalusia crowded into Madrid, jamming public facilities, sports fields, and the like. As the Franco forces approached the capital in the winter of 1936, the Republican government evacuated a large part of the population, sending them to Catalonia and towns on the Mediterranean below Valencia. When Málaga fell to the insurgents early in 1937, an estimated 150,000 refugees streamed along the coastal road in the south, occasionally strafed by Nationalist aircraft. This human tide surged north to Alicante and finally to Catalonia. On both sides the numbers of destitute fugitives increased catastrophically, imposing a hopeless burden on authorities already strained to the limit by the exigencies of war. In the summer of 1938 the Catalan government estimated that there were two million refugees on the Republican side, about half of whom were in Catalonia itself. By January 1939, following the government retreat on the Ebro River, Sir John Hope Simpson raised this estimate to about three million.[113]

Early 1939 saw the Republican forces crumble, leading to the panic-stricken flight of refugees. Franco's troops entered Barcelona at the end of January and took Madrid in March. In the wake of these disasters, masses of refugees surged toward the French frontier. By the beginning of February, 300,000 had gathered at various border points. The French admitted about 80,000, but sent back many of military age. Pressure mounted. Some of the refugees suffered severely from exposure, malnutrition, and typhoid fever. After hurried meetings with Republican officers, border guards allowed defeated soldiers and civilians to cross into France. Within days about 350,000 entered, over half of whom were once Republican soldiers.

Given the deep divisions over the war in French society, this gesture was bitterly resented by the French right. The latter accused the government of welcoming mobs of radicals and subversives and lamented the spectacular financial costs, said to reach 7 million francs per day. In what has been called the "Great Fear" of 1939, people responded to the Spaniards much as townspeople during the French Revolution of 1789 reacted to mysterious and menacing "brigands" from the countryside. "Terrorist hordes," announced the progovernment *La Dépêche de Toulouse;* according to the deputy mayor of one southern town, awed by the enormous sea of people, "one was astonished that the earth did not follow [after them]. . . . It was grandiose, enormous, frightening."[114]

By the end of April, when the Nationalist forces had finally triumphed everywhere, over 450,000 Spanish refugees had reached France. Many were sick or wounded, and over a third were old people, women, and children.[115] The French had expected a small fraction of this amount and were totally ill equipped to deal with the emergency. Most of the refugees were housed in makeshift camps, hastily constructed and widely dispersed throughout the country. In the remaining months of winter, conditions were exceptionally difficult, especially for soldiers. A report prepared for the Council of the League of Nations in May noted that food was insufficient, amounting to only 1,600 calories a day; shelter was poor or nonexistent; discipline was harsh, mail was censored, and the men were treated as hardened criminals. Civilian camps varied considerably; local authorities were sometimes extremely hostile to the refugees and sometimes not. Huge agglomerations such as Bacarès or Argelès housed tens of thousands of inmates; around these centers the French Red Cross, interested political groups, and a variety of other voluntary agencies clustered to supplement the meager levels of government assistance.

For several reasons, the French were desperately concerned to rid the country of the Spaniards. Threatened now by fascist regimes on all three frontiers—the Rhine, the Alps, and the Pyrenees—the government of Edouard Daladier sought a modus vivendi with France's neighbors and strove to distance itself from the anti-fascist crusade associated with the defeated Republicans. In sharp reaction against Léon Blum and the Popular Front, French leaders were eager to conclude an agreement with Franco, hoping to secure his neutrality in the event of the Franco-German conflict that was so widely feared. In addition, the government wanted to avoid the heavy economic burden of refugees. France, by this point, had the largest refugee concentration in the world, for which the government was paying about two hundred million francs a month. Firebrands on the right bitterly denounced this expenditure, fanning the xenophobic and antileftist hysteria.

These circumstances pushed Paris toward an amiable settlement of the issue with Franco, followed by prompt repatriation. Almost as soon as the last Loyalists straggled across the frontier, indeed, some were on their way back to Spain. At first, the Nationalist government in Burgos dragged its feet on formal repatriation arrangements, refusing to receive refugees until France returned the gold from the Bank of Spain that the Republicans had sent out of the country. To Franco's apparent surprise, Daladier eagerly complied, and the two sides settled the issue. As the fighting began in Poland, therefore, the French were busily sending the Spaniards home. Although repatriation was voluntary, the French applied considerable pressure on the refugees to leave. Between 150,000 and 200,000 did so before the end of the year.[116]

Those who did not soon suffered the Nazi occupation in France, with the exception of about 20,000 or so who went to Mexico.

Thousands of Spanish refugees enlisted in the French army and fought against Germany, many ended up in the French Resistance during the Nazi occupation, and thousands were caught by the Germans to be deported as forced laborers, murdered in concentration camps, or summarily shot. Almost 10,000 are said to have perished during the Second World War at the hands of the Nazis either in battle or captivity. In 1951, according to the French Ministry of the Interior, there were still 112,000 Spanish refugees in France.[117]

Among those who did not seek asylum in France a small group of Republicans, mainly Communists, managed to enter the Soviet Union. Because the Russians had been so prominently committed to the Republican cause, at least until 1938, the Communist state appeared a most reasonable place for the defeated revolutionaries. Valentín González, known by his nom de guerre, El Campesino, scarcely thought twice about his destination: "It was decided that I should go to Russia. Where else? I was an exile and a Communist. I had nowhere else to go. I had no business anywhere on the face of the earth."[118]

Predictably, the Soviets and the Spaniards detested each other from the start. Politically and temperamentally, the two could never get along. El Campesino, a rough-hewn, bearded warrior from the Estremadura, was one of the most prominent Republicans to reach the Soviet Union. He instantly disliked what he found—bureaucracy, boundless corruption, social inequalities, and a bourgeois style of life for civil servants. In his first encounter, an official gave him a form to fill out. The Spaniard refused: "The Russians knew me well enough. I don't have to answer all these questions." As Raymond Carr puts it, the Soviets "instinctively distrusted revolutionary enthusiasm and believed in discipline."[119] Brooding on internal problems and fearful of outside ideas, Stalin singled out his own generals who had fought in Spain for particular repression during the great purges. The Soviet leader was not well disposed to flamboyant Latin revolutionaries. Furthermore, by the time the Spaniards arrived, the Soviets had long since abandoned their cause and were concentrating instead on improving their relations with the Reich. Within months, Stalin was to sign a pact with Hitler. In this sense, too, the Republicans were an embarrassment. So far as one can tell, their presence was even more awkward for their hosts than was the case with German Communists, who were treated badly enough.

The Soviets received about 5,000 Spanish children, evacuated in 1936 and 1937, and about 4,000 Communists in 1939. The children were placed in schools and orphanages in order to obliterate their national identity. According to one disgruntled Spanish Communist who reported on their suffering in the Soviet Union, 2,000 of these youngsters died during the war, often owing to overwork and undernourishment.[120] Adults ended up in forced labor camps and eventually perished, as no doubt they were intended to do; they were kept isolated from the rest of Soviet society and often dispatched to the most distant places of internal exile. El Campesino spent two years in Uzbekistan and eventually joined a band of robbers in order to survive. Most were not so durable and frequently fell into the hands of the NKVD. It is not known how many survived the war or whether any managed to leave at its end.

Refugees from Poland and Eastern Europe

The Germans shocked the world in 1939 by the speed of their advance in Poland and by their overwhelming technical superiority. Despite a gallant resistance, the Poles were hopelessly outclassed in numbers and equipment and especially in tactics. German armored columns, acting in close coordination with the Luftwaffe, drove deep into Poland, leaving the defenders stunned and disorganized. Crucially, the Poles fought alone, unaided by their allies Britain and France. Then, on 17 September, the Red Army entered Poland and dealt the country a final blow. Throughout this short campaign, the fighting decimated the civilian population. Most of the Polish air force was destroyed on the ground in the first hours, permitting the Luftwaffe to operate virtually at will.

Refugees proved to be a powerful weapon on the German side. In panic, hundreds of thousands burst out of endangered cities, jamming the roads and throwing Polish military communications into chaos. From above, Stuka dive-bombers swooped down on the crowds, hurling terrified inhabitants into the Poles' defensive positions. The result was a great uprooting of the Polish population, limited only by the suddenness and completeness of the Nazi victory. It is impossible to say exactly how many civilians were thrown out of their homes. The Polish army suffered 200,000 casualties in a few weeks of fighting, and civilian losses were many more. Alexander Donat was living in Warsaw at the time and heard a government radio broadcast announcing a German breakthrough on 6 September. "Hundreds of thousands left Warsaw that night," he later wrote.[121] The government abandoned the city the next day although Warsaw was to hold on for a few weeks. A torrent of escapees fled south in confusion, joined by military units in various stages of disintegration. A substantial proportion managed to cross the "Green Frontier" into Hungary or Rumania; those of military age attempted and often managed to reach France, where the Polish government and army set up shop in the provincial town of Angers. Polish authorities counted about 140,000 evacuees taken through Iran to Palestine and then often to England and France. Even after the fighting ceased, some refugees continued to flow. Many simply returned to their homes. As word of Nazi antisemitic atrocities spread, an estimated 300,000 Jews, almost 10 percent of the entire Jewish population, fled German-held territory in western Poland and crossed into parts occupied by the Russians. For about a month fugitives from the Nazis moved freely into the Soviet zone, and the Nazis often permitted people to leave. Then, on 30 October, the frontier dividing the Russian and the German zone was officially closed, although some managed to cross until January 1940.[122]

The Soviet role in these events was the political shock that accompanied the military surprise. Despite their nonaggression pact with Poland, the Russians joined the Germans in overrunning the country. On 29 September the two invading powers formally partitioned Poland, giving small shares to Slovakia and Lithuania. The entire eastern half of Poland, with its fourteen million people, now lay under Soviet rule. In the West, there was much incomprehension about the Soviets. As Winston Churchill put it, Russian policy was "a riddle wrapped in a mystery inside an enigma." The key, Churchill said, was Russian self-interest. Self-interest had nudged Stalin toward the pact with Hitler, signed on 23 August 1939, ensuring that their two countries would cooperate in the coming conflict. Self-interest, rather than humanitarian concern, similarly guided Soviet refugee policy.

Among the first victims of Stalin's new relationship with Hitler were a handful of anti-Nazi refugees who still lived in Russia, having managed extraordinarily to survive the purges of the late 1930s. For those still outside prison, the most striking shift occurred following the Nazi-Soviet agreement. Immediately, all criticism of Nazi Germany disappeared from the Soviet press, stage, and screen. Anti-fascist opposition, the lifeblood of the exiles still at liberty, was now to be eliminated from Soviet life. The NKVD rounded up hundreds of German Communists. According to one witness, Soviet agents forced some to request their own return to the Reich. At the beginning of 1940, the Russians simply handed over to the Gestapo several hundred German Communist refugees, including a number of Jews, almost certainly destined to be killed by the Nazis. One of these, Margarete Buber, in Russia since 1933, even managed to survive the ordeal, though only after five years in the Ravensbrück concentration camp. Officially, the Comintern denounced these exiles as enemies of German-Soviet understanding. They were certainly that, but they were also a pathetically weak and helpless lot, ideologically stranded in the shifting seas of Stalinist policy.[123]

Consolidating their newly acquired territory in Poland, the Russians wanted to bind the new subject peoples to the Soviet state and pursue the supreme goal of national defense. In Moscow the official view was that the U.S.S.R. had not really attacked Poland at all: the Red Army had moved east simply to protect Ukrainian and Belorussian "blood brothers" in a previously Polish-dominated region and to unite them with the Soviet motherland. In practice, the Soviets worked to break up the institutions of Polish national identity, to intimidate the Poles, and to subdue any opposition to merging into the Soviet state. No independent Polish voice was permitted. Even Polish Communists like Wladyslaw Gomulka found themselves under a cloud. Refugees from Nazi-held territory were treated with great suspicion, probably in an effort to reassure Berlin that Moscow would honor the new friendship with the Reich.

To facilitate their annexation of newly conquered Polish territory, the Russians deported hundreds of thousands of Poles eastwards. Before the winter of 1939–40 had ended, the Soviets began a massive uprooting of the Polish population, adding to the extraordinary confusion and dislocation that had followed the German campaign. The idea was to send important segments of the Polish population deep into the Soviet interior, far away from formerly independent Poland. Families were split apart, and many hundreds of thousands were dispatched to remote settlements and camps for "corrective labor" in Siberia and elsewhere. Shipments included as many as 30,000 people a day in unspeakable conditions—sealed into cattle cars for journeys as long as 6,000 miles. From London, the Polish government in exile said that there were 1.8 million Poles in the Soviet Union in 1941, but they had no way to test the accuracy of these totals. Nicolai Tolstoy estimates that between a million and a million and a half Poles, including prisoners of war, were deported eastward before the Nazi invasion of June 1941.[124]

The deliberate uprooting of great masses of people widened a few months later, when the Russians turned to the independent Baltic states—Latvia, Estonia, and Lithuania—having decided some time before to incorporate them into the Soviet Union. Moscow moved against the Baltic countries in the summer of 1940 as the Western world was distracted by the spectacular Nazi victories in the West. Following ultimatums, the Russians simply sent troops across the frontiers. "In the first year of Soviet occupation," according to Tolstoy, "34,250 Latvians vanished without a trace—more than 2 percent of the population." Shortly after their invasion in 1941, the Nazis published photocopies of previous Soviet plans, credited by Lithuanian exiles in Switzerland, that 24 percent of the entire Lithuanian population was earmarked for deportation. Baltic peoples by the tens of thousands now followed the Poles into exile. Deportation trains sent them deep into Soviet Asia and Siberia during the next year. Tolstoy concludes that 61,000 Balts were thus packed off to the east.[125]

Jews living in all these territories were frequently turned into refugees—the second time for many of them, having first been driven east by the Nazi invasion of 1939. By the middle of the next year, after the annexation of the Baltic states plus some Rumanian territory, about two million Jews had come under Soviet rule.[126] In eastern Poland, close to 20 percent of the Jews were refugees, and officials experienced difficulties in absorbing them into the local economy. The Soviets regarded these fugitives from Nazi-held Poland with particular suspicion and distaste, subjecting them to systematic harassment by the NKVD. Many of these Jews had relatives still living in Nazi-occupied territory, and many were reluctant to become Soviet citizens as the Russians had hoped. In the eyes of Soviet officials, these continuing links with the West made Jewish refugees especially grave security risks. Consequently, almost all who had fled to Russian-held Poland were deported further east. The Soviets were somewhat more lenient toward local Jews, who, together with Ukrainian and Belorussians in formerly Polish territory, were declared Soviet citizens in Nobember 1939. The Russians apparently wanted to boost these nationalities as a way of further undermining the Poles; Soviet propaganda regularly stressed the non-Polish character of the recently annexed land. Stanislaw Kot, who became Polish ambassador to Moscow, estimated that there were 400,000 Jews among the Poles in the Soviet Union, their proportion among the exiles being more than triple their proportion in prewar Poland.[127]

Until the Nazi invasion, refugees trickled in both directions across the new Russo-German frontier. Although the border closed in the autumn of 1939, hundreds continued to sneak across, occasionally with the connivance of frontier patrols. Nazi massacres in western Poland determined some to stake their future on the Soviet Union. The Soviets occasionally welcomed these refugees and occasionally turned them back. Until early 1940, the Germans continued to expel thousands of Jews into Soviet-held Poland—sometimes using the terms of an agreement of November 1939 between the two countries providing for exchanges of Ukrainians and Belorussians living under Nazi rule for ethnic Germans in Russian territory. Following stiff Russian objections, this practice ceased. Meanwhile, some Jewish refugees were fleeing westward to Nazi-held territory. The reasons, of course, were various: family ties in the West, ignorance about conditions under the Germans, severe economic conditions in eastern Poland, and political repression by the Communists. As conditions worsened on both sides of the border, many assumed that life had to be better somewhere else. One author recounts an incident to illustrate: in one Polish town, just inside the German border, a trainload of Jewish refugees moving east crossed a similar train heading west. When the Jews leaving the German zone passed their coreligionists from Russian-held territory, they shouted, "You are insane. Where are you going?" From the other train, however, came the same cry: "You are insane. Where are you going?!"[128]

Among those refugees trapped by the clash of Germany and Russia, thousands of Jews in the city of Vilna drew particular attention in the West because unusual circumstances permitted a small number to escape.[129] Once the capital of the Grand Duchy of Lithuania, Vilna lay within Polish frontiers during the interwar period. When the Red Army moved west in September 1939, the Russians captured the city and permitted its integration into the still-independent Lithuania. For some eight months—until the Soviets overran all three Baltic countries in June 1940—Vilna remained under Lithuanian control and became an important crossroads for refugees, especially Jews. Renowned as a seat of

Jewish learning, Vilna had about 50,000 Jews at the beginning of the war, to whom were added thousands of fugitives both from Nazi- and, to a lesser extent, Soviet-held territory. Between 15,000 and 20,000 Jews amplified the Jewish population. After the first influx, however, the Lithuanians sealed their borders, and few were permitted to enter. Within the city, thousands of refugees were kept alive through the winter by funds provided by the Lithuanian government, the Joint, and the World Jewish Congress. Meanwhile, Western agencies funneled some 4,000 Jews out of Lithuania, mainly passing through Vilna. The refugees traversed Soviet territory to Odessa or to Istanbul and then went on to Palestine; others rode the Trans-Siberian Railway to the port of Vladivostok, eventually reaching Kobe, Japan. Even after the Soviet takeover, refugees moved along the increasingly precarious lifelines. At the end of 1940, however, the Russians broke off relations with the western refugee organizations, and the fate of Lithuanian Jews, including 14,000 remaining refugees, was sealed. Finally, in the last chapter of Vilna's refugee story before the Nazi occupation, the Soviets began mass deportations from Lithuania in mid-June 1941, only a week before the German attack. About 30,000 people were expelled, including thousands of Jews and refugees.

Nearby Finland generated several hundred thousand refugees during this period, following two wars in succession against the Soviet Union. Hostilities began in subzero temperatures at the end of November 1939, when the Russians attacked their small northern neighbor suddenly and without warning. Despite stubborn resistance, during which the Finns inflicted enormous casualties, the Soviets triumphed in the Winter War at the end of February 1940. In the ensuing Treaty of Moscow, the Russians forced their opponents to cede the entire Karelian Isthmus and the region around Lake Ladoga. Local inhabitants were given a short period to leave their homes in order to remain under Finnish rule. According to Joseph Schechtman, about 415,000 moved of the 420,000 living there—97,000 evacuated during the fighting and the rest following the signing of the treaty.[130] Within months, 11 percent of the entire Finnish nation became refugees. Simultaneously, according to Nicolai Tolstoy, the Russians turned on the Finns in their own midst living in Leningrad and Soviet Karelia, dispatching them to camps north of the Arctic Circle.[131]

War between the two countries resumed in June 1941, when Finland took advantage of the Nazi invasion of Russia to win back lost territory. The Finns did well at first, moving deeply into what had become Soviet territory. By the autumn, the evacuees of the previous year spontaneously began to return home. At the end of 1941, 77,000 Karelians were back, and the numbers continued to rise. Eventually, well over half of the refugees ended their temporary displacement. These people experienced a new disaster during 1944, however, when the tide turned once again, and the Soviets entered Finnish territory. This time a quarter of a million Finns took to the roads, now definitively leaving their homes to escape the Red Army. Over 100,000 more fled from other parts of the country, imposing staggering problems of resettlement on Finland in the immediate post-war period.[132]

To complete this survey of refugee movements in Eastern Europe in the first phase of the Second World War, one should briefly note the territorial changes on the borders of Rumania, forced on Bucharest in the summer and autumn of 1940. Balanced awkwardly in the late 1930s between the Western powers and the Axis, Rumania attempted to protect herself from three revisionist neighbors—Hungary, Bulgaria, and the Soviet Union. Everything depended on her ability to keep an increasingly powerful Third Reich from lending too much support to her enemies. The Nazi-Soviet pact and the collapse of Poland revealed how exposed was Rumania's position. The Nazis now had no reason to favor Bucharest, and her foes could strike with impunity. In June 1940 the Russians forced Rumania to surrender Bessarabia and northern Bukovina. Shortly thereafter the Hungarians exacted the greater part of Transylvania, and the Bulgarians annexed southern Dobruja. Each of these losses, totaling more than one-third of prewar Rumanian territory, prompted movements of refugees. Disgruntled Rumanians from all three regions made their way to the capital, adding to the numbers of uprooted who had previously come from Poland. Bucharest became an important refugee center, boiling with intrigue and violent opposition to the unsuccessful government of King Carol.

Flight in the West, 1940

Following the Nazi breakthrough in the West in the spring of 1940, refugees poured into France, and close to a fifth of the entire French population surged south in the greatest single upheaval of the entire war in Europe. This colossal avalanche was caused by the rapidity of the German advance. Beginning on 9 April, the Germans burst into Denmark and Norway and then raced through the Low Countries. In the second week of May, General Heinz Guderian's panzer divisions drove unexpectedly through the wooded Ardennes hills in Belgium and emerged in open country around Sedan, shattering defensive formations to the north and south. In only a few more days than it had taken to conquer Poland, the Blitzkrieg broke the backs of the French and British forces.

The French government, mindful of the experience of the First World War, had prepared a careful plan for withdrawing the civilian population from frontier areas at the outbreak of war. Overwhelmed in this as in every other respect in 1940, officials watched helplessly as waves of Belgian, Dutch, and Luxembourg refugees swept into the country in early May. Among these were tens of thousands of Jewish fugitives from the Reich, many of whom had found refuge in the Low Countries in the years before. All were trying desperately to keep out of the Nazis' hands and ahead of the frightful destruction accompanying the German advance. At the end of the month, Red Cross authorities telegraphed Red Cross societies in other countries that about "two million French, two million Belgian, 70,000 Luxembourg and 50,000 Dutch refugees or evacuees are in a state of serious destitution in France."[133] German fighters, with complete domination of the air, strafed the highways, paralyzing communications and scattering the fugitives. Even worse followed. In June the government abandoned Paris, starting a cascade of millions more refugees. In any vehicle that could move, on bicycle and on foot, frightened civilians struggled west or southwest to escape the Germans. Historians differ on the numbers involved, but the total was almost certainly between six and eight million people. Some 90,000 children, according to one report, were temporarily lost in the confusion. Some areas were almost deserted: only one-tenth of the inhabitants of the industrial regions around Lille, Roubaix, and Tourcoing remained; at the other end of the refugee chain, quiet villages like Beaune-la-Rolande, with 1,700 inhabitants, mushroomed to 40,000 people.[134]

It took many months to sort out this confusion, to allow hundreds of thousands to find friends, relatives, and resume normal life. A year later, close to a million people were still uprooted. Complicating the return were stringent German controls along the demarcation line dividing unoccupied France, known as the "Free Zone," from the Nazi-occupied territory in the northern three-fifths of the country, including a strip along the Atlantic coast going south to the Spanish border. The French government, now established in the sleepy resort town of Vichy, was eager to collaborate with the Germans and sought to curry favor with the occupation forces in order to facilitate this process. Unfortunately for Vichy, the Germans proved uncooperative in two respects that affected the refugees. First, the Nazis refused to permit tens of thousands of Jews who had fled south to cross the demarcation line and enter German-occupied territory. Many wished to do so, having no inkling of what lay in store for them under Nazi rule. The German authorities proved adamant despite Vichy's official adoption of an antisemitic policy and the government's obvious desire to get rid of the Jews. Indeed, Vichy repeatedly tried to bundle Jews across the demarcation line, only to find the refugees sent back by the Nazis. Second, the Germans had their own expulsion program, dispatching unwanted persons to the Unoccupied Zone. After July 1940, pursuing the integration of Alsace and Lorraine into the Reich, the Nazis began to expel francophones and pro-French elements, unceremoniously thrusting about 100,000 of these people on Vichy. Nazi planners considered Jews especially undesirable and planned eventually to force over a quarter of a million into Vichy territory. In October 1940, in a foretaste of the brutal character of this scheme, the Nazis sent a shipment of 6,500 Jews from Baden and the Palatinate across the demarcation line to Lyons. Vichy protested vigorously, but was forced to accept the unwanted Jews.[135]

Vichy now had to live with the refugees as best it could. Even without Nazi pressure, the government of Pierre Laval, working under the authority of the aged head of state Philippe Pétain, moved boldly against all Jews, singling out the refugees for special attention. A Vichy decree of 4 October 1940 authorized prefects (the state's most powerful representative on the local level) to intern foreign Jews in "special camps" or to assign them to live under police surveillance in remote villages. Thousands of Spanish Republicans were locked once again in French camps. At the end of October, Vichy established a national identity card to be carried by all French citizens over sixteen; this enabled the police to scrutinize the population regularly and to search out foreigners deemed to be in the country illegally.

Formally, Vichy encouraged the emigration of Jews and other

undesirables, including thousands of refugees. It removed some of the remaining Spanish exiles by sending them to Mexico. About 10,000 Jews reached Portugal from France in the autumn of 1940, clearly without opposition from Vichy. The difficulty was that shipping for refugees was extremely scarce, and there were few countries overseas willing to accept them. Throughout the Occupied Zone the French permitted extensive operations of HICEM, the main Jewish emigration agency operating in Europe. With offices in Marseilles, Lisbon, and Casablanca, this organization worked in close contact with Vichy officials even after the doors began to close everywhere in Western countries by the autumn of 1941. In March 1943, when its formal activity ceased, HICEM claimed to have assisted about 24,000 Jews to leave France.[136]

On a practical level, however, French bureaucrats often hindered emigration efforts despite repeated indications that official policy favored the departure of refugees. Those in charge harassed Jewish applicants and multiplied the hurdles of paperwork in the path of Jews trying to escape. Refugees had to run a complex gauntlet defined by currency restrictions, exit permits, property restrictions, police certificates and the like. In the end, bureaucratic obstruction and antisemitic prejudice sometimes proved stronger than official policy. Even fewer Jews managed to escape than the numbers of available visas permitted.

Prospects for Escape in the Years 1939–41

More than 350,000 refugees managed to escape the Nazis before war broke out in 1939. The overwhelming majority of these were Jews, and the greater part left Europe shortly afler they crossed the German frontier. Nevertheless, about 110,000 Jewish refugees were spread across Europe as the fighting began. Many more were in the process of attempting to emigrate—awaiting immigration visas or some other crucial link in the chain of paper required for departure. For the next two years, refugees continued to flow from Europe although at a vastly reduced rate. New restrictions both on the Nazi and the Allied sides made emigration a bureaucratic nightmare. Corruption, inequities, and bureaucratic absurdities slowed the machinery—fatally in the case of tens of thousands. Much of refugee work slipped into some degree of illegality. Escape networks ran on mixtures of greed and benevolence, intrigue and idealism, concocted on the edges of Europe—in places like Lisbon, Casablanca, and Istanbul.

To many refugees from the Reich who found temporary asylum in Western Europe, the great menace was internment, a loss of liberty that reduced substantially the prospects of escape from the Continent. For, with internment, refugees frequently lost the ability to gather documents, to wait in queues, to do all those things necessary to secure passage overseas. Even in the best of circumstances, as we have seen, fugitives from Hitler's Germany were sometimes viewed with suspicion and antipathy. The outbreak of war triggered the internment of enemy aliens, and in the emergency situation it was seldom considered possible to distinguish between genuine refugees and other Germans living abroad.

Of the refugees from Nazi Germany who were still in Western Europe when war broke out, most were in England or France. Understandably, perhaps, these exiles immediately came under suspicion for being Germans; in countries rapidly mobilizing for war, bureaucracies had some difficulty distinguishing between secret Nazi agents and helpless victims of persecution. The British began the war with a relatively liberal policy toward enemy aliens that attempted to assess how serious was the menace posed by various categories of people. One hundred and twenty tribunals across Britain decided whether suspected foreigners really posed a danger and whether internment was necessary. Within months three-quarters of the aliens, over 55,000 people, were declared authentic refugees from Nazism and either released from camps or protected from internment. Much the same was true in France, where early and often indiscriminate internment was followed by prompt release. This trend was reversed, however, when the Nazis struck in the West. In Britain, the catastrophic defeats in the spring of 1940 were sometimes attributed to the nefarious plotting of pro-German fifth columnists, a conviction that fed a deep-seated anxiety about enemy aliens and prompted another round of internments. In France, directly threatened with Nazi invasion, the fear bordered on panic. At least 25,000 Germans and Austrians, most of them Jews—and many only recently released from camps—suddenly became prisoners once again.[137]

Defeat brought new problems for the refugees. With unprecedented disregard for the anti-Nazi exiles in France, the Vichy government agreed to hand over to the Reich any German national designated by Berlin. German inspection teams roamed the

Unoccupied Zone and pored over lists of close to 10,000 internees supplied by the French. The Nazi government expressed no interest in Jews and most refugees, but did request over 1,500 people, some of whom had no desire to return home. The Germans further indicated that those who had entry visas for neutral and overseas countries could leave for Spain, but the French Ministry of the Interior apparently thought otherwise at the time and did not act on this suggestion.

In Britain, 27,000 enemy aliens (including some merchant seamen) found themselves behind barbed wire. British intelligence officers and other officials then began sometimes clumsy efforts to ascertain whether the internees were, in fact, dangerous enemy agents. Just under 20,000 were kept for a time in camps, the largest number being sent to the Isle of Man. The British also began a controversial policy of deportation, dispatching over 7,000 overseas to Canada, Australia, and other locations. Then, early in July 1940, disaster struck. At the beginning of that month, the Blue Star liner *Arandora Star* left Liverpool bound for Halifax with 712 Italian and 478 German enemy aliens, hastily selected for deportation. After only a day at sea the ship was struck by a German torpedo and promptly sank. Six hundred internees were drowned. Critics promptly alleged that those deported posed no threat whatever to Britain, but were simply refugees caught up in an indiscriminate internment net. Many charged that the internees had been treated unfairly and that classification errors had been widespread. In the end, the loss of the *Arandora Star* helped shift British policy toward a more highly selective internment policy and an ending of deportations. Gradually, officials released interned aliens throughout Britain. By the following year, only 1,300 remained in camps, and the numbers fell steadily thereafter.[138]

Britain received the last wave of refugees from the Continent during the German invasion in the West: in addition to the civilians of many nations who fled from France and the Low Countries, some 30,000 Channel Islanders escaped the Germans, joining 14,000 British refugees who quit Gibraltar about that time. Remarkably, however, refugees continued to emerge from Nazi-held Europe and from countries either controlled or influenced by the Third Reich. From the beginning of the war until the end of 1941, it is reckoned, a total of 71,500 Jews managed to flee Germany and Austria. Before the spring of 1940, many of these sailed from Belgium, Holland, France, and Italy. After the fall of

France, ships continued to carry refugees to the Western Hemisphere from Lisbon, Casablanca, Tangier, Oran, and Marseilles. From ports on the Aegean or on the Black Sea, ships chartered by the Jewish Agency painstakingly made their way to Palestine with their cargo of legal and illegal immigrants. Between April 1939 and the end of 1942, nearly 40,000 Jews reached Palestine, half of them illegally. According to Yehuda Bauer, some 25,000 Jewish refugees passed through Spain and Portugal in 1940 and another 15,000 in 1941. Other refugees entered Switzerland and later emigrated to North or South America. Jewish emigration agencies managed, despite great difficulties, to smuggle many refugees across frontiers and to secure their departure overseas. Even with restrictions, U.S. visas enabled significant numbers of fugitives to leave the Reich itself. During 1939, for the first time since Hitler took power, American officials permitted the entire German-Austrian immigration quota to be met—23,370 immigrants, almost all refugees. In 1940, 26,000 entered; and in 1941, 13,000.[139] In June of that year, however, the American Congress passed the Russell Bill, which severely curtailed the flow of refugees to the United States. American consulates throughout Nazi-occupied Europe closed that summer, and prices of the few available visas to Latin American countries skyrocketed. By the autumn, the new, even more harsh restrictions were everywhere in force. At the same time, the Nazis redefined their own policies and refused to allow Jews to emigrate. In December, according to a count made by HICEM, there were 50,000 refugees in France still trying to leave. Jews still approached Jewish emigration offices in the Netherlands and within the Reich itself. By now, however, the gates had definitively closed. Except for a tiny trickle, the refugees were locked in the European continent.

Even for refugees who qualified for the few available visas, the difficulties involved in emigration were legion. Many problems arose because Nazi policy, which theoretically favored the departure of Jews, left them virtually penniless. In 1939 foreign ships transporting Jews refused to accept German currency in payment. Because American consulates demanded proof that passage had been made before granting visas, some applicants failed for this reason to obtain entry permits.[140]

Moreover, as war spread to the North Atlantic, shipping space became extremely scarce. One exporting company official told William L. Shirer in December 1940 that there were thousands of desperate refugees in Lisbon awaiting passage and begging for any available berth. The ships carried only 150 passengers, however, and there was only one such vessel per week. Emigration officials in France noted at the end of 1941 that there were only twenty ships *anywhere* taking civilian passengers from Europe. One must be cautious in concluding that refugee traffic had reached the upper limit of possibility even while accepting the fact that shipping was scarce. Ships were found later in the war to send over 400,000 German prisoners of war to camps in the United States. The refugee advocate Eleanor Rathbone claimed in 1943 that many ships bringing troops and supplies from the New World to the war zone returned only lightly loaded. Scarcity of shipping seems largely to have been a reflection of wartime priorities. And these, in virtually every sphere, told against the refugees.[141]

4

Under the Heel of Nazism

THE most important goal of Nazism, the aggrandizement of the Third Reich, involved the eventual resettlement of millions of Europeans, to be moved about the Continent in accordance with Hitler's geopolitical ideas. In this, as in other respects, ideology was a serious business for the Nazis, and their broad views remained consistent over time despite important tactical shifts and alterations of policy. Based on racial preconceptions, Hitler's long-range plan was to build a vast, Pan-German empire, intended to achieve the purest expression of Aryan civilization and to last for a thousand years. There were two strategies to achieve this end. First, Germany had to absorb extensive territory in Eastern Europe to set the empire properly on its economic and biological foundations, thus carving out the Lebensraum or "living space" described in the Führer's *Mein Kampf*. Eventually, this would mean a massive confrontation with Soviet Russia. Second, to protect the racial fabric of the new order, Hitler encouraged vast population movements: non-Germans had to be ruthlessly excluded from the territory of the Reich; at the same time, pure Germans, wherever they lived, were to be brought within the fold, particularly in the new eastern marches.[1]

Until the outbreak of war, Hitler's racial politics riveted on his supreme enemy, the Jews. Nazism defined the Jews as the fundamental opponent of the German people, and the drive to isolate and remove them from German society began almost immediately after Hitler seized power, continuing with remarkable ferocity until the last hours of the Reich. In 1941, as we know, his objectives hardened into genocide: the Jews were simply all to be murdered. To his dying breath, Hitler fulminated against Jews, admonished his followers to preserve faithfully his racist credo of the 1920s.

During the first six years under Hitler, therefore, as the Nazis attempted to rid the Reich of what they called "non-Aryans," refugees from Nazism were overwhelmingly Jews. But once they controlled significant parts of Poland, the Nazis began the construction of the New Order in earnest. In the autumn of 1939, Heinrich Himmler, head of the SS and master of the Gargantuan police apparatus known as the Central Office for Reich Security (*Reichssicherheitshauptamt,* or RSHA), brought under his control a series of agencies devoted to racial and settlement matters. In October, as soon as the guns were silent, Hitler authorized Himmler to institute a Reich Commission for the Consolidation of Germandom (*Reichskommissariat für die Festigung deutschen Volkstums,* or RKFDV), a powerful bureaucracy to coordinate the Nazis' population schemes. Under Himmler's direction, vast numbers of people, Jews and non-Jews, began to move in Nazi-held Eastern Europe. One result was the creation of hundreds of thousands of new refugees.

NAZI POLICY, 1933–44

Jews in Central and Western Europe, 1933–41

Historians have been unable to determine any consistent Nazi program for the Jews during the struggle for power or in the first period of Nazi rule. From his earliest writings we know that Hitler's hatred of Jews infused his thinking about geopolitical questions to the point that his antisemitism and his international vision could never be disentangled. The more abstract and "profound" his discussions of world affairs and Germany's future, the more likely he was to bring up the Jews. With his thoughts drifting on this level of abstraction, the Führer invariably preferred vehemence to concrete plans. To be sure, Hitler envisaged a radical solution to what he called the "Jewish problem"—much as he envisaged radical solutions to every problem presented to him. There can be no denying his passionate commitment to settling accounts with the Jews or his desire for comprehensive, thoroughgoing action against them. In an oft-quoted

letter of 16 September 1919, Hitler opposed "antisemitism based
on purely emotional grounds," which "will always find its ultimate expression in the form of pogroms." The Führer argued
rather for a "rational antisemitism" and noted that its "ultimate
goal . . . must unalterably be the removal [*Entfernung*] of the Jews
altogether."[2] But however determined he was on this score, Hitler made no effort to specify what he meant by "removal" or what
path would be followed to achieve that objective. The official
program of the National Socialist German Worker's Party, proclaimed in 1920, remained silent on the issue; and throughout
the decade, as the Nazis campaigned for support, there was no
further clarification. Even after 1933, Hitler does not seem to have
urged any specific anti-Jewish strategy for the first five years of
his dictatorship. Instead, Jewish policy was initiated on a variety
of lower levels, and anti-Jewish moves lacked both consistency
and coordination.

  Two contradictory tendencies appeared from the outset. One
was a grass-roots lashing out against the Jews sparked by brownshirted Storm Troopers and Nazi party activists. In the euphoria
of their victory in early 1933, Nazi zealots brutally assaulted Jews;
a boycott of Jewish businesses followed at the beginning of April,
accompanied by public denunciations and random dismissals from
public positions. Second, there were signs of caution in government circles and even criticism of anti-Jewish stalwarts. In the
still desperate economic climate of 1933, many worried, for example, that a boycott of Jewish businesses would have severe repercussions for the German economy. Bullying and dismissals of
Jewish war veterans, to take another case, prompted an urgent
letter to Hitler by the elderly president Paul von Hindenburg, on
4 April 1933, warning the chancellor that the abuse of these Jewish notables was "intolerable" and that an "honorable" and
comprehensive solution had to be found. Hitler received this ad-
monition like a delinquent schoolboy still not entirely brought to
heel; he acknowledged the president's superior authority, but still
seethed with anger and self-righteousness. During the early days
of their government, as David Schoenbaum has said, the Nazis
were like a small group of stowaways taking over an ocean liner:
the task before them was awesome, the risks of disaster were
substantial, and they needed time and experience to grope their
way forward.[3] Hitler apparently concluded that he could not
proceed too far and too fast with the Jewish issue.

  Restraint gradually got the upper hand. After a flurry of laws against Jews in the civil service, the legal and medical professions, and the schools, the worst appeared over. More anti-Jewish laws followed, but these seemed part of a general pattern of Nazi vindictiveness rather than an anti-Jewish program moving to a terrible climax. Reviewing the situation in a book published in 1934, Arthur Ruppin was by no means completely pessimistic. "The position created by the Nazi attitude and legislation is untenable for either side," he wrote. "The violation of the principle of equal rights for citizens is certain to raise difficulties for Germany herself. Nor is the hostility of sixteen million Jews, which Germany has incurred, a neglible factor, either politically or economically." Everything depended, Ruppin believed, on the resolution of economic and political tensions peculiar to the Depression era. If these related crises could be settled, anti-Jewish feeling would diminish, even in Germany.[4]

During the next four years or so, Nazi long-range objectives toward the Jews seemed to have been guided by the Nuremberg Laws of 1935 and by the concerted campaign of Aryanization that followed. The first provided the basis for a systematic discrimination against Jews and the removal of citizenship rights; the second enabled the state to confiscate most of their property. Beyond these moves, however, there was still no publicly articulated goal for Nazi Jewish policy. One branch of the Nazi movement, the SS, periodically pressed for clearer thinking on the issue. As the elite corps of Nazism, the SS strove for ideological as well as racial purity, binding its members to the will of Adolf Hitler and providing the cutting edge for the enforcement of Nazi goals against any possible opposition. Unlike other groups, which might have been satisfied with robbery, public humiliation, or periodic violence, the SS promoted the emigration of Jews from Germany. An SS report of the late spring of 1934 was one of the first Nazi documents seriously to envisage the mass departure of Jews, although this was not seen as the exclusive answer to the Jewish question. At the time of the report, there were severe practical difficulties impeding such a scheme. Abroad, worldwide depression generated severe unemployment and encouraged immigration restrictions. As the SS knew well, most German Jews were of the middle class and were ill-equipped for the pioneering and development required in some regions. Moreover, an unusually large percentage of German Jews was over fifty years of age, too old to begin a new life in some distant country, and unlikely to emigrate under most circumstances imaginable. In view of these obstacles, therefore, no official weight was thrown behind a comprehensive emigration plan, however desirable this was seen to be.[5]

The Nazis did support one particular emigration project in 1933, precisely because it did not generate homeless refugees and because it seemed consistent with other Nazi objectives. This was the *Ha'avara* or Transfer Agreement worked out with the Zionist movement, enabling some Jews to migrate to Palestine and take a fraction of their property with them. According to emigration rules inherited from the Weimar Republic, German Jews quitting the Reich faced the prospect of leaving much of their wealth behind. To hasten the departure of affluent Jews, the Germans negotiated an arrangement with Jewish authorities in Palestine acting through a Palestinian bank. According to this plan, the confiscated assets of emigrating Jews would be deposited in special accounts in Germany; this capital could then be used to purchase German goods to be exported to Palestine. Jewish representatives there established a trust company, which reimbursed immigrants for a fraction of the value of their German accounts. This arrangement assured emigrating Jews that they could retain at least some of their fortune, and it assured British immigration officials that the German Jews entering Palestine under the scheme would not be penniless indigents. The Germans believed the plan would facilitate the emigration of unwanted Jews. Palestine circumvented the bottleneck caused by scarce entry visas elsewhere and the cumbersome bureaucratic processes of other governments. In addition, the Reich Economics Ministry calculated that the agreement would stimulate the export of German products and help generate much-needed foreign exchange. As a political bonus, the subsequent trade would help break the boycott of German goods announced by various Jewish organizations outside the Reich.

Although extremely controversial within the Jewish world and among some Nazis, the *Ha'avara* agreement began operation at the close of 1933 and functioned thereafter until the outbreak of war. For several years it worked without difficulty, notably because the British imposed few restrictions on German entry into Palestine. The year 1937 was the peak year of operation, when the capital withdrawn from Germany under the Transfer Agreement amounted to thirty-one million marks—a substantial sum though very small in comparison with the total Jewish wealth in the Reich. By the autumn of that year, according to German calculations, about 120,000 Jews had left the Reich since 1933: of these, about one-third remained in Europe, one-third had migrated overseas, and one-third had gone to Palestine. *Ha'avara* played its part in assisting the latter group. Moreover, the agreement served the interests of Nazi propaganda. Still sensitive to charges that Germany was creating a refugee problem at the expense of other states, Nazi spokesmen were able to point to the Transfer Agreement as an instance of "orderly" departure in the interests of everyone concerned. Nazi leaders, therefore, considered that *Ha'avara* served them well. Toward the end of 1937, interministerial meetings in Berlin reviewed the arrangement and reaffirmed it. According to Nazi Party experts on foreign affairs, Hitler wanted to encourage Jewish emigration by all possible means. Although foreign policy specialists worried increasingly about the prospect of a Jewish state in Palestine, the Führer's priority was that the Jews should leave.[6]

Emigration emerged even more sharply as a focus of Nazi intentions for the Jews in 1938, when the inhibitions of the previous period became far less important and when so many more Jews came under Nazi rule. After four years of Nazism, indications of German economic recovery were everywhere—the practical elimination of unemployment, the increasing solidity of Germany's international trade position, and the gradual rise in the standard of living. These successes meant less preoccupation with the risks of Aryanization, permitting a new drive to liquidate Jewish property in Germany. At the end of 1937, major changes in the German government signaled an abandonment of many conservative policies that derived from the previous insecurity of the Nazi regime. Rearmament accelerated. Economics Minister Hjalmar Schacht resigned his post in November, giving way to Hermann Göring, a man prepared to assault established interests in German society and press forward with Nazi goals. Within weeks, Göring's staff prepared the legal basis for the expropriation of Jewish businesses. Within a few months, the "voluntary" transfer of property from Jews was replaced by forcible seizures of Jewish concerns. In October 1938, Göring announced that Jews were to be completely "removed from the economy." Immediately after Kristallnacht, he issued a Decree on Eliminating Jews from German Economic Life, which promised to do just that.

Throughout 1938, German foreign policy achievements encouraged Nazi leaders to disregard the mounting European con-

cerns about the creation of Jewish refugees. Hitler stood trium-
phant in Europe: his spectacular diplomatic moves in Austria and
Czechoslovakia utterly transformed the face that Nazism presented to European statesmen. Repeatedly, in the mid-1930s,
German diplomats heard objections from European governments about the expulsions of Jews, about the burden that this
policy imposed on Germany's neighbors. Invigorated now by
success, the German Foreign Office became increasingly haughty
over the issue of Jewish refugees. As we have seen, the Wilhelmstrasse disdained the Evian conference of 1938 and proved
consistently uncooperative over refugee issues. In response to a
British inquiry, Ernst von Weizsäcker of the Foreign Office reiterated that no Jewish capital could be removed from the Reich.
As a matter of principle, Foreign Minister Joachim von Ribbentrop indicated, "Germany refuses to participate with other countries in the question of German Jews." Generally, German
spokesmen were far less interested after Anschluss in the appearance of "orderly departure." Indeed, it has been suggested
that the ruinous Nazi policies worked against the stated Nazi objectives of Jewish emigration by making it far more difficult for
the impoverished Jews who remained to leave the country.
Henceforth, Germany thrust Jews naked and destitute on a world
unwilling to receive them. Western protests were ignored.[7]

At the same time, however, Anschluss undercut the work of
more than five years of Jewish emigration from the Reich. Although about 150,000 had left by 1938, the German Foreign Office calculated that union with Austria brought 200,000 more Jews
within the Nazi domain. The prospective incorporation of
Czechoslovakia promised to include many more. Unless the drive
to eliminate the Jews from German life were to languish, the situation called for a much more radical policy than ever before.

At the end of 1938, the Nazi leadership placed a new emphasis on forcing Jews out of the Reich. *Kristallnacht* provided part
of the impetus, for in the wake of the riots it became clear that
random, uncoordinated action tended to undermine Nazi interests. With the glass still littering the streets and smoke still rising
from the gutted synagogues, Nazi leaders surveyed the serious
damage: widespread, wanton property losses, millions of marks
in insurance claims, and the insidious example of public disorder. And despite all the destruction, the solution to the Jewish
issue was no closer. A few days later, a meeting in Göring's office criticized the haphazard character of Nazi Jewish policy and

received Hitler's oracular statement "that the Jewish question be now, once and for all, coordinated or solved in one way or another."[8] Those present accepted the call for a more centralized, disciplined approach to Jewish matters. Acting on Hitler's wishes, Göring assumed responsibility for the orchestration of persecution and warned his subordinates not to act independently against the Jews. Defining the new course, Göring underscored the importance of emigration, which was then succeeding so impressively under Adolf Eichmann's authority in Vienna. In one of his first moves, Göring told RSHA chief Reinhard Heydrich to set up a Reich Central Office for Jewish Emigration. Heydrich's orders were to accelerate emigration, using all possible technical means; poor as well as rich were to go—and all under the auspices of the SS.

Surveying the Jewish question for all German diplomatic missions and consulates in January 1939, a Foreign Office official noted that the "ultimate aim of Germany's Jewish policy is the emigration of all Jews living on German territory." Never before had the goal been expressed so clearly. The "year of destiny," 1938, was seen to have brought the Jewish question close to solution. Economic measures, which had impoverished the Jews, would now bludgeon them into flight. All signs within Germany now were positive. Unfortunately, the official noted, serious external obstacles had to be overcome. The United States and the European powers wanted Germany to release Jewish assets in order to help finance Jewish emigration. Germany, of course, still refused to assist by sacrificing even a small fraction of Jewish wealth. In the official view, the Jews' wealth really belonged to the Reich: the Jews had become rich by exploitation; having arrived in Germany penniless, they would leave in a similar fashion. The real problem, of course, was where the Jews could go. The official noted that "almost all the countries of the world have hermetically sealed their boundaries against the undesirable Jewish intruders." Practically speaking, this scarcity of havens slowed the Jews' departure. Nevertheless, the official view was not pessimistic. Criticism of Nazi policy was sure to diminish; eventually, the world would recognize the Jewish danger and act on it. The main thing was to keep up the pressure: the problems of destination and the financing of the exodus would ultimately take care of themselves. The future would see an "international solution of the Jewish question," hammered out by Germany and inspired by German anti-Jewish ideology.[9]

In the few months that remained before the outbreak of war, Nazi policymakers pursued this double-edged strategy: throughout the newly expanded Reich, furious persecution forced some Jews to flee even without destinations, provoking a refugee crisis; meanwhile, German negotiators kept the lines of discussion open, engaging in a remarkable series of discussions with Western representatives over the Jewish question. Ostensibly, these negotiations involved the impact of Germany's policies on her neighbors; in reality, the Germans were groping toward precisely that "international solution" that the Foreign Office was promoting. On Göring's orders, the Germans began several approaches to George Rublee's Inter-Governmental Committee on Refugees (IGCR), the Evian Committee previously boycotted by Nazi diplomats. The United States, having fathered the IGCR the previous summer, became principally involved in these contacts although in an indirect fashion; the British, never keen on the organization from its inception, were drawn in, in the wake of the Americans, and took no initiative.

In mid-December 1938, former Economics Minister Schacht, then head of the Reichsbank, met Rublee and two British representatives. Operating with Hitler's approval, Schacht proposed a mechanism for letting 150,000 Jewish wage earners and their dependents leave Germany over a three-year period. Schacht was not concerned about destinations. He suggested a complex scheme by which an ill-defined "international Jewry" would finance the emigration, eventually to be repaid from the proceeds of Jewish wealth impounded in Germany; reimbursement, however, would depend on increasing German export revenue, thus assuring that the Reich would not lose in the bargain.

Despite the extortionate character of Schacht's proposal and despite the violent opposition of American Jewish leaders, Rublee's committee decided to continue discussions. In Washington the view was that such meetings might ease the refugee crisis and were, therefore, worth pursuing. London went along, allowing the United States to take the lead and anxious not to assume any new burden of assistance to the refugees. In Berlin, however, the climate shifted as Heydrich's authority in emigration matters was increasingly felt. Schacht fell from grace in February 1939, and more radical elements, such as Ribbentrop, clamored for a speedier resolution of the Jewish problem.

Nevertheless, the Germans maintained their double-edged approach. On the one hand, Göring pursued the earlier line of negotiations, with Schacht's place being taken by Helmut Wohlthat, a foreign exchange expert working for the Four-Year Plan. German negotiators now favored a variation of the *Ha'avara* agreement, with its mechanism for increasing German exports. More conferences with Rublee followed, reaching a tentative arrangement without exploring the crucial matter of finding havens for the Jewish refugees. On the other hand, Heydrich's Central Reich Office for Jewish Emigration pressed ahead, believing it could force Jewish departures by streamlining bureaucratic procedures in the Reich itself and by adopting the brutal methods of Eichmann's Vienna machine. Gestapo teams now literally dumped Jews across frontiers, forcing them at gunpoint into wooded areas at the French or Swiss borders, or cramming them onto leaky vessels sailing down the Danube. German and Italian shipping companies assisted, carrying refugees without papers to distant ports in the West. Meanwhile, in the Reich, bureaucrats squeezed the last remaining resources out of the Jews. In February the Ministry of Economic Affairs ordered them to surrender all gold, jewelry, and other valuables they still had in their possession. Heydrich and his people appeared little concerned about the scarcity of immigration opportunities. Somehow, it was believed, space could be found. On 11 February, Wohlthat reported to Heydrich a rumor he heard in London that Palestine could still absorb between 800,000 and a million Jews. In conversation with Czech Foreign Minister Chvalkovsky a few days earlier, Hitler refered to "unlimited spaces" at the disposal of Britain or the United States. According to the Germans, there was no lack of territory to which the Jews could be sent. What really counted was to keep up the pressure of refugees.[10]

During the spring of 1939, Wohlthat pursued negotiations with Rublee's successor, Sir Herbert Emerson. The Germans' continued persecution of Jews and their obvious unwillingness to relax their expulsion policies strained relations between Washington and London. Under Roosevelt, the Americans still believed that some alleviation of the refugee crisis was possible. The British, feeling the pressure of a new wave of Czech emigrants after Hitler's absorption of Czechoslovakia and smarting from Jewish denunciations of the Palestine White Paper, feared that thousands more impoverished refugees might land on their doorstep. Discussions were still under way when Hitler's panzer divisions moved into Poland, and only then did the IGCR negotiations with Germany finally cease.

War did not foreclose the "international solution" originally proposed by German leaders. On the contrary, the spectacular achievements of the Wehrmacht in Poland convinced Nazi hierarchs that Hitler retained the initiative in this as in other questions of European importance. Several weeks after the fighting began, Berlin signaled to London and Washington that the Reich still favored Jewish emigration and still wanted to deal with the IGCR. Washington and London, of course, broke off discussions. Nevertheless, for well over a year, Nazi leaders declared their support for Jewish departures from Europe. War did prompt a new German interest in territorial solutions—the notion of specific places being assigned to Jews in remote parts of the world. Conquest, indeed, stirred the imaginations of Nazi planners and ideologues. Those in charge of Jewish affairs occasionally speculated on how a peace settlement, to be worked out sooner or later, would see a German-conceived resolution of the Jewish question. In the end, none of these plans came to fruition, and Hitler imposed his genocidal Final Solution.

For two more years, however, Jewish refugees trickled out of the German Reich. Nazi authorities commissioned a Viennese Jewish businessman, Berthold Storfer, to organize emigration and illegal entry into Palestine, hounding and pressuring him constantly to hasten departures. Over one thousand entered Italy with visas for Western countries, but this route closed when Mussolini entered the war in June 1940. Others went first to Belgium or Holland before proceeding further west. Despite the priority given to German Jews, the Nazis occasionally even permitted a small number of Polish Jews to emigrate provided they had the necessary papers. These Jews traveled through German or Italian territory with transit visas obtained by a shipping company engaged as an intermediary by the World Zionist Organization. Several thousand left via this route before the spring of 1940, when fighting resumed in Western Europe and the Germans blocked Jewish departures from the east. According to Yehuda Bauer, some 71,500 Jews left the Greater German Reich between September 1939 and the end of 1941, when the exits were finally sealed. These represented about a fifth of the Jews remaining in the expanded German state. Most went to Britain, to the western hemisphere, to Shanghai, or to Palestine.[11] The least fortunate among the émigrés remained in Western Europe. When

Hitler struck in the west in the spring of 1940, many refugees from Nazism found themselves living once again under German rule. In their case, too, Nazi policy favored Jewish emigration although the top priority was to rid the Reich of its Jews. Occasionally, as we have seen, the Nazis even dumped Jews from newly annexed territory into the Vichy-controlled zone of France. Similarly, the Gestapo forced transports of Austrian Jews into neighboring Yugoslavia. Given the scarcity of transport and visas, the Germans soon obstructed the path of refugees in Nazi-occupied Western Europe in order to accelerate emigration from the Nazi heartland. As late as May 1941, Göring ordered Jewish departures from Bohemia and Moravia speeded up. Then, as the invasion of Soviet Russia triggered great changes in Nazi policy toward Jews, a signal went out to close the gates to Jewish refugees. On 23 October 1941, Gestapo Chief Heinrich Müller passed along an order from Himmler that henceforth, apart from a few exceptions judged to be in the German interest, no more Jews were to emigrate from Germany or German-occupied Europe. This order was slow to percolate through the far-flung bureaucracy of the Nazi empire, but by this time few refugees were moving in any case. Locked up in Nazi-controlled Europe, Jewish refugees from Nazism were among the very first to be deported and murdered in Hitler's Final Solution.[12]

Lebensraum in Eastern Europe, 1939–41

Throughout the Weimar era, Hitler and the Nazis championed the cause of the *Auslandsdeutsche,* the millions of ethnic Germans living outside the boundaries of Weimar Germany. These were often descendants of the German colonists who since the Middle Ages had settled in the Baltic region, eastward beyond the Vistula into Russia, southeastward along the Danube, throughout the Balkans, and southward to the Alps. Following the First World War, when additional millions found their links with Germany or Austria severed by the territorial settlements, these Germans formed substantial minority groups in the Baltic states, the Soviet Union, Poland, Hungary, Rumania, Yugoslavia, and Czechoslovakia. No one knew precisely how many people were involved. German authorities had preconceived notions of who these Germans were and how one defined *Auslandsdeutschtum.*

To nationalists eager to assuage the wounded pride of a defeated nation, these were the "lost" Germans who might one day be brought back to the fold. A highly charged issue within Germany, the fate of these people preoccupied right-wing politicians and inevitably concerned the Nazi movement, which presented itself as the most energetic champion of the national cause. Hitler imposed a racialist framework on the issue. He referred to the ethnic Germans as Volksdeutsche and, in the very first plank of the Nazi Party program, called for their incorporation in a Greater Germany.

Following the Nazis' seizure of power, the question of the Volksdeutsche remained in abeyance for some years. The Nazis' first priority was to secure the Reich, then to restore the boundaries of Imperial Germany, and only then to conquer the Lebensraum needed for the future. In view of this long-range program, the issue of Germans living abroad arose primarily in the context of specific diplomatic crises—notably over Austria and Czechoslovakia. Meanwhile, German propaganda for the union of individual Volksdeutsche with the Reich intensified with each of the major German diplomatic successes from Anschluss to the outbreak of war in 1939. This, plus the enhanced power and prestige of the Hitlerian state, stimulated the flow toward Germany. Government aid helped install Germans from across Europe after 1938. A special organization known as Heim ins Reich cared for the new immigrants, who numbered 2,500 a month in 1938 and about 4,000 a month in the following year.[13] Up to the outbreak of war, the incorporation of the Volksdeutsche went smoothly.

The conquest of Poland in 1939, however, posed the most serious challenge to Nazi policymakers. Some two million Germans lived in the recently conquered territories. Hundreds of thousands more lay further east, in Polish land occupied by the Soviet Union. Germans in the Baltic states, soon to be occupied by the Russians, were also a matter of concern. Should the Reich continue to promote the unification of all Germans by voluntary migration as before? Should Germany annex new territory and thus incorporate the Volksdeutsche? Or should the Reich forcibly uproot the Germans in the east, obliging them to resettle westward in the Greater German Reich?

Perhaps by temperament as well as ideology, the Nazis adopted the most grandiose design, guided by racism and the urge to dominate other peoples. They began by dividing their con-

quered Polish territory into two parts: northern and western Poland, including Danzig, West Prussia, Posen, and Eastern Upper
Silesia, were incorporated into the Reich (the bulk of these regions forming the new *Reichsgaue* of Danzig-Westpreussen and
the Wartheland). The rest, known as the *Generalgouvernement*, was
placed under the authority of a German governor, Hans Frank,
responsible directly to Hitler. The incorporated provinces would
be subject to the most intense Germanization to eliminate all im-
pure racial elements; the *Generalgouvernement*, to which the latter
would be sent, was to become a vast work camp, an immense
repository of unskilled slave labor to serve the needs of the enlarged German state.

This grand design called for vast shifts of population. Taking
charge of this effort was Heinrich Himmler, the Nazi leader with
the dominant voice in the incorporated territory. Given the task
by Hitler of "the consolidation of the German people" and
equipped with a new bureaucracy for the purpose (the RKFDV),
the Reichsführer SS began a vast campaign to integrate the new
Reichsgaue into the Reich. His deputy and head of the RSHA,
Reinhard Heydrich, drafted a blueprint for the New Order before the end of September. A party agency, the Volksdeutsche
Mittelstelle (VOMI), the Liaison Office for Ethnic Germans, was
assigned the task of gathering and caring for the Volksdeutsche
prior to their resettlement.[14]

Resettlement took place through agreements between the Nazis and their neighbors. Nominally "voluntary" transfers, these
were intended to strengthen Hitler's alliance system by removing a potential source of friction on Germany's borders and also
strengthen the Nazi hold on the newly incorporated territories,
where many of the Volksdeutsche were to be settled as agriculturalists. Understandably, the various countries with large German minorities showed a keen interest in getting rid of a potentially troublesome mass of ethnic Germans. In October 1939, Berlin
negotiated agreements with Lithuania, Latvia, and Estonia in
which Volksdeutsche living in these countries would be released
from their Baltic citizenship to proceed immediately to the Reich.
At the end of the month, an agreement was signed in Rome, intended to bring to Germany over 9,000 Germans from South Ti-
rol. A few weeks later the Germans worked out a much more
important transfer with the Soviet Union concerning their respective zones of what had once been Poland. The Russians permitted Volksdeutsche from Ukrainian and Belorussian regions,

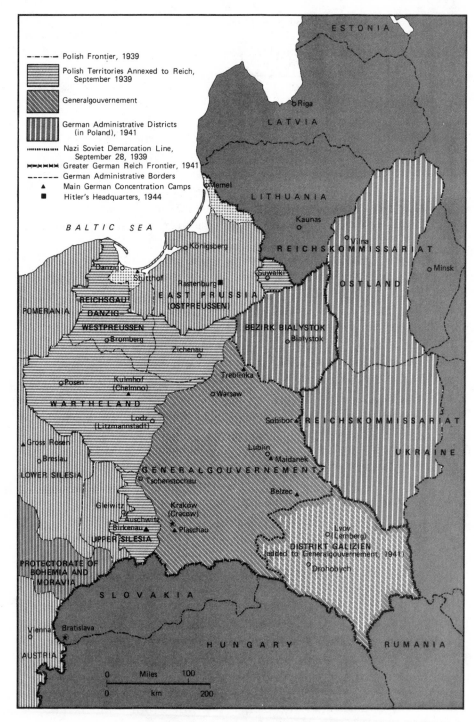

Legend:

- — · — · — Polish Frontier, 1939
- Polish Territories Annexed to Reich, September 1939
- Generalgouvernement
- German Administrative Districts (in Poland), 1941
- ·········· Nazi Soviet Demarcation Line, September 28, 1939
- ▶◀▶◀▶◀ Greater German Reich Frontier, 1941
- ------- German Administrative Borders
- ▲ Main German Concentration Camps
- ■ Hitler's Headquarters, 1944

ESTONIA

LATVIA

○ Riga

LITHUANIA

○ Memel

Kaunas ○

BALTIC SEA

○ Königsberg

REICHSKOMMISSARIAT

○ Vilna

○ Minsk

○ Danzig

▲ Stutthof

Rasterburg ■

EAST PRUSSIA (OSTPREUSSEN)

OSTLAND

REICHSGAU DANZIG WESTPREUSSEN

POMERANIA

○ Suwalki

BEZIRK BIALYSTOK

○ Bialystok

○ Bromberg

Zichenau ○

▲ Treblinka

WARTHELAND

Kulmhof (Chelmno) ○

○ Warsaw

○ Posen

Lodz (Litzmannstadt) ○

Sobibor ▲

REICHSKOMMISSARIAT

▲ Gross Rosen

Lublin ○

○ Breslau

GENERALGOUVERNEMENT

▲ Maidanek

UKRAINE

LOWER SILESIA

○ Tschenstochau

Belzec ▲

Gleiwitz ○

Kraków (Cracow) ○

▲ Auschwitz

▲ Birkenau

▲ Plaschau

Lvov ○ (Lemberg)

UPPER SILESIA

DISTRIKT GALIZIEN (added to Generalgouvernement, 1941)

PROTECTORATE OF BOHEMIA AND MORAVIA

○ Drohobych

SLOVAKIA

Vienna ○

Bratislava ★

AUSTRIA

HUNGARY

RUMANIA

Miles 0 ——— 100

km 0 ——— 200

POLAND UNDER NAZI OCCUPATION

After Norman Davies, *God's Playground: A History of Poland*, Vol. II, *1795 to the Present* (Oxford, 1981).

formally absorbed into the Soviet Union on 3 November, to enter Nazi-held Poland. At the same time, the Nazis determined to oust Germans living in the *Generalgouvernement*, obliging them to resettle in the newly annexed territories. The following months saw additional transfer treaties—with the Russians on 5 September 1940, regarding Germans from Soviet-incorporated Bessarabia and Northern Bukovina; with the Rumanians on 22 October 1940, covering Rumanian-held Southern Bukovina and northern Dobruja; and again with the Soviets on 10 January 1941, dealing once again with the Baltic countries, now absorbed by the Russians. After the Nazi invasion of the Soviet Union, tens of thousands more Volksdeutsche were brought from the former Baltic states, Belorussia, the Ukraine, and the Caucasus—all to be deposited in the incorporated territories.

These supposedly voluntary evacuations of Germans westward from Eastern Europe were often little different from the Nazis' forcing of Poles in the other direction. Foreign correspondents were banned from the scene. Resettlement squads, or *Umsiedlungskommandos*, sent villagers on long treks to railway junctions; even the German press reported severe hardships along the way.[15] Peasants from Wolhynia and Galicia, often without the slightest desire to "return" to a Reich they had never seen, suddenly lost all their possessions save those they could carry. Evacuees trudged along the highways and piled their worldly goods on primitive carts. One witness saw the passage of a group of German peasants from Bessarabia in 1940: "a spectacle which reminded me of engravings of the American frontier era: a long line of wagons, covered with white canvas, oxen-drawn, sometimes with a colt or horse running alongside."[16] To be sure, circumstances varied considerably. Some of the Volksdeutsche may indeed have been eager to leave their homes. Many feared the Sovietization of Russian-incorporated areas, an apprehension deepened by sometimes desperate economic circumstances during the first winter of the war. Nazi propaganda beamed its messages to the German population, generating rumors of massacre together with visions of an idyllic life in the Reich. Joseph Schechtman describes a widespread sense among Baltic Germans that they could no longer continue as a minority. "Our historic mission in the Baltic has come to an end," one of them reported, echoing a typically apocalyptic Nazi theme. "It is no longer a historic mission if one has to fight for one's mere existence, for a barely sufficient loaf of bread. It is no longer a historic mission if one faces slow but certain death."[17] In one case, Yugoslavia, the program manifestly failed. Not only did the local Germans object to their emigration, but the Nazi supporters in Yugoslavia bombarded the Wilhelmstrasse with protests—charging that the evacuation harmed valuable German stock and removed a potentially useful element for future imperial expansion. Faced with a barrage of criticism, Berlin canceled the Yugoslav transfer—but not before some 36,000 Volksdeutsche had been shifted to various parts of Nazi Europe. Himmler's RKFDV indicated nearly 500,000 German evacuees uprooted in the year after the conquest of Poland—70,000 from the Baltic states, 130,000 from Soviet-occupied Poland, 90,000 from the *Generalgouvernement*, 90,000 from Bessarabia, 90,000 from Northern and Southern Bukovina, and 14,000 from Dobruja. And the process was still under way. During the entire course of the war, the RKFDV moved about 1.25 million Germans, of whom fewer than 500,000 were ever settled on farms—mainly in the incorporated territories of Poland; the rest were simply moved from one camp to the next until the end of the war.[18]

Resettling the German evacuees proved a far more difficult and complex process than Nazi fantasies suggested. After hundreds of thousands were removed from their homes at short notice, it was often impossible to install them immediately on a neat *völkisch*-style farmstead. Instead, the German evacuees often languished in camps run by the SS. The Nazis soon ran into difficulties assessing the racial credentials of the new settlers. Were they, in fact, of pure German stock? To determine the answer, the Reichsführer SS decided on a cumbersome screening of every individual—a slow procedure that collapsed into hopeless muddle and confusion. Once planted in the Polish countryside, the Volksdeutsche were exposed to violent attack and systematic sabotage—from the resident Polish population, which detested the intruders, and from guerrilla bands who fought the German occupation. Even when secure in their new homes, moreover, the settlers could not keep up agricultural production; the Germans were simply unable to maintain the property of Poles pushed off to the east, the guerrillas exacted their toll, and the countryside was thrown into turmoil. Local German military and civilian administrators were frequently outraged at this example of racialist social engineering run riot. "We are not a settlement team but refugee commissioners," complained one RKFDV employee in the Wartegau. Hundreds of thousands of evacuees remained unsettled in 1943, and the entire program faced catastrophic breakdown. Oblivious to reality on such matters as was his wont, Himmler became even more convinced of the need to consolidate the Volksdeutsche. Even those in the western hemisphere, he believed, would have to "return." During 1943 the Nazis transported additional German stock from Croatia as well as the other centers of *Auslandsdeutschtum*. Every drop of German blood, Himmler told a conference of naval officers at the end of that year, had either to belong to Germany or face eradication." [19]

Accompanying this fantastic project was the forcible removal of "undesirable elements" from the incorporated territories, largely to be deposited in Hans Frank's *Generalgouvernement*. This was, as Himmler explained in October 1939, "one of the most essential goals to be established in the German East." "We either win over the good blood we can use for ourselves . . . or else we destroy that blood. For us, the end of this war will mean an open road to the East. . . . It means we shall push the borders of our German race 500 kilometers to the East." [20] Despite the vehemence of this tirade, however, the results often lacked consistency; Nazi activity in this, as in other, spheres frequently emerged from competing power interests of Nazi hierarchs and from rival military and civilian bureaucracies. Deportations sometimes followed a haphazard pattern in which groups of non-Germans were moved about to serve very different ends. In broad outline, however, the task Himmler set for the newly annexed eastern provinces was the deportation eastward of all Jews, all Poles who had moved into the region after 1918, and all Polish leaders and intellectuals. During the so-called Operation Tannenberg, the Nazis drove at least 750,000 Polish peasants from their homes in the New *Reichsgaue* in order to accommodate Volksdeutsche from the Baltic region and elsewhere. Several weeks after war ended in Poland, trainloads of deportees began to move, meeting railway targets of about 10,000 people daily. [21]

The SS conducted the deportations with extreme brutality. Evacuees were given between ten minutes' and one hour's notice before "resettlement." The Germans permitted them only hand luggage—at first, 12.5 kilograms per person, and later, 25 to 30 kilograms. Poles were allowed 200 zlotys; and Jews, half that amount. Later, everyone could take only 20 zlotys. The Nazis sealed their human cargo into freight cars—one thousand per convoy. Deportees suffered terribly en route, and many died of cold, hunger, and illness. At their destination the victims were sometimes abandoned, with the local population expected to care for them; sometimes the men were separated and forced into work camps. Sometimes, too, the children were taken away, to be brought to the Reich for "Germanization."[22]

In theory, the non-German deportees were to be moved about like prisoners, with their destinations carefully controlled by Nazi occupation authorities and racial planners. In practice, differing visions of the future of Eastern Europe emerged from various quarters, and the attempt to implement them resulted in a patchwork of resettlement efforts—sometimes working at cross-purposes and sometimes leaving masses of evacuees to their own devices or forced to and fro by various German authorities. As the numbers of deportees climbed into the hundreds of thousands, many looked more and more like refugees—still moved about at the point of a gun, but with their destinations less clearly fixed, with plans for their ultimate disposition less definite, and occasionally with rival branches of the Nazi establishment responding spontaneously to the hordes of victims.

For the Germans, a basic problem was the uncertain, shifting status of the *Generalgouvernement*. At first, Himmler's dynamic vision for the Germanization of the incorporated territories seemed to confirm the dumping-ground status of Hans Frank's preserve. Unannounced, train after train brought evacuees into the *Generalgouvernement*, causing immense logistical problems. Convoys arrived in freezing weather, without food, and without any preparation at their point of termination. By the end of the first year of occupation, no less than 10 percent of the entire population of the *Generalgouvernement* was made up of refugees. German officials simply did not know what to do with the flood of newcomers, deemed by Himmler to be the refuse of the Third Reich. In his castle in Cracow, conceived as the glittering capital of his "kingdom," Frank felt swamped. It was "absolutely intolerable," he declared, "that thousands upon thousands of Jews should go slinking around and occupy apartments in the city which the Führer has granted the great honour of becoming the seat of a high Reich authority."[23] Frank's pretensions grew—to the point that he longed for a more elevated status within the German Lebensraum. Urgently, he protested the arrival of so many refugees. In February, Frank personally delivered his complaint in Berlin, apparently to no avail. Jews, Poles, and other "undesirables" continued to arrive on his doorstep. At the end of March, however, Frank joyfully announced the Führer's promise that his domain would one day be made *Judenrein* (free of Jews) and eventually cleansed also of Poles. Within a year, relief of a sort finally came. During the course of the invasion of Russia, the Nazis began their Final Solution of the Jewish question, thus eliminating a substantial portion of Frank's refugee problem.

Jewish Refugees and the Final Solution

To the world outside Nazi-occupied Eastern Europe and even to many Nazi officials, the Jewish question in the first two years of the war was largely a question of refugees. Staggering numbers were uprooted, and even more were supposed to follow. Initial plans called for one million Jews to be sent into Hans Frank's domain—600,000 from the incorporated territories and the rest from the Reich, rendering the latter almost entirely free of Jews.[24] Occasionally, the Nazis spoke of an as yet ill-defined "final aim," usually implying that its formal definition would await the end of the war. As a stock formulation the term "final solution" may have first appeared in June 1940, as a "territorial final solution," and it was increasingly used in the spring of 1941. By "final solution" the Nazis implied a vast process of deportation and emigration, in which Jews would leave Europe en masse.[25] The top priority was to eliminate German Jews; then the others would follow. While awaiting a peace settlement in which details would be worked out, the Germans pressed forward with necessary preparations. This helps explain the ferocity of Nazi policy toward Jews in Poland. Apart from periodic massacres, robbery, and humiliation, the Nazis were laying the groundwork for the gigantic expulsions of the future. The *Generalgouvernement*, the great population center of world Jewry, was to be the concentration point for further deportations. Thus, as we have seen, the Nazis began immediately to ship Polish Jews eastward from the incorporated territories. In mid-October 1941, before the killing centers were in place, mass deportations of Jews started from the Reich. The first goal was Lodz, which the Germans called Litzmannstadt, in the middle of the Wartheland. In the spring, the Nazis believed, the Jews would be sent further east.

During the two years following the German attack on Poland, the uprooting process was an unprecedented calamity for the

Jewish population—overshadowed subsequently by the horrific mass murder in Nazi death camps. To facilitate future movement, refugees were packed into teeming ghettos in the poorest and least adequate portions of cities in both incorporated territories and the *Generalgouvernement*. Everywhere, the Nazis cleared Jews from the countryside and forced them into towns where ghettos were established. Evidence suggests that at least a million of Poland's three million Jews were torn loose from their homes by the effects of war and Nazi persecution during this period. Even the smallest ghettos contained huge numbers. Kutno, for example, had 5,239 Jews, of whom one quarter were refugees. Nearly half of the population of the Piotrków Trybunalski ghetto population in the *Generalgouvernment* were refugees in July 1940, and other ghettos in the Lublin and Radom districts reported a majority coming from elsewhere. Substantial proportions of the ghetto inmates of Cracow, Czestochowa, Lodz, and many others were similarly homeless refugees. Warsaw housed 90,000 Jewish refugees when its ghetto gates were closed in November 1940; the Nazis forced even more into the city, however, so that they numbered 130,000 in the spring of 1941–about one-third of the entire Jewish population.[26] To Adam Czerniakow, the Jewish head of the Warsaw *Judenrat*—the Nazi-imposed Jewish council set up to run the affairs of the ghetto—accommodating refugees was among the most urgent problems he faced. In desperate circumstances, with everything scarce—food, fuel, medical supplies, accommodation, even burial places—the Nazis continued to force Jews on the ghetto administration. "Is another swarm of refugees going to descend on our shoulders?" Czerniakow asked his diary on 3 April 1942. Czerniakow continually reported on the arrival of refugees: he seldom knew until the last moment when or where they would come or how many to expect; each arrival was a crisis; each time he approached the German authorities to lighten his burden, they refused to budge.[27]

Within the ghettos, overcrowding contributed to spectacular mortality rates. Typhus, dysentery, tuberculosis—all took their toll. The Jewish Councils' elaborately organized public welfare operations constantly broke down because of inadequate resources and the endlessly rising tide of need. The arrival of new refugees constantly exacerbated the situation. Not infrequently newcomers quarreled with residents of longer standing. Invariably, the refugees were at a disadvantage. Newcomers camped in schools, synagogues, and the few other public buildings within the ghettos. Along with their periodic raids on the ghettos for laborers, the Germans proceeded systematically to starve the Jewish population—a task made easier by their concentration in tightly enclosed areas. For a time, aid from the Joint Distribution Committee trickled into the ghettos of the *Generalgouvernement*, hopelessly inadequate in amount. At the end of 1940, according to a report of the Joint Distribution Committee, 1.25 million Jews needed public relief in German-held Poland. Raul Hilberg calculates that between 500,000 and 600,000 Jews died in ghettos and work camps as a result of Nazi policies—about one-fifth of Polish Jewry.[28] And this was before the Nazis' Final Solution.

For a short time in 1939 and 1940, Nazi strategists focused on the Lublin area in the southeastern corner of the *Generalgouvernement* as one concentration point for Jewish refugees who could not be accommodated elsewhere. This effort, subsequently known as the "Lublin Plan," fitted into the broader effort to force Jews and other "undesirables" out of the incorporated territories into the *Generalgouvernement*. It stands out from the general pattern of population movement, however, because it also reflected a recurrent Nazi interest in "autonomous" Jewish reservations (such as the "Madagascar Plan") and because it gave rise to myths about Jewish deportations to "work camps" in the East.

The inspiration for this scheme may well have come from Alfred Rosenberg, one of the top Nazi Party ideologists and eventually Reich minister for the occupied Eastern territories. Months before the war, Rosenberg explained to representatives of the foreign press that millions of Jews needed resettlement and that their destination should be a colony somewhere, ruled by "administrators trained in police work." During the autumn of 1939 this idea blossomed into a plan for a Jewish reservation, a supposedly autonomous area that, said Nazi propaganda, would become a "Jewish state." Adolf Eichmann, mastermind of the Jewish expulsions from Austria, directed deportations to the region. In reality, the Germans seem to have had in mind an experimental concentration of Jews as a buffer zone between the Russians and the *Generalgouvernement*. The site chosen was a desolate, marshy place, south of the city of Radom near the town of Nisko and west of the San River. The Nazis likely intended this "reservation" as a temporary camp, a *"Durchgangslager"* as Eichmann put it, from which survivors would eventually be pushed even further to the east. Whatever the long-range plans, however, it seems clear that the project had but a small place in the Nazis' overall vision.[29]

Unlike the Madagascar Plan, revived by the Germans in 1940, following the defeat of France, the Lublin reservation passed into an operational stage. Beginning in October 1939, convoys arrived from the Reich, Bohemia and Moravia, and the newly incorporated territories. In November, a newspaper in Luxembourg reported that 45,000 people of all ages were dumped at the reservation after a dreadful journey; the bewildered refugees staggered from the trains and were told to build their homeland. "There are no dwellings," an SS officer announced. "There are no houses. If you build, there will be a roof over your heads. There is no water; the wells all around carry disease. There is cholera, dysentery and typhoid. If you bore and find water, you will have water." On other occasions, the Jews were spared the charade of such a briefing. One transport from the Reich was simply ordered to march east to the San River, which marked the beginning of the Soviet zone. "Go to Red Palestine," was what they were told.[30] For the deportees to the Lublin region, the result was catastrophic mortality, details of which were soon reported in the Western press. Journalists looked into the so-called reservation and reported the shocking truth: as many as 60,000 people were being sent to the area with no preparation whatever having been made for them. Even from the German viewpoint the plan raised difficulties, apparently pitting the civil administration against the local SS emissary, Odilo Globocnik. Frank wanted desperately to check the latter, a zealous champion of deportations into the *Generalgouvernement*. German administrators nagged Berlin about disease, transportation difficulties, food supplies, and so forth. Finally, the Nazis lost interest in the project by the end of March 1940, and the deportation trains to the region finally stopped.

Nazi Jewish policy pursued the chimera of mass migration and expulsion until the latter part of 1941. Then, during the course of the Russian campaign, code-named Barbarossa, a new "final solution" took shape: the Nazis determined to deport Jews from everywhere in Europe to specially designed killing centers in Poland, where they would all be murdered. What accounts for the change? Some historians have seen mass murder as implicit in Nazi ideology, and some have even read it into Hitler's intentions at a very early date. But these scholars have sometimes failed to appreciate the zeal with which the Germans pursued their earlier emigration objectives and the radical redirection implied by the genocidal program begun in the midst of the Barbarossa campaign. According to the new directives, no Jew would be permitted to escape—not the old, not women, not even tiny infants. Emigration, far from being encouraged, was now to be blocked at every opportunity. Eventually, Jews were to be hunted down everywhere to be murdered—not only in the Reich, but in every corner of Europe, including countries like Ireland, Finland, Turkey, and Norway, where the size of the communities was relatively insignificant. It is difficult to assess fully the reasons for a shift in the Nazis' stated objective, given the paucity of written directives and plain language addressing a crime of this magnitude. Certainly, the change was part of a general radicalization of Nazism during Barbarossa and conformed to the pattern that Hitler had defined for that campaign: it was a *Vernichtungskrieg*, a war of destruction, conceived as a struggle to eradicate once and for all the entire "Jewish-Bolshevik system," seen as fundamentally at odds with Nazism. But, in addition, this shift to mass murder stemmed from problems posed by accumulating masses of Jewish refugees in Nazi-occupied Poland and Russia, problems that reached a critical point under the impact of war.[31]

Ironically, Nazi victories in the first two years of the war rendered even more difficult the great task of ridding the Reich of its Jews. Territorial conquest in the east and the west brought more millions of Jews under German rule without advancing the prospects for Jewish emigration. Meanwhile, the Nazis increasingly fell back on vaguely defined settlement schemes as war conditions reduced the actual numbers of Jewish departures. From 1939 to 1941, the beginning of the "final solution" was regularly postponed until peacetime. Within Poland the eastward expulsions of Jews created severe problems for the German administration and continuing political difficulties for the authorities in Berlin, now hearing loud local protests from occupation officials in the field. The latter complained about the overcrowding, supply and transport troubles, as well as impediments to maintaining order. Hans Frank, as we have seen, strenuously objected to the notion of the *Generalgouvernement* being used as a dumping ground.

Barbarossa merely highlighted the impasse. Calling for a climactic struggle against a rival Kultur, a decisive contest in which all the rules of war could be abandoned, Hitler sent his divisions against Soviet Russia on 22 June 1941, expecting sudden, dramatic success. German military commanders predicted victory within between eight and ten weeks. As the army drove into Russia, mobile killing units of the SS, the so-called *Einsatzgruppen*, swept through newly conquered territory, massacring hundreds of thousands, mainly Jews, in an unprecedented slaughter. Consistent with the apocalyptic expectations for the outcome of the war, the Führer seems to have advised his followers that the decisive moment had come for the resolution of the Jewish question. At the end of July, buoyed up by the first successes of the Wehrmacht, Göring issued his famous order to Heydrich to prepare "a total solution [*Gesamtlösung*] of the Jewish question in the German sphere of influence in Europe." We cannot be certain what precisely was understood by "total solution" at that point, but the instructions should be seen in the context of the anticipated early conclusion of hostilities. Quite possibly, emigration or expulsion of large masses of Jews still remained the overall conception.

Before long, however, things went wrong. The war did not end in a few months as had been expected. Rather, the Wehrmacht failed to meet its objectives. The fighting became more difficult with the autumn rains, the siege of Moscow, and the well-ordered Soviet retreat across the freezing Russian countryside. The Jews, under these circumstances, became more bothersome than ever. Even with killings on a spectacular scale, the Nazis now had more Jews on their hands. Their concentration in Polish ghettos continued, and large numbers were being assembled at railway junctions ready for the long-awaited expulsions—probably to the east, deep into the Soviet Union. But the Soviets were not prostrate, and the expected territory to which Jews could be dispatched remained a battle zone. Madagascar, meanwhile, even if considered seriously for more than a few hundred thousand, remained an abstract prospect so long as there was no treaty with France and so long as British sea power controlled access to the island. Most important, it became evident in the autumn of 1941 that the war would continue into the following year.

About that time, as the news from the battlefields was becoming worse and as the Nazis faced the prospect of yet another winter with the Jews, directives to prepare for mass killing on a Europewide scale seem to have gone out. On 23 October, in a striking reversal of emigration policy, Himmler ordered the exits closed, even for German Jews. Deportations from Germany to the east, in fact, began a few days before. Within days, a team of technicians visited the village of Chelmno (Kulmhof) in the Wartheland to begin work on one of the first of the death camps. The killing center at Belzec was launched at about the same time. At the end of November, Heydrich sent invitations to Nazi Jewish experts across Europe to participate in a conference at the Berlin suburb of Wannsee on the "total solution." When the Wannsee meeting finally assembled on 20 January 1942, Heydrich told those present that emigration was finished. Up to 30 October 1941, he reported, approximately 537,000 Jews had left since 1933, 360,000 of whom were from the *Altreich*, the Germany defined by the pre-Anschluss frontiers. Now, however, Himmler had forbidden such departures "in view of the dangers of emigration during wartime and in view of the possibilities in the East." "Emigration," he said, "has now been replaced by evacuation of the Jews to the East as a further possible solution, in accordance with previous authorization by the Führer." Although Heydrich's words veiled aspects of this "coming final solution," the minutes of the meeting, kept by Adolf Eichmann, leave little doubt as to the murderous intentions of the Nazi leadership. The deportees were to be grouped into huge labor gangs, in which "a large part will undoubtedly disappear through natural diminution." After this, only a remnant of the Jews would remain, and those "will have to be appropriately dealt with" lest the survivors "be regarded as the germ cell of a new Jewish revival."

The decisive change had now occurred. All along the many chains of command, Nazi officials now explicitly rejected expulsion or emigration schemes they had championed so long. Crucially, the timetable for the Final Solution had also altered. It was not to be put off until peacetime. On the contrary, it was extremely urgent to work quickly, *before* the fighting ceased. Franz Rademacher of the German Foreign Office made this crystal clear: "The Jewish question must be resolved in the course of the war, for only so can it be solved without a worldwide outcry."[32] Refugees would no longer be a problem for German officials. The refugees, like the other Jews, would be no more.

Rise and Fall of Lebensraum

At the height of the Nazi empire in Eastern Europe, Hitler's New Order reached as far as the Caucasus in the south and the outskirts of Leningrad in the north. Germany carved out two enormous Reich commissariats in occupied Soviet territory—the Ostland (the former Baltic states and Belorussia) and the Ukraine. All Poland now came under Nazi rule, with Polish territory formerly held by the Red Army becoming the German province of Bialystok. As many as sixty million Soviet citizens, including many Soviet nationalities, found themselves part of this empire. Because Soviet resistance continued, much of the conquered territory remained a war zone and never left military rule. Conditions varied substantially from place to place; the character of German administration shifted significantly over time with the ebb and flow of battle, the attacks of partisans, and the exigencies of the Nazi war machine.

The earliest Nazi plans for the new land in the East preceded the Barbarossa campaign and meshed with the Nazi racial program begun in Poland. The quest for Lebensraum involved the annexation of the Baltic states and much of the Ukraine plus the Crimean peninsula. By 1941 the Germans showed none of the caution exhibited in 1939, when they carefully divided their part of Poland into two and restricted the territory they would absorb. Now no limit was set on the land that Germany would one day acquire: eventually, the Nazis thought, natural growth would propel the empire even further east. To prepare for annexation, Hitler had in mind a vast program of Germanization: uprooted Germans would be brought from across all the Reich to develop the conquered spaces, preferably in the individual family homesteads the Nazis found so congenial. Germanic populations from the Low Countries and Scandinavia were included in these grandiose designs, and Rosenberg even believed that Englishmen would one day join this tide—or at least the pure Aryans among them. Settlement was the key, and high-ranking Nazis always spoke of millions to be planted in what had once been Russia. With settlement went its unmistakable corollary, the expulsion of all racially inferior elements. According to the General Plan East of May 1942, these were to be driven across the Urals into Siberia or Central Asia. As Himmler put it in mid-1941, "Our duty in the East is not Germanization in the former sense of the term, that is, imposing German language and laws upon the population, but to ensure that only people of pure German blood inhabit the East."[33]

Relentless in logic, Nazi settlement policies were hesitating, confused, and disorganized in practice. The Führer himself had other things on his mind. Increasingly isolated from the world in the *Wolfschanze,* his underground command post called "the Wolf's Lair," deep in the Masurian forests, the Nazi leader became totally engrossed in military strategy. Meanwhile, warring bureaucracies struggled within the Reich to impose their versions of Hitler's grand design. Essentially, there were two approaches to the Russian occupation. One group adhered fanatically to the concepts of Lebensraum set forth by Nazi ideologists: to them, the eastern territories were indeed, as Hitler once said, "our Africa," to be repopulated by Aryans, with the local inhabitants ("our Negroes") reduced to a slavelike existence. Another group, driven by the need to mobilize for the long, protracted war that Hitler always had wanted to avoid, sought more pragmatically to utilize the recently acquired land for the German war effort. Their priorities sometimes involved working with indigenous peoples, particularly when the latter showed signs of popular resistance to Soviet domination of their nationalities. Variations on these themes existed in profusion, and, in the long run, according to Alexander Dallin, Berlin followed many different courses at once. The contradictions within German policy, he suggests, ultimately subverted Nazi policies in the east and ruined the ability of the Reich to benefit from its occupation.[34]

During the next four years, Nazi imperial fantasies shattered on the hard realities of war in Russia. In the field, Germans proved much better at creating chaos and brutalizing the civilian population than resettling ethnic Germans. On a much smaller scale than had been envisioned, villages were sometimes cleared of their inhabitants, to be replaced by Volksdeutsche brought from the west. Throughout Nazi-occupied Russia, many thousands were turned into refugees in this way. Sometimes the locals were massacred, and sometimes they managed to escape behind Russian lines. Jews were liquidated by the hundreds of thousands. Suffering great hardships, German settlers were installed in Estonia, Lithuania, and the Ukraine. Throughout German-held territory, Nazi policies produced mounting terror, the decimation of local populations, and widespread resistance in partisan movements. As the tide of war shifted, moreover, requiring an ever more strenuous economic effort from the Reich, the insatiable German demand for labor checked Nazi policies that had created so many refugees within their own territory. Unforeseen by racial theorists, it turned out that the most important practical resource from Eastern Europe was the almost unlimited supply of unskilled workers, desperately required to serve German industry and agriculture. Even if it had been possible to drive people off to the east, to scatter the natives like wild animals from the new German settlements, the need for workers by 1942 proved too great to permit the indulgence of the original plans. The only broad exception was with the Jews, where Nazi fanaticism triumphed over all rational calculation. Throughout 1942 and 1943, with extraordinary strains being put upon German communications, the SS marshaled the Ostbahn Railway network to carry nearly two million Jews to their deaths in Polish camps. Labor boss Fritz Sauckel, meanwhile, scoured the Russian countryside for ablebodied workers, sending hundreds of thousands to Germany to work in factories and fields. Some 2.8 million *Ostarbeiter* from the eastern territories, mainly the Ukraine, were deported as slave laborers to Germany in the last few years of the war. To these were added millions more Soviet prisoners of war, herded together in unspeakable conditions. Altogether close to seven million foreign workers toiled in the Third Reich in 1944, about one-fifth of the entire labor force. Instead of being driven off to the east as refugees, ironically enough, Soviet citizens were sent west to the Reich.[35]

On a smaller scale, similar population movements occurred in Yugoslav territory, with the process complicated by the presence of several expansionist states. Following her defeat in 1941, Yugoslavia was divided between the Reich and her three European allies—Italy, Hungary, and Bulgaria. All, with the exception of Italy, expelled particular Yugoslav minorities from their zones of occupation. The Nazis sent at least 20,000 Slovenians from their slice, known as the Südsteiermark, to their puppet state of Croatia. Croatia, in turn, dispatched some 18,000 ethnic Germans from its territory, mainly to the German-incorporated parts of Poland. The Croats and Hungarians expelled tens of thousands of Serbs to Serbia, the former core of Yugoslavia, now under German military administration. Bulgaria, which took most of the Yugoslav regions of Macedonia, engaged in an aggressive Bulgarianization of the region, involving the expulsion of tens of thousands of Macedonian refugees. Toward the last days of the war, as the German armies fell back from Yugoslav territory, many ethnic German or German-sympathizing refugees followed the path of the retreating forces heading for the Reich. In all, according to one researcher, "the flight and evacuation, initially from one Yugoslav region to another, later to Austria and Germany proper, may have approached 300,000."[36]

Rumania, Germany's erstwhile dependent vassal, took advantage of the Barbarossa campaign to reinstate Bukovina and her lost Bessarabian province, creating a Rumanian version of Lebensraum, complete with several hundred thousand refugees. Driving eastward, the Rumanians reached the Dniester River, the easterly boundary of Bessarabia, and then proceeded even deeper into Soviet territory, entering the Black Sea port of Odessa in mid-October. The very next day the government of Marshal Ion Antonescu proclaimed a major part of the southeastern Ukraine, between the Dniester and the Bug rivers, to be its province of Transnistria (the lands beyond the Dniester), with Odessa as its capital.

Like the Germans with the incorporated Polish provinces, Antonescu intended to settle the region with his own people wherever they lived in Soviet lands, hoping to create there a territorial expression of Rumanian domination and superiority. In practice, little colonization took place, and the province became instead a dumping ground for Rumanian Jews, then being cleared out of formerly Soviet-held country. During the heat of the fighting in 1941 and immediately after, the Rumanians pushed the Jews across the Dniester into the waiting arms of *Einsatzgruppe* D, which was busy murdering all the Jews it could find in the Ukraine. Later, when the Germans moved further east, across the Bug, the Rumanians devised their own Final Solution for the Jews. From everywhere in the newly expanded Rumania, the unwanted Jews were deported to Transnistria, where the bulk of them perished. In 1943 even Jews from the Regat, the heartland of the old Rumania, were sent to the new province. In all, the region received between 250,000 and 300,000 Jewish refugees. Sometimes the Jews were sent to ghettos in the new territories, sometimes they were quartered among local residents and organized into work teams, and sometimes they were kept in special camps. Massed together in appalling conditions of filth, disease, and near-starvation, the Jews suffered staggering mortality. Of the 200,000 deportees, according to the Jewish relief agency organized at the behest of the Germans, almost two-thirds died by March 1943. The International Red Cross found 82,000 Jews in

Transnistria in September 1943, over 60 percent of whom were deportees. Of the total number deported to the province, only 50,000 survived at the time. Another 10,000 were to perish before the Rumanians finally left the region, driven westward in the beginning of 1944 by the Red Army.[37]

Visions of Lebensraum dimmed almost everywhere in the second half of 1943, with a convenient turning point being the gigantic tank battle of Kursk-Orel in July. Following this, the Russians moved steadily westward, reaching the former Polish frontier in February 1944. As they fell back, the Nazis determined to make their former "living space" uninhabitable. One result was hundreds of thousands of new refugees caught in the juggernaut of the Third Reich.

In the autumn of 1943, following the Germans' loss of Kharkov, Himmler issued a terrible order that helped to set the tone for subsequent fighting in Europe: "Not a human being, not a single head of cattle, not a hundredweight of crops and not a railway line is to be left behind. Not a horse is to remain standing, not a mine is to be available which is not destroyed for years to come and not a well which is not poisoned."[38] Although occasionally the Germans were more lenient, the reality was ghastly enough. The extent of destruction was astounding, and the suffering of civilians as great as at any time during the war. A substantial part of the local population was shot or starved to death. After rapine and murder, the Nazis dragged westward a substantial part of the adult population of certain regions to prevent their labor from ever serving the Communist side. By this point the Reich had developed a Gargantuan appetite for workers, and Sauckel's conscription program sucked up huge masses of people. Traveling with the advancing Soviet forces, Alexander Werth witnessed some of the effects of Himmler's decree. West of Moscow, he reported, the Germans produced "a harrowing catalogue of mass shootings, murders and hangings, rape, the killing or starving to death of Russian war prisoners, and the deportation of thousands as slave labor to Germany." Enduring their third winter in Russia, the Germans regularly faced attack from partisan units, now galvanized into action. The Nazis' brutal countermeasures, including periodic massacres, drove away civilians still in occupied territory. In 1944 the Germans undertook the expulsion of civilians as a systematic policy. Anticipating a retreat from Belorussia, the Wehrmacht ordered winter crops destroyed and tried to prevent spring planting. Practically all the livestock was killed, villages were set afire, and the survivors were scattered to the four winds.[39]

With the deportees, masses of Soviet citizens in one way or another cooperating with the German occupation fled westward, fearing to fall into the hands of the Red Army. These included the Volksdeutsche, seen by the Russians as collaborators with the SS and the Wehrmacht. Over 300,000 men, women, and children were deemed Volksdeutsche—a designation many were eager now to escape. Once celebrated as the pride of the Third Reich, standard-bearers of Aryan civilization in the savage East, the Germans now trudged toward Poland in often primitive circumstances. The refugees covered hundreds of miles on foot on journeys only rarely interrupted by rail transport. One of them described his westward march through the Ukraine with a Mennonite group in the autumn of 1943:

> In the September heat and in a thick cloud of dust the crowded trek made its way toward the Dnieper as rapidly as possible for the thunder of cannons and the flare of rockets signaled the approaching front. It was a reminder of an earlier trek. Nearly one and one-half centuries before our fathers had moved from Prussia in a similar way. We were repeating the trek, only somewhat poorer, less certain about the future, and in the opposite direction.[40]

As Dallin points out, the SS deeply distrusted many of these so-called pure Germans; in the enveloping mood of recrimination, it was assumed that these onetime Soviet citizens had "absorbed to a large extent Bolshevik and Russian doctrine." Himmler decided that although there was no question of their "return" to Germany proper, they should never again be settled in Russian territory. In what proved an overly sanguine move, the Reichs-führer SS approved their installation in Polish territories incorporated into the Reich. Joining the Volksdeutsche were other Soviet peoples who had in one way or another cooperated with the occupation in order to win a measure of self-determination. From the Caucasus, for example, came thousands of Cossacks who had commited themselves against the Soviets and who were subject to severe reprisals if caught by the Red Army. Field Marshal von Kleist, commander of military forces in the Caucasus and the Ukraine, authorized their exodus in 1943. By the end of that year, long columns of refugees formed caravans of horse-drawn wagons, together with their cattle and pathetic belongings, heading west.[41]

Evacuations of this sort continued for the next year and a half from Russia, the Baltic states, Poland, Rumania, Hungary, and Yugoslavia, as the Nazis determined to remove the Volksdeutsche from the path of the Soviets. In the late summer of 1944, for example, it was the turn of Hungary, where anti-Soviet propaganda was particularly intense and the devastating Russian invasion of 1849 still lived in popular memory. Following Rumania's surrender to the Soviets in August 1944, Berlin ordered a systematic transfer of about 50,000 ethnic Germans. As in Russia, the Germans set out with horses and wagons. While modern aircraft screamed overhead and German armor rolled east, the refugees plodded west at a snail's pace with the most primitive conveyances. Most of the wagons were driven by old men, women, and children—almost all ablebodied men having been mobilized by the armed forces. There was little doubt, near the end, as the Russians approached, that the New Order was in ruins and that local Volksdeutsche were paying some of the price. An elderly German farmer, Michael Schinko, told how he left his farm in the autumn of 1944. The Hungarian villagers, including the local gendarmerie, gathered to see him off: "These strong men of the law pressed my hands hard, the silver pearls from their eyes were falling on my hand that reached out to bid a last farewell. Their last words were [in Hungarian]: 'Dear Uncle Schinko, you know at least where you are going, but what about us? God bless you and keep you on your way!' "42

WARTIME ESCAPE ROUTES

How effective were the Germans in sealing escape routes from Nazi Europe? Wartime security measures often looked more formidable than they were in practice. For example, although the demarcation line between occupied and unoccupied France was studded by checkpoints and patrolled by military police, people regularly sneaked across, even for recreational purposes. The Germans did their best to block exits from the Continental coastline by using small patrol vessels, but it was impossible to watch everywhere at once, and very small craft could slip past the Nazi blockade. Sizable ships still left Aegean ports heading for Turkey and the Middle East during the height of the war. Within Europe, too, there was movement. Couriers for the Polish underground crisscrossed Europe carrying secret messages, and many thousands of Allied airmen escaped capture after being shot down over the Continent. Refugees slipped into Spain until 1944, and some also managed to enter Switzerland. Despite the Germans' efforts to control all movements of persons, they were never completely able to do so.

Although individuals and small groups could sometimes breach German security, it was otherwise with large masses of civilians, the refugee movements discussed in this book. At the outer edges of the Nazi empire, hundreds of thousands of potential victims did escape although, for some, freedom lasted only until the next Nazi advance. Successful flight in these circumstances needed organization—normally both on the part of the refugees and their rescuers. It usually required a friendly haven, such as was sometimes provided by neutral countries or allies of the Reich like Italy that diverged in some way from Nazi policy. Finally, escape was possible when the Germans were distracted by battle—as occasionally in Eastern Europe, when the Soviets believed it in their interest to keep people out of German hands.

The Soviet Union and Eastern Europe, 1941–44

Suspicious of potential subversives in the best of times, the Soviets verged on panic in the opening phase of Barbarossa as Nazi armored divisions swept everything before them, paralyzing the country's political leadership as well as its military machine. We have seen in the last chapter how, in the year and a half before the German attack, the Russians forced large numbers of newly subjugated nationalities from the Baltic states and what had been Poland deep into the Soviet interior. This process accelerated rapidly during the invasion, and to the extent that other traffic permitted, the Russians strove to remove anyone of doubtful loyalty to the regime. Citing fears of subversion and sabotage, NKVD agents ousted masses of people, often with murderous ferocity. Nicolai Tolstoy presents the evacuation of entire communities in the context of the rughless suppression by the Soviets of any element seen to be a potential source of opposition to the regime. Stalin, he argues, was fighting a "war on two fronts"—against the Germans on the one hand and against a substantial proportion of the Soviet people on the other. The deportation of Volga Germans, in this view, was little short of a massacre, with survivors being carted off to camps as slave laborers.[43] Evacuation, however, was not only for enemies of the regime. The Soviets were also anxious to remove Communist officials, skilled workers, and others deemed essential for national defense. In Estonia a special Committee of the Republic for Evacuation was set up to screen the populace, allowing only the most privileged and the most useful among them to escape.[44]

In practice, "dangerous social elements" went eastward together with valued Party officials and other notables. During the supreme crisis, the Russians had neither the time nor the means to distinguish carefully between the two. The Germans advanced with extraordinary rapidity, swallowing up huge areas of western Russia in the summer and autumn of 1941. Those uprooted had a few hours or even less time before they left. They then traveled enormous distances—often on foot for part of the way and normally by rail. Careful surveillance by police or militiamen was a constant reminder of the authorities' dark suspicions. Even in Soviet Central Asia they were kept under close watch. A Czech-Jewish lady, a refugee in Russia since 1939, later described her experience:

> Early one Wednesday morning, we were awakened by the NKVD and ordered to pack our most essential belongings. We were taken to the train station and from there we traveled together with masses of people to the city of Ajaz, in the Semipalatinsk district in Kazakhstan. Exiled together with us and for the same reasons were Czechoslovak citizens and the wives of Polish officers who had fought in the Polish army abroad. In Ajaz we were kept under guard; we lived in collective barracks and were permitted to move around only in a limited area.[45]

Other evacuees were left alone to fend for themselves. Lost in hopeless confusion and panic, entire families and individuals tried to find their own way to a place of refuge. The Russians they met seemed equally confused and unable to help. A young woman from Estonia, who fled from Tartu, later reported:

> Everything was on fire, and we didn't know where to run. Father, an old soldier who had already been through one war, was wearing two suits, one on top of the other, with a short fur jacket on top of the lot [and this in summer, in June!] . . . He had a little money, which he had saved over the previous year. He gave each of his children a few hundred rubles, in case we would be separated, and we hid the money in our belts. We got onto a train which we discovered was going to the front, so we got off and boarded another train . . . .[46]

Stalin determined to salvage what he could from the territory about to fall to the Nazis and resolved to leave a barren desert to the occupiers. Soviet planners worked feverishly to shift much of Russian industry, heavily concentrated in western regions, to more secure locations in the east. This required the forcible transfer of large masses of people, destined to work in the newly established industrial centres. Millions were uprooted for this purpose and sent to Siberia, the Caucasus, or Central Asia. In general, says Alexander Werth, "the transplantation of industry in the second half of 1941 and the beginning of 1942 and its 'rehousing' in the east, must rank among the most stupendous organizational and human achievements of the Soviet Union during the war."[47] The evacuees, however, paid a heavy price for this remarkable accomplishment. Everything was in extremely short supply, housing was primitive, and working conditions abominable. The death toll from hunger, overwork, accidents, and disease was extremely high. Many who had escaped the Germans and who had likely thought of themselves as refugees turned out in reality to be prisoners, officially classified as such and designated to spend the rest of the war confined to remote regions of the country.

More properly, refugees were those who eluded deportation, at least for a time, but were cast adrift on their own, rushing headlong from the panzer divisions. Twelve days after Barbarossa began, Stalin decreed a scorched-earth policy for the Red Army in retreat. Peasants were ordered to drive away their livestock, destroy all their property, including fuel and seed. After they had done all this, there was little left but to take to the roads. People also poured out of cities immediately threatened by the German invasion. Just before its encirclement in August 1941, the Russians claimed to have evacuated 350,000 civilians from Odessa. German aircraft and submarines on the Black Sea took a heavy toll of defenseless refugees. On 25 August, ten ships left the Estonian port of Tallin (Reval), carrying evacuees to Leningrad; the Germans immediately attacked these vessels, sinking four of them. Altogether, the Germans sank seventy ships removing people from Tallinn. In mid-October, when news of the government's decision to move to Kuibyshev raced through Moscow, the result was a stampede of refugees out of the city. A few months later, half-starved survivors of the Germans' siege of Leningrad slipped through enemy lines by crossing the frozen waters of Lake Ladoga; close to 500,000 managed to escape, with priority given to women, children, and the elderly.[48]

EUROPE UNDER NAZI OCCUPATION (BEFORE 22 JUNE 1941)

After C. E. Black and E. C. Helmreich, *Twentieth Century Europe: A History* (New York, 1966).

244

With Barbarossa, the Soviets attuned their policy on nationalities to meet the emergency, favoring some groups and penalizing others, driving them into internal exile. The Russians were extremely suspicious of Soviet citizens of German descent, numbering about 1,425,000 according to the census of 1939. Nearly 500,000 ethnic Germans lived in the Autonomous Soviet Socialist Republic of the Volga Germans, in territory that ultimately escaped German occupation. At the end of August 1941, however, having lost much of the Ukraine, the Soviets took extreme measures against these Germans. Moscow ordered the Republic dissolved and its entire German population to be moved to Siberia and Central Asia. The expulsion decree spoke of "thousands and ten-thousands of diversionists and spies" among the local population, against whom punitive measures were to be taken. The decree mainly affected dependents because most of the men between eighteen and forty-five were at the front or in various military support units. A survivor recounted the extraordinary sight: "A mournful procession of refugees filled the roads leading to the railway stations of the Middle Volga, 400,000 of them, carrying bedding, dragging domestic animals, the women weeping, all with the bitterness on their faces of those who have been driven from their homes . . . ."[49] The same decree ordered other German communities evacuated, especially along the Black Sea. Around 200,000 of these Germans, it has been estimated, were herded across the Urals at the time. More expulsions of Germans occurred the following year, this time from the Transcaucasian region between the Black and Caspian seas.

As conditions worsened catastrophically for the Volga Germans and other elements deemed suspicious in 1941, they improved in some respects for Polish refugees in the U.S.S.R. The latter, it will be remembered, bore the weight of Russian repression after September 1939, with Moscow attempting to extirpate all expressions of independent Polish nationhood. Barbarossa, however, forced the Soviets to reverse their policy toward the Poles. With the Red Army staggering back in retreat, the Russians needed military and political support from non-Communist Polish exiles who would help, in turn, to improve Soviet relations with the West. Moscow signaled accordingly to its former enemy, the Polish government in exile, producing an accord, signed in London in July 1941. The Russians agreed now to work with the Poles to defeat the Nazi aggressor. An important focus of this document was the estimated 1.8 million Poles living within the Soviet borders of 1941, of whom a large majority languished in camps or distant places of exile. The Russians promised an amnesty to all interned or imprisoned Poles and agreed to permit a Polish relief agency to assist them on Soviet soil. The Poles sent an ambassador, Stanislaw Kot, to the Soviet Union, charged with the task of overseeing the fate of their exiled compatriots. To assist the Polish relief effort, the Russians granted the Poles a million rubles, conceding to the exiles in London the determination of regional representatives who would, among other things, decide who was, in reality, a refugee of Polish citizenship. Remarkably, the Soviets also allowed the Poles to recruit and organize an important military force under the Polish general Wladislaw Anders, which would eventually leave Soviet territory. After the force was trained and organized, some of them actually did quit Russia in mid-1942. These were mainly soldiers, but among them were about 40,000 civilian refugees, the first sizable group to leave the Soviet Union since the German invasion. In striking contrast to what was claimed about the Jews, namely, that there was no room anywhere for them, space was somehow found for the Polish civilian exiles, seen as cobelligerents against Hitler: the British admitted some 16,000 to East Africa, 3,000 went to Rhodesia, 4,000 went to India (of whom half continued on to Mexico), and 7,000 were located in various places in the Middle East.[50]

Barbarossa decimated the Jewish communities that lay in its path, throwing hundreds of thousands of escapees into Soviet hands. The Jews were heavily urbanized in western Russia and, therefore, were caught in a relatively small number of centers; the Nazis moved so rapidly that major concentrations were overrun before anyone had a chance to flee. The northernmost killing unit, *Einsatzgruppe* A, swept through the Baltic states to the outskirts of Leningrad. On 15 October 1941 it reported to Berlin that it had already murdered 125,000 Jews.[51] Horrific massacres, mainly of Jews but also of Communist officials, partisans, and hostages from the local populations, sent shock waves of terror through civilian communities that lay in the path of the Nazi advance. At the end of August, SS units slaughtered some 23,600 Jewish refugees at Kamenets Podolski. On 29 and 30 September, in the largest massacre of its kind during the war, squads from *Einsatzgruppe* C, with Wehrmacht and local assistance, killed over 33,000 Jews in the ravine of Babi Yar, outside of Kiev. Fleeing these horrors, Jews from central and eastern Ukraine and from

Belorussia surged eastward, sometimes with the help of the Soviets. As many as 10,000, according to one researcher, found
sanctuary by hiding in the so-called "family camps"—clusters of
refugees who gathered in wooded areas of Belorussia, Polesye,
and northern Volhynia and who managed to survive on their own,
despite extraordinary hardships and the frequent hostility of the
local peasants. Most Jews, however, could save themselves only
by reaching Russian-held territory to the east. Raul Hilberg estimates that as many as 1.5 million managed to elude the Nazis in
this way.[52]

Once past the first shock of the invasion, the Soviets geared
their refugee policy to a protracted struggle with the Nazi enemy. Although they managed to stop the Germans at the gates
of Moscow in December 1941, the Russians faced the winter of
1941–42 with the loss of their industrial heartland, a decrease of
more than half of their industrial production, ammunition supplies nearly exhausted, and their armed forces in tatters. At that
point, when the Nazis were still flushed with victory, the Soviets
mobilized every ounce of energy for what they were to call "the
Great Patriotic War." Soviet citizens applied backbreaking effort
to the task, eventually to achieve remarkable results by 1943.
Without hesitation, the Russians enlisted hundreds of thousands
of refugees on their soil in this effort. Along with prisoners of
war, refugees were sent to the most primitive locations, to perform the most strenuous tasks, frequently under conditions of
military discipline. They worked in the frozen wastes beyond the
Arctic Circle and in the republics of Turkistan. They cut timber
in the Siberian tundra and did agricultural tasks in collective farms
in the Ukraine. Those whom the Russians deemed citizens could
enroll in labor divisions attached to the Red Army; but others who
had refused Soviet citizenship in the first two years of the war
were often treated as prisoners, kept alive only to serve the
motherland they had rejected. The death toll among these refugees was staggering, though probably no higher than that for
millions of internal deportees. In the second half of 1942, Jewish
agencies attempting to monitor refugees in Russia sent an investigator to Iran who reported on the "two to three hundred and
fifty thousand Jewish refugees scattered throughout the length
and breadth of Russia." His informants mentioned "outposts so
remote from any community or center that they do not even bear
place names but are simply designated Labor Station No. So-and-
So." It took a journey of several months overland from the near-

est railway station to reach such labor camps in northeastern Siberia. Work stations were isolated from one another as well as from the major cities; this made a general evaluation of their situation extremely difficult. Attempting to ascertain the total number of Jews in such camps, the reporter heard guesses of between 50,000 and three times that figure.[53]

El Campesino, the legendary Spanish Republican exile in the Soviet Union, described what life was like during the war for a veritable army of refugees who ended up in Uzbekistan, deep in Soviet Central Asia. He met all sorts of uprooted people, citizens and noncitizens, some of them apparently free to roam about on their own:

> The Spaniards were only a small part of the army of refugees in Uzbekistan. The whole territory of the Republic and the neighboring Republics were at the time crammed with masses of people of all sorts and all breeds—and all uprooted. They had come following a rumor that the regime had lost control of those regions. There were those who wanted to get out of the way of the authorities for a particular reason, and those who had no personal reason but only shared the general fear of the police and wished to enjoy something like freedom thanks to the chaos. There were deserters, evacuees from factories and *kolkhozes*, Poles who had been released from internment camps when the German attack on Russia had turned them from enemies into allies. There were public servants who left their posts and had taken great sums of money with them and prisoners of war who had escaped from German-occupied countries. There were political refugees of all nationalities, Yugoslavs, Poles, Czechs, Austrians, Germans, Italians, Frenchmen, Spaniards and so forth; and there were professional bandits.[54]

El Campesino's cheerful account should be read in the context of the grim realities of exile in Uzbekistan and notably the phenomenal mortality among deportees who were sent there. They died of hunger, overwork, disease, exposure—every conceivable fate for a huge population transplanted without adequate preparation and material means. The former Soviet historian Aleksandr Nekrich cites statistical evidence on the fate of the Tatar "special settlers" deported to Uzbekistan from the Crimea: from May 1944 to 1 January 1945, 13,592 persons died, 9.1 percent of the total; nearly 18 percent of all Tatar deportees eventually perished before the beginning of 1946 in their new Central Asian homeland.[55]

Emigrés who could not appeal from Soviet internment to a

government in exile friendly to the Russians suffered severely as a result of the crisis that opened in mid-1941. Czechoslovak refugees, for example, who carried passports from the pro-Nazi Protectorate, were branded enemy nationals in June 1941 and immediately interned. Months later, when the Soviets signed an agreement with the Czech government in exile, they were finally released. Ukrainian internees, who had no one to whom they could appeal, remained behind barbed wire, as did the Balts, Yugoslavs, and others. With the Nazis in control of so much Soviet territory, the Soviet leadership feared that occupation would stimulate centrifugal forces of nationalism among national groups deemed hostile to Russian rule. Since the summer of 1941, when some Soviet regions actually welcomed the Nazi invasion, the Germans beamed nationalistic appeals to various peoples behind Russian lines. Experts in Berlin believed they had particularly good prospects in the Crimea and the Caucasus, having taking the trouble to establish links with émigrés from those regions who had left Russia during the Revolution. Moslem nationalities were considered particularly vulnerable. The Nazis promised much. In the words of one slogan: "Long live the free Caucasians in alliance with and under the protection of the Great German Empire of Adolf Hitler!" Seeking to shore up their battered state, the Soviets fell back on the deportation of large masses of people, entire small nations in some cases, a procedure first used during the forced collectivization of agriculture in the 1930s. Eventually, nearly a million were removed from the Crimea, Transcaucasia, and the Caspian steppes—Tatars, Chechens, Ingush, Balkars, Karachai, and Kalmucks. These deportees followed the many hundreds of thousands who had gone before to Siberia and Central Asia.[56] Many years later, at the Twentieth Party Congress in 1956, Nikita Khrushchev admitted that the deportations had been a great injustice—"rude violations of basic Leninist principles," as he called them. Notably, the process began in 1943, *after* the tide of battle shifted in the Soviets' favor. The deportations seem to have been dictated by a combination of factors, including the desire to punish certain "unreliable" peoples and the urge to settle border regions with nationalities believed to be more loyal.

With some military success and a greater confidence in an ultimate victory, the Russians hardened their attitudes toward many refugees within the Soviet Union, preparatory to the Red Army's drive westward into the Baltic countries and Poland. As early as the spring of 1942, Moscow amended its agreement with the Poles over refugees, withdrawing Polish jurisdiction over former citizens of Jewish, Ukrainian, and Belorussian nationality. The Russians now deemed these to be Soviet citizens. During the latter part of the war, the Soviets drew into increasingly sharp confrontation with the Polish exiles in London. The breaking point came in the spring of 1943, with the Germans' revelation of a horrendous massacre of thousands of Polish officers in the Katyn Forest near Smolensk, apparently committed by the Soviets three years before. Following an anguished Polish appeal to the International Red Cross to investigate, Moscow abruptly severed relations with the London Poles. Foreseeing an assault on German positions in Poland, the Russians had their own plans for that country and were increasingly concerned to have a controlling voice in Polish affairs. As a result of the new strategic shift, conditions for most Polish exiles in the Soviet Union reverted to what existed before Barbarossa. Now, however, the Soviets began to court the relatively small numbers of Communists among them. In May, Moscow began recruiting its own Polish force, the Polish Union of Patriots, to assist in the liberation of the country and the determination of its political future. Much the same reversion to earlier times occurred in the Baltic states in 1944, when the Red Army threw out the Wehrmacht and reestablished Soviet control. Police and political authorities began a new purge of the local population intended to remove all anti-Communists, nationalists, and those suspected of collaborating with the Germans. Close to 500,000 were deported to the east. Soviet victories over the Germans, therefore, saw a major new uprooting of the Soviet population.[57]

Minor currents of refugees flowed elsewhere in Eastern Europe, forming pathetic eddies of humanity sometimes forgotten in the great torrent of deportations and expulsions. A few anti-Nazis headed for Hungary or Rumania instead of going west in 1938 and 1939, following the Anschluss with Austria, the dismemberment of Czechoslovakia, and the attack on Poland. A substantial number, between 15,000 and 35,000 in one rough estimate, were Jews. More refugees probably managed to flee southward from Poland during the early days of Soviet and German occupation. Hungary later claimed to have received 140,000 Poles, including part of the Polish army. The Hungarians confiscated the Poles' weapons but allowed them to proceed to the West via Yugoslavia. They also permitted the establishment of a refugee-aid committee for Poles, which was in touch with the Polish government in exile and the Polish underground. The committee co-operated with the Hungarian authorities, and its members were on close terms with leading officials in Budapest. To the north, in the puppet state of Slovakia, the local government had a certain independence in internal affairs, and some of its senior officials treated with a Jewish underground agency assisting refugees and in contact with the West. This rescue committee, known as the Working Group, operated secretly in Bratislava within the framework of an officially sponsored Jewish Council and smuggled refugees from Poland into the relative safety of Hungary.[58]

From early 1942, when ghettos across Poland were being emptied for the Final Solution and as cattle cars began to bring their miserable human cargo of Jews from across Europe, the Hungarian Axis ally oddly provided a sanctuary for persecuted Jews. From both Poland and Slovakia, thousands of refugees slipped into Hungary to escape deportation and murder. More than ten thousand, it has been estimated, came from Slovakia alone in 1942. Under the aristocratic and conciliating Miklós Kállay, prime minister from March 1942 until March 1944, Hungary sheltered fugitive Jews, former members of the Polish army, and Allied airmen or escaped prisoners of war. In Budapest, activist local Jews established an elaborate assistance network for refugees, known as the *Vaada*—taken from its Hebrew designation, the Jewish Committee of Mutual Assistance. This body, apparently with the passive consent of Hungarian officials, hid thousands of escapees and managed to smuggle some of them out of the country to Palestine. The Nazis knew about this activity, but until 1944 were unwilling to bludgeon their Hungarian ally on the issue. Edmund Veesenmayer, the regional expert from the Wilhelmstrasse, reported the existence of a Hungarian haven in April 1943 in his correspondence with Berlin. He explained reactions in Budapest as a supposed Hungarian effort to curry favor with the West through the Jews and thereby to be spared Allied air raids. There is no doubt that the Hungarians were trying desperately to extricate themselves from the German embrace at this point, and a permissive policy toward refugees was merely one way of signaling their preference to the Allies. This was why government officials agreed, for example, to include limited numbers of Jewish escapees among shipments of Allied prisoners of war being sent to Turkey via Yugoslavia. Everything, of course, depended on the Nazis' remaining at arm's length, too preoccupied with the Russians and Allied air bombardment to coerce their Hungarian ally. These circumstances finally changed in the spring of 1944, when the Germans occupied Hungary, ending the Kállay era and closing the door to any further escapees.[59]

Neutrals

SWITZERLAND

When the war began in 1939, Swiss authorities were profoundly committed to their country's neutrality, which had proved its worth in more than a century of European conflict. Swiss politicians spoke proudly of their country's vocation for freedom and independence, her commitment to humanitarian service. But woven into their rhetoric of neutrality were many threads of insecurity, a sense of the great vulnerability of this nation of only four millions, which was seeking to escape the scourge of a ferocious conflict. In 1940, when Germany attacked in the west and Italy entered the war, the Swiss found themselves for the first time in contemporary history completely surrounded by an ascendant, predatory, and bellicose military bloc. Nazi or Fascist forces stood on every frontier. Expenditure on defense skyrocketed, food supplies were cut, and rationing was strictly enforced. The Swiss ate much less than ever before, sinking in 1944 below 2,000 daily calories per person. Every hectare was under cultivation. Many thousands of young men were called to the colors, women and girls worked the fields, and draft animals were taken away by the military. Government officials constantly pictured themselves as strained to the limit, desperately warding off catastrophe, struggling with vastly inadequate means.[60]

Hardly anyone seems to have thought that refugees might be an asset rather than a liability. The general assumption was rather that Switzerland was dangerously overpopulated. In 1939 federal authorities ventured that the country had an absorptive capacity of only 6,000 refugees—a tiny number, given that the Republic had harbored about 75,000 civilians and military personnel during the First World War. No one of importance seems to have proposed that refugees could work the fields, where they were obviously needed, or fortify the will to resist Nazi or Fascist aggression. Paralyzed by their sense of the country's weakness, Swiss leaders saw foreigners as posing the gravest threat to the entire society. Federal Councillor Edouard von Steiger, heading the country's Ministry of Justice and Police, popularized an image of the Swiss Confederation that was widely accepted—"a lifeboat in a great sea disaster, with only very limited space and even more limited provisions." Steiger, in one heated public debate, saw himself as the captain:

> When . . . the captain of this lifeboat must choose whom he shall embark and for whom he wishes to reserve the space and supplies that are available, while thousands and thousands have the same right and while he should and would like to help thousands and thousands, he faces the great spiritual torment of choosing whom he shall and shall not take aboard. Shall he take the women first, or the children, or the ailing, or the married couples, and say to himself: "Perhaps the young and strong still have a choice to survive in spite of all dangers; but I have to make a choice!"[61]

The choice, however, was conditioned by the knowledge that some refugees would leave the lifeboat after the war and others might not. Like other neutrals, the Swiss government viewed rescue with an eye to an eventual postwar situation in which some refugees would likely be able to return home whereas the stateless among them might become a permanent burden. Stateless Jews, in particular, were considered a liability. In an effort to improve the situation of its refugees, the Polish government in exile publicly declared its intention to facilitate the return of its nationals; it is unclear, however, whether this promise impressed the Swiss captain. Above all, the government wanted to limit the numbers granted asylum and to clear every obstacle in the path of those already in the country to travel abroad.

The result was a policy stringently restricting the acceptance of refugees, continuing the exclusionism of the 1930s described in the last chapter. When hostilities began, there were about 7,300 fugitives in Switzerland, the great majority of them Jews. Most of these had been accepted on the understanding that they would emigrate; indeed, this had been practically the sole basis on which refugees could cross Swiss border posts in the preceding year. Small numbers continued to leave the country after September 1939, having successfully forged, bought, or otherwise procured the precious visas to the western hemisphere. From Geneva, 1,563 refugees managed before the summer of 1942 to travel abroad, embarking by ship from Marseilles. The largest group, 638 people, went to the United States; 214, to Cuba; 190, to Brazil; 156, to Santo Domingo; and 132, to Uruguay. In addition to these emigrants, the Swiss announced tough measures in October 1939 against new illegal immigrants: federal police authorities ordered them expelled immediately.[62]

Refugee pressure mounted precipitously with the collapse of France in June 1940. As millions of French took to the roads and their army was thrown into chaos, thousands attempted to enter Switzerland. Close to 45,000 French soldiers crossed Swiss frontiers in the hour of defeat, along with 7,500 civilians. In response, the federal authorities acted quickly: all refugees, instructed a circular of 18 June 1940, were to be turned back, except women, children under 16, men older than 60, and invalids. Frontier guards were reinforced, and the civilians who fled at the time of the German invasion were promptly sent home. Thereafter the Swiss borders were practically closed. Until the beginning of 1942, refugees entered legally only in driblets—twenty-six during the rest of 1940, and 124 throughout 1941.[63] Federal authorities now refused even to admit refugees with valid entry visas for other countries, fearing the emigrants might be trapped in transit and forced to remain. Fugitives caught illegally crossing the Swiss frontier were summarily expelled. Sometimes cantonal officials mitigated the effects of this tough policy by permitting a very brief sojourn or allowing expellees to return secretly so as to escape capture. Sometimes, if Vichy did not object, the Swiss allowed fugitives from the Reich to enter the unoccupied zone of France. But Bern was extremely solicitous of Vichy's wishes on such matters, not wanting to jeopardize the passage of emigrants from Switzerland through Marseilles.

The roundups of Jews throughout both zones of France for the Final Solution in the summer and early autumn of 1942 sent new waves of refugees battering on the Swiss frontiers. More came in November, when the Germans entered the formerly unoccupied zone. On the French side, professional guides or *passeurs* helped organize a massive effort to evade Swiss controls. What was to be done? Police Chief Heinrich Rothmund issued tough instructions in August, September, and October, explicitly excluding Jews from the category of acceptable political refugees. Military units were sent to patrol certain border areas, and barbed wire was strung along vulnerable entry points. Over a thousand fugitives were turned back between August and December 1942, according to police records. From various sectors of the Swiss public, however, there was a sharp reaction. Swiss Protestant churches, labor organizations, and liberal politicians brought the issue into national prominence; Swiss newspapers trumpeted the refugee cause, and many of them raised funds for refugee relief. Strong protest over the *refoulement* of Jews at the frontiers induced authorities to relent, and thousands were granted asylum. Much depended on officials at the cantonal level, for the latter managed on some occasions to ignore or to soften federal policy. During the second half of 1942, nearly 9,000 refugees came into the country. Henceforth policies altered according to the attitude of local authorities, the periodic instructions from Bern, and the shifting tide of war. *Refoulements* remained common. Federal authorities recorded 3,324 turned away at the frontiers in 1943, 3,998 in 1944, and 2,282 in 1945.

Bern gradually agreed to accept more refugees as the Allies made military headway, particularly after their landings in Italy. To house the refugees, the Swiss government built a network of special camps in 1943 and increased expenditures on refugees sixfold. By June of that year, the Swiss had received 19,000 refugees. In September, when Italy surrendered suddenly to the British and Americans, the Swiss accepted record numbers of fugitives, notably thousands of Jews attempting to escape from what had been an important sanctuary in the zone of France formerly occupied by Mussolini's army. At the end of the year the total number of refugees stood at over 26,000. During 1944, with the end in sight, masses of new refugees were accepted—deserters, wounded soldiers, escaped prisoners of war, political refugees, orphaned children, and Jews who had somehow managed to escape the Nazi dragnet. Only now did the proportion of Jews sink to less than half of the total number of refugees. With the liberation of France, many who had found asylum returned home; yet, even so, the number of refugees continued to rise. In May 1945, when the fighting finally ceased, the total was the highest ever—115,000 of all categories, about half of whom were civilians.

The Swiss later calculated that, at one time or another, they sheltered 295,381 people—of whom civilians numbered 191,512. Some of these were emigrants passing quickly through the country, inhabitants of frontier areas who briefly sought shelter from the fighting, and 60,000 children, rescued near the end of the war. There were 55,000 civilian refugees, of whom 21,600 were Jews. No one can say how many refugees were turned back. One recent estimate puts that total at 20,000, but this remains conjecture. Any such calculation, moreover, can only be a pale reflection of the thousands more who were dissuaded by the reality of expulsions from even trying to reach Switzerland.[64]

The Swiss never faced direct Nazi pressure to refuse refugees or to send back specific fugitives who did find asylum in the Swiss Confederation. True, the Axis military presence was a grave menace, and the Nazi press vociferously attacked Bern for harboring Jews. The Germans were in a position to tighten the economic noose around the Confederation and actually did so in the summer of 1942. But Berlin and Rome seem to have been largely indifferent to Swiss actions on behalf of refugees, and their wishes can hardly be said to have played a direct role in the policy of restriction. After the war, some Swiss officials claimed that if they had only known the realities of the Final Solution, they might have acted otherwise. But there can be little doubt that, even without a full grasp of the details, every literate Swiss knew that the Jews faced a terrible, mortal threat under Nazi occupation. Bern and Geneva were important collection points for all sorts of information about wartime atrocities; articles on the murders of Jews appeared regularly in the Swiss press from mid-1942. Even if one accepts that a measure of skepticism existed about the extent of Nazi killings, the plainly reported facts were chilling enough. "Those who wanted to inform themselves were able to do so" seems a reasonable enough conclusion.[65]

Yet the Swiss feared that the lifeboat could be swamped at any moment. As Heinrich Walther, a national councillor from Lucerne, put it, "there exists a *sacro egoismo* that at certain times must be taken into account, and a certain influence on the exercise of the right of asylum must be permitted to this *sacro egoismo*."[66] The Swiss military command harped on the dangers that excessive numbers of foreigners posed to national defense. Eventually, they feared, Switzerland could be drawn into the fighting and overrun by the Wehrmacht. Others, preoccupied with the administration of refugee policy, shared the police officer's wartime view of outsiders as a menace to public order. In an economic climate of extreme penury, particularly mid-1942, these arguments had considerable force. Interestingly, many working on behalf of refugees at the time had some sympathy with the Swiss position. Donald Lowrie, a refugee worker acting for the World Alliance of the YMCA in France, found Bern's refusal to grant visas quite understandable. Arieh Tartakower and Kurt Grossmann, well-informed Jewish researchers writing in 1944, offered a remark-

ably positive judgment: "While, owing to extremely urgent considerations of a political and economic nature, the Swiss government could not open the doors of its country more widely to the
thousands of unfortunates knocking for admittance, Switzerland
has certainly made a considerable contribution to the refugee
problem."[67]

These observers may well have underestimated the strength of
anti-Jewish feeling in high quarters. But if antisemitism was, in
part, responsible, it was seldom openly expressed, and its effects
diminished in the latter part of the war. Throughout this period,
government spokesmen who dealt with the issue certainly rejected Nazi-style racism. There was talk about how unassimilated Jews could not be permitted to swamp Swiss society. Heinrich Rothmund, who was most immediately involved in these
matters, observed, after a visit to the Oranienburg concentration
camp in 1938, that "the people and government of Switzerland
had long been fully cognizant of the danger of Judaization."[68]
Rothmund denounced the Germans' form of Jew hatred and their
methods of dealing with the "Jewish problem," but he also believed that the best way to prevent the implementation of this
kind of antisemitism in the Swiss Confederation was to hold the
line against the Jewish outsiders. Relentlessly, he walled himself
off from the desperate calls for help, the accounts of massacre,
and strong humanitarian instincts of his own countrymen. Less
than two weeks after the Allied declaration on Nazi war crimes
against Jews on 17 December 1942, Rothmund moved to reinforce restrictionism once again at Swiss frontier posts. Rigid,
blinkered, and utterly convinced of his own rectitude, Rothmund saw the helpless Jews as a great source of danger. He was
no ideologue, and, like most of his compatriots, he never saw
the Jewish question in the obsessive manner of anti-Jewish ideologists. When pressure mounted on the Jews' behalf, he was
prepared to make concessions. In December 1943, for example,
he telephoned an order to receive all Jewish refugees from Italy;
his police instructions of mid-1944 finally threw open the doors
to all threatened Jewish refugees.[69] From his own pen comes an
account of a discussion he had with German State Secretary von
Weizsäcker in Berlin in October 1942, at the very height of Nazi
persecutions of Jews in neighboring France. Rothmund claims to
have insisted on the Swiss commitment to the policy of asylum.
There is reason to be skeptical about this version, given the real-

ity on the Swiss borders at the time; one of his comments, however, rings true and may stand as an epitaph for Swiss policy: "For me, the essential thing is that we be left alone . . . ."[70]

SPAIN AND PORTUGAL

Despite their superficial affinities with Hitlerian Germany, neither Spain nor Portugal had any real sympathies with Nazism, and both remained neutral throughout the war. Both countries were ruled by conservative autocrats, eager to steer their old-fashioned vessels safely away from adventure. Both were linked historically to persecutions of Jews via the Inquisition and the calamitous expulsions of 1492. Both were seen in the 1930s as producers, rather than receivers, of refugees: although fugitives from the Salazar regime were never numerous, the flood of Spanish refugees after the collapse of the Republic in March 1939 surpassed anything that Europe had seen since the Russian Revolution. There was some irony, therefore, in the role of the two countries in the rescue of thousands of refugees from Nazism during the course of the war.

Geography placed Spain in a critical position for refugees in the spring of 1940. In addition to Frenchmen fleeing their own country, a ragged assortment of Belgians, Dutch, Luxembourgers, and Central Europeans looked toward the Iberian Peninsula as the Allied armies reeled back from the onslaught of the Wehrmacht. During May, June, and July, 50,000 refugees crossed the Spanish frontier, some of them swept up in the panic of the *Exode*. Many returned to France following the stabilization of the military situation and the Franco-German armistice. Generally, the other fugitives sought to embark at Lisbon or other ports, emigrating to America. First, however, the refugees had to cross into Spain. In Madrid there was never any question of harboring large numbers of foreigners. Spain emerged devastated from the Civil War in 1939, facing enormous problems of demobilization and reconstruction. The fighting killed nearly 600,000 Spaniards; tens of thousands more died of disease and malnutrition. Half a million Spanish workers sat idle, agriculture and industry were in shambles, and internal communications were crippled. Spain lost one-third of her livestock and at least 250,000 homes. Food was scarce, rationing was imperfect, and most of the population was reduced to a primitive level of existence. Moreover, as the end of the conflict coincided with the beginning of a European war, it was practically impossible for Spain to make up the losses abroad. Food and medicine were in even shorter supply in the years 1939–41 than during the Civil War itself.[71]

Following the triumph of Franco, his government drew close to its Nazi benefactor, quitting the League of Nations; signing the Anti-Comintern Pact with Germany, Italy, and Japan; and forging military and economic links with the Reich. Franco was, therefore, hardly in a position to welcome with open arms the armies of fugitives from Nazi persecution. But neither was the Caudillo prepared to mobilize his prostrate society on behalf of Hitler's war aims. From the end of the first year of fighting, indeed, Spain began to assert its independence from the Reich, at the very moment when the Führer appeared invincible in the West and the Wehrmacht stood at the Spanish frontier. On 23 October 1940, Franco met Hitler at the Spanish border town of Hendaye and bluntly refused the Nazi effort to get Spain into the war. Hitler later said that he would rather have four teeth pulled than face another session with Franco. The latter set his terms so impossibly high that Germany could never possibly meet them. Spain, therefore, excused herself from the contest. Extravagant with his expressions of loyalty to the Axis, Franco kept up a profitable trade in strategic materials with Britain and the United States. He was playing a delicate and dangerous game, which succeeded only because Hitler ran into difficulties elsewhere.

Shortly after the establishment of the Vichy government, the Spanish consul in Marseilles suddenly became a key player in the refugee drama of Western Europe. German policy actively encouraged Jewish emigration at this point, particularly from the Reich but also from the recently defeated countries. As we have seen, visas and shipping space were extremely scarce. Some of the luckiest managed to sail directly from Marseilles and from ports in North Africa. Two Spanish ships occasionally carried refugees to America from Seville; most emigrants, however, went by Portuguese ships from Lisbon. For thousands of desperate Jews seeking to escape Nazi Europe, therefore, crossing through Spain became an urgent necessity. The Spanish consul received and processed the transit visa applications. HICEM, the principal Jewish emigration agency, maintained offices in Marseilles, Lisbon, and Casablanca. With its assistance, emigration machinery began to turn, and the Spanish authorities were bombarded with requests to traverse Spanish territory.

The Spaniards made it clear at the outset that their country was not a haven for refugees. Although Franco's Falangist government did not regard the Jewish issue as did the Nazis, it appeared distinctly unfriendly to Jews who were seeking escape from occupied Europe. Jewish organizations could not operate openly in Spain, and Jewish representatives had great difficulty communicating with the Spanish authorities. The Spanish-Jewish community was tiny, insignificant, and in any event associated with the recently defeated and despised Republicans. After three years of bitter internal upheaval, Spaniards seemed eager to go their own way and to have as little as possible to do with outsiders. The ideal, as Foreign Minister Jordana y Sousa once put it, was to have foreigners "passing through our country as light passes through a glass, leaving no trace."[72] Jordana and his government wanted no expense, no headlines, and no confrontation with the Germans over the issue. But they had no objection to the passage of refugees and exercised no discrimination against Jews among them. Instructions issued in November 1940 stipulated the exclusion of men of military age who were fit for service and set elaborate currency regulations. Madrid tinkered with the rules from time to time, but basically kept its gates open to those able to proceed elsewhere. One experienced refugee worker, Varian Fry of the American-based Emergency Rescue Committee, burrowed his way into emigration bureaucracies and became an expert in circumventing obstructive regulations. He was struck by the permissiveness of the Spanish so long as the precious transit visa was affixed to the refugee's passport. Spanish officials did not care if the refugee had left France illegally or whether his or her other papers were in order. Their sole concern was the continuation of the voyage through Spain and conformity with local currency regulations. Once this was settled, all was well. Spaniards may well have profited from the passage of many thousands of refugees in the year or so after the fall of France. Among the least reliable were the professional smugglers, who were known to leave parties of illegal refugees stranded in the mountainous regions they were paid to cross. At the frontier it was common to bribe Spanish border guards, many of whom simply sold entry permits. Spanish shipowners charged extortionate rates for berths on the few ships leaving the country for North and South America. The fugitives were packed on board crudely refitted vessels to venture across the submarine-infested Atlantic. One notorious case, the freighter *Navemar*, with 1,200 bunks in its hold, took forty-eight days to go from Seville to New

York in the summer of 1941, arriving with most of its passengers in a pitiful condition; the latter eventually filed damage claims against the Spanish owners totaling more than 3.5 million dollars.[73]

Illegal entries posed a different problem, but here, too, the Spaniards were not particularly hostile to the refugees. Those caught without the appropriate visas were generally imprisoned and sometimes ended up in the grim Miranda de Ebro concentration camp outside Burgos. But very seldom were they sent back to Vichy or German-occupied territory. One reason may well have been the strong interest shown by the British and American embassies in the care of refugees in Spain. The Allies were primarily concerned with their military personnel who managed to cross the Spanish frontier after having engaged in underground operations, escaped from camps for prisoners of war, or been shot down from the skies over Nazi territory. Carlton Hayes, the American representative in Madrid, blocked the efforts of Jewish refugee workers, arguing that their activity hindered work on behalf of Allied soldiers; his intercessions, however, may well have nudged the Spaniards toward decent treatment for illegal refugees. Allied pressure may also have helped keep the Spanish frontier open.

Spanish policy was severely tested in the summer and autumn of 1942, when the number of illegal entries jumped spectacularly. First came a wave of Jews fleeing the roundups by Vichy police and the deportations to the "East." Then came additional fugitives after the German occupation of the southern zone of France, followed by thousands of young French nationals threatened with conscription into German factories early in 1943. In addition, there were political refugees—part of a rising tide of opposition to Vichy by previous supporters of the regime, including many now seeking to join the Resistance movement in England or North Africa. In February 1943 there were some 12,000 French refugees in Spain. Both Vichy and the Germans actively opposed Jewish emigration at this point and strenuously denounced the flight of anticollaborationist elements. German border troops and French *gardes mobiles* patrolled the frontier with Spain, and Berlin applied serious pressure on the Spaniards to block entry to the fugitives. Two German divisions stationed nearby were meant to impress Madrid and force its compliance. For a brief moment, Spain wavered. On 25 March 1943 the Pyrenees border stations closed to all entrants except those with

Spanish entry visas. The Germans, however, were distracted by their agonizing experience in Russia, and they were awkwardly placed to coerce the Spaniards. Counterpressures mounted. For reasons that are difficult fully to understand, the Vichy ambassador in Madrid, François Piétri, took up the cause of the fugitives and pleaded their case to the Spanish foreign ministry. More important, the British and American ambassadors urged Spanish authorities not to send back the illegals. Mindful of the political and even military importance of some of the refugee traffic through Spain, the Allies signaled to Madrid how seriously they took the issue. British Prime Minister Churchill himself sternly lectured the Spanish ambassador in London, warning him that good relations depended on humane treatment of fugitives. The wind, by this point, was beginning to blow against the Axis side. Finally, deferring to Allied wishes, Spain accepted even illegal entrants on its soil. At first, the government forbade the use of its ports to ferry refugees to Algeria or Morocco, permitting only passage to Portugal. Carlton Hayes worked out an arrangement whereby French refugees, many of them openly intending to join the Allied armed forces, embarked for North Africa from the Portuguese port of Setúbal. Then, in September, following the Italian surrender, Jordana allowed sailings from Spanish ports. During the course of 1943 more than 20,300 refugees left Spain, including 16,000 Frenchmen, 800 American airmen, and about 3,500 stateless Jews.[74]

Through the latter part of the war the Spaniards extended a protective mantle outside their own country to cover several hundred Sephardic Jews in a curious gesture combining nationalistic pride and cultural pretension. Behind it lay an unstated sense of obligation to Spain's "lost children"—the descendants of Jews expelled from Spain at the end of the fifteenth century. Many of these Jews still spoke the Spanish-based Ladino language and were scattered throughout the Mediterranean world, especially former parts of the Ottoman Empire. The Germans were prepared to exempt small numbers of Jews from neutral countries from the death factories in Poland provided they were "repatriated" in short order. The Spanish, in turn, were willing to receive a limited number of refugees who could somehow demonstrate their Spanish "citizenship." In this largely fanciful exercise, much depended on consular officials, some of whom were willing to bend the rather stringent formal requirements set in Madrid. Spanish representatives in Sofia, Athens, Salonika, and

Budapest brought about 800 Jewish refugees out of countries where they were in great danger. In addition, Spanish diplomats, taking advantage of their relatively favorable standing with the Germans, extended consular protection to an additional 3,235 persons, mostly in Hungary.[75]

In close alliance with Spain since March 1939, Portugal was, if anything, even more committed than her Iberian neighbor to keeping out of the European conflict. Balancing a pact with England with support for Nazi Germany as an anti-Communist bastion, the wily Portuguese dictator António de Oliveira Salazar worked assiduously throughout the war to keep on good terms with everyone and to extract the maximum economic benefit from the neutral's role. Both the Germans and the Allies found the Portuguese stance useful; neither wished to risk this utility by pressing Lisbon too hard. As a result of careful management of her political and economic resources, Portugal prospered during Europe's darkest days. Red Cross aid to prisoners of war and civilian internees passed through the port of Lisbon before distribution, much of it carried by Portuguese vessels. The Portuguese capital swarmed with spies and shady characters; wealthy refugees settled into hotels, and their impoverished counterparts slept in makeshift shelters.[76]

Lisbon soon became the refugee capital of Europe, the nerve center of various relief agencies and the principal port of embarkation on the European continent. The first wave of arrivals came in 1940, including as many as 10,000 Jews who had been formally ineligible to receive Portuguese visas. Most of these received these documents anyway, thanks to the intervention of a remarkable diplomat, Aristides de Sousa Mendes, then Portuguese consul in Bordeaux. A devout Catholic who claimed to be of Marrano descent, de Sousa Mendes singlehandedly rescued thousands of fugitives following the French defeat. Against the express instructions of his superiors, Mendes issued visas apparently on demand in the spring and summer of 1940. "My government has rejected the refugees' requests for visas," he observed at the time, "but I cannot let these people die. Many of them are Jews and our constitution states that the religion and political views shall not constitute grounds for their being refused asylum in Portugal. . . . Even if I am dismissed from my post I cannot act but as a Christian, faithful to the dictates of my conscience." Other Portuguese consulates seem similarly to have issued visas freely shortly after the collapse of France. De Sousa

Mendes was soon recalled, however, and Portuguese policy tightened.[77]

Refugees continued to trickle into the country throughout the war, with the active or tacit approval of the government. As in Spain, the policy was to require reemigration. In practice, however, given the scarcity and uncertainty of transport, refugees gathered in the Portuguese capital for increasingly long periods. The small Portuguese Jewish community, headed by an associate of Salazar, Moses Amzalak, assisted many of the refugees, about 4,000 of whom were always present in Lisbon during the war. Representatives from the Joint Distribution Committee funneled in financial help and received cooperation from the Portuguese police. As late as October 1941, the Nazis were actively encouraging Jewish emigration from Portugal, and the SS even commissioned a travel agent to accompany Jewish refugees leaving the Reich that month and heading for Lisbon.[78]

It is difficult to say how many people passed through the Iberian Peninsula during the course of the war. Relatively few remained in Spain or Portugal for the duration—several thousand at most. Because so many refugees were assisted by Jewish emigration agencies, we have the clearest idea of refugees who were Jews. HICEM materials suggest that as many as 30,000 Jews traversed Spain in the first half of the war until September 1942: some of them embarked from Lisbon; and others, from Spanish ports. Fewer and fewer of the new arrivals were Jews; the roundups throughout France gradually reduced the numbers of Jewish refugees who might have been able to escape. As many as 40,000 Jewish refugees may have been saved by flight into Spain. Within Spain, the refugee problem became acute by the spring of 1943 after a new wave crossed the Pyrenees from France. By this point Jews constituted less than half of the refugee total. Refugees in Spain were one of the most important problems discussed by delegates to the Allied Conference in Bermuda at the end of April that year. Spain had to be relieved of its refugee burden, it was suggested, lest the entire Iberian escape route be jeopardized. At the time, the delegates were told, the country sheltered the following: about 14,000 French refugees, hoping to go to North Africa; about 800 Allied nationals, mainly Poles, who had been accepted into the armed forces of their countries and whom the Allies wished to take to Britain; and finally between 6,000 and 8,000 other refugees, mainly Central European Jews. No mention was made of Allied military personnel, whose presence on

Spanish or Portuguese soil was not a matter for public discussion. In the end, the borders were not closed, and thousands more escaped Nazi-held territory via this route. Including the first wave of refugees, in the late spring and summer of 1940, the total number rescued through Spain and Portugal probably approaches 100,000 persons.[79]

THE VATICAN

To no one's surprise, the Vatican also emphasized its neutrality in the war against Hitler—a posture that was to earn it much criticism after the war and that conditioned its action on behalf of refugees. Neutrality, as the Vatican understood it, continued the policy of conciliation that was fundamental to Church diplomacy in the interwar period. During the political storms of the Great Depression years, this course was charted by Eugenio Pacelli, the cardinal secretary of state under Pius XI and later to become the wartime pope, Pius XII. Pacelli shared with his predecessor a profound sense of the spiritual and pastoral mission of the Holy See; his goal was to avoid association with power blocs and to create an environment in which the Church could operate as freely and openly as possible. As fascism extended its influence in Europe during the 1930s, therefore, the Vatican remained aloof, occasionally challenging fascist ideology when it touched on important matters of Catholic doctrine or the legal standing of the Church, but unwilling to interfere with what it considered to be purely secular concerns. Undoubtedly, the Vatican also found certain aspects of right-wing authoritarian regimes congenial: their opposition to Marxism, their patronage of the Church in several countries, and their frequent championship of traditional social structures.

But the Vatican quarreled with both Hitler and Mussolini on race, eventually drawing Church officials into the refugee crisis of the 1930s and 1940s. At issue in this controversy was certainly not the welfare of Jews or the programs to discriminate against them. The Church did not generally oppose such persecutions and rarely denounced them; when it did so, it usually admonished governments vaguely to act with "justice and charity." The problem was rather that the pseudobiological justification for racism directly challenged Catholic doctrines and Church authority. The claims of fascist movements, especially Nazism, to use race as a foundation of their regimes, hit squarely at the

Church's insistence on its own rules for marriage, baptism, and, more broadly, the definition of who was and who was not a Catholic. Although the Holy See always worked within the well-worn channels of diplomatic and pastoral communication, the Vatican, nevertheless, staked out its opposition to racist doctrine in unmistakable terms. In March 1937 the papal encyclical *Mit brennender Sorge* (With a Burning Heart) condemned the false and heretical teachings of Nazism; related protests by the Vatican of Mussolini's racist policies the next year highlighted Church concern over the issue. Yet at the same time the Holy See carefully avoided an open breach. As always, the goal was political neutrality, the safeguarding of Church interests in a perilous political world.

The Vatican first became involved in refugee matters in the late 1930s, when racial persecutions intensified throughout the Reich and when large numbers of Catholics of Jewish origin were forced to emigrate. From the start, however, the Holy See was cautious. The pope made no public pronouncement against the governments that refused to take in refugees. Although Catholic emigration societies existed throughout the Reich, the Vatican also opposed suggestions for the establishment of a centralized refugee organization with headquarters in Rome. The Holy See then believed it possible to dissuade the Mussolini regime from pursuing a path of racism and did not want to disrupt its diplomatic demarches by announcing forthrightly its assistance to the Catholic victims of racial persecution. Instead, papal officials maintained their private communication with German emigration organizations like the *Raphaelsverein*, run by a clerical order in Hamburg, and related Jewish agencies in Italy that assisted Catholics of Jewish origin. Through correspondence and in private discussion, Vatican diplomats tried to clear the path for these refugees, attempting to use their influence with Catholic countries such as Ireland and Chile. Generally, they had little success. In 1939 and 1940 the Holy See pressured the Brazilian government into granting visas to 3,000 "German Catholic non-Aryans"; the Brazilians later withdrew their offer, complaining about the refugees' conduct and questioning the authenticity of their Catholicism. At the same time, the Vatican remained extremely skeptical about schemes for mass migration to places like Angola or Madagascar. Prudent and cautious to a fault, the Holy See adapted its diplomacy to a world that was closing its doors to refugees.[80]

The outbreak of war in 1939 signaled the failure of the Vatican's efforts to achieve conciliation and increased pressure for action. But now, more than ever, officials in the Holy See weighed every word and measured every gesture to ensure that their neutrality would be respected. Against a hailstorm of appeals for humanitarian gestures of all sorts, Pius XII adhered stubbornly to what he believed to be the higher interests of the church and, ultimately, mankind itself. Carefully schooled in the ways of the Vatican bureaucracy, Pius was a scrupulous, anxious, methodical diplomat. The published documents of the Vatican reveal a pope horrified by the war and passionately committed to a negotiated peace. Preoccupied with the vulnerability of the Church in a conflict that pitted Catholics against each other and that, he feared, could end with either Nazi or Communist hegemony, Pius clung to the wreckage of prewar policy—"a kind of anxiously preserved virginity in the midst of torn souls and bodies," as one sympathetic observer puts it.[81] In practical terms, this meant no ringing denunciations of one side or the other; a continuing and exaggerated hope to limit the spread of war and encourage peace negotiations; and an effort, strictly within the framework of wider Vatican objectives, to alleviate the suffering caused by war.

From the beginning of the conflict, this approach clashed with political realities. The Vatican wanted to remain above the melee, able to intervene and excercise its moral authority. In reality, the pope disappointed most of those who appealed to him without significantly affecting the conduct of the war. Thus, for example, the pope refused to denounce the Reich as the aggressor in Poland, and, despite the appeals of Polish clerics, he would not explicitly protest the Nazi occupation. While reiterating his compassion for the Poles, his detestation of particular methods of war, his hatred of injustice, he turned a deaf ear to these and all other calls for ringing moral pronouncements against any of the belligerents. And having maintained this neutrality, the pope did not thereby win the ability to deliver more humanitarian assistance. Neither the Nazis nor the Soviets allowed Vatican officials to dispense aid to the Polish population, and the pope's charitable activity, therefore, mainly reached the Poles least in need—those refugees who had escaped the occupation of 1939. To be sure, neutrality facilitated the Vatican's support of countless victims on both sides of the conflict. The Vatican Information Service served refugee movements as well as prisoners of war, missing persons, orphans, and deportees. Vatican radio broadcast thousands of messages monthly helping individuals to communicate across the abyss. Throughout, the Vatican's refugee work focused on Catholics of Jewish origin who were directly threatened by Nazi racial persecution. Many of these Catholics, it was claimed, could not receive assistance from Jewish charitable agencies. Opportunities for diplomatic intervention on their behalf narrowed drastically at the end of 1941, when the Nazis foreclosed the possibility of emigration from anywhere in occupied Europe.[82]

The most controversial aspect of Vatican policy remains its reticence on the persecution of Jews. Throughout the war Jewish agencies and countless individuals called in desperation to the Holy See, demanding some public gesture on behalf of European Jewry. Almost invariably, they were disappointed: Vatican officials constantly repeated that they were doing all they could, that more should not be expected. Nuncios, bishops, and other clerics were allowed to go their own way to assist the Jews in flight, and many did so with the Vatican's blessing and encouragement. Thousands of Jewish refugees found shelter in Church institutions, received unstinting aid from priests and laypersons, and were brought to safety by Catholic underground workers. But the Holy See itself took little leadership in such activity and remained bound by its reluctance to take sides openly in the conflict. The risks of forthright statement were great, and the likely returns deemed extremely uncertain.

Using traditional diplomatic channels, the Holy See occasionally helped Jewish refugees in Nazi-occupied Europe. The Vatican was unsuccessful in its effort to halt deportations from Slovakia in the spring of 1942, but its widely communicated disapproval may have strengthened popular resistance to these measures, which were suspended during the summer. Later that year, the Vatican more successfully intervened with Italian authorities, urging them not to surrender 2,000 or 3,000 Jewish fugitives in Croatia to the Germans. Papal nuncios or apostolic delegates in Catholic countries could sometimes be decisive so long as the Nazis did not force a confrontation. Thus, in the case of Slovakia, local church officials, with the approval of Rome and Giuseppe Burzio, the Vatican representative in Bratislava, were able to assist some Jews to escape. The nuncio in Madrid helped keep the Spanish frontier open in 1943; and Monsignor Bernardini, the nuncio in Switzerland, intervened with Bulgarian authorities to rescue Jewish refugees in 1944. Interestingly, Vatican representatives persistently opposed the Zionist project in Palestine and throughout the war objected to sending the pathetic handful of Jewish escapees to the Holy Land. Cardinal Secretary of State Maglione, who must have known better, even claimed in 1943 that "it does not seem more difficult to find other territories that might be more suitable." "If Palestine fell under the power of the Jews," he went on, "it would create new and grave international problems, upset all the Catholics in the whole world, provoke the justified complaints of the Holy See, and be a poor reply to the charitable concern which the Holy See has shown and is showing toward the non-Aryans."[83]

The Vatican had more opportunity to assist refugees during the latter part of the war, when the Nazis' resolve weakened in certain areas and when some of Germany's allies were more amenable to pressure from the Holy See. In the autumn of 1943, just after the Italians surrendered to the Allies, the pope found himself with Jewish refugees almost literally on his doorstep. After the armistice the Germans occupied Rome, then filling with large numbers of fugitives, including Jews. Abandoning all concerns about Italian opposition, the Nazis intensified their persecution of Jews and prepared their deportation convoys for Auschwitz. Many Jewish refugees were of French or Polish origin, knew no Italian, and were easy prey to the Gestapo, which was then aggressively rounding them up. After an initial hesitation, the pope ordered the Vatican City and its institutions open to Jewish refugees, having received some assurances from the German representative, Ernst von Weizsäcker, that the Nazis would observe the inviolability of the Holy See and its territory. On a knife's edge, the pope seems to have balanced carefully, fearful at any moment that the SS might descend on the Vatican itself. The SS succeeded in deporting to Auschwitz about a thousand Jews from Rome, captured without the help of the Italians on 16 October. According to one authority, 477 Jews found refuge in the Vatican City and its enclaves; and 4,238 others, in convents and monasteries throughout Rome.[84]

In the final year of the European war, rescue activity intensified in Hungary, where the Nazis had only begun the deportation of Jews in May 1944. That summer international protest rained down on Admiral Horthy and the Hungarian authorities, who were actively assisting the Germans. The pope's call on Horthy to cease the deportation of Jews was particularly telling, and after its receipt, together with other statements and threats from the

Allies, the Hungarians halted the transports to Auschwitz. Later that year, the overthrow of the Horthy regime saw renewed deportations and further efforts to save the Jews who had thus far escaped massacre. The nunciature in Budapest was one of several diplomatic posts that issued safe conduct documents, enabling about 2,500 Jews and Catholics of Jewish origin to escape the rampaging of the fascist Arrow Cross and death marches to the Reich. Some 15,000 people in all, including fugitives from all over Hungary and elsewhere in Eastern Europe, were saved by miscellaneous protective passes issued in this way.[85]

SWEDEN AND TURKEY

On the periphery of occupied Europe, Sweden and Turkey were conveniently situated to provide escape routes and sanctuary for refugees from Nazism and the victims of war. Of the two, Sweden seems the most likely refuge, having a close cultural affinity with large numbers of refugees from neighboring countries engulfed in the conflict—mainly Norway, Denmark, and Finland. Sweden also shared liberal and democratic traditions with the forces arrayed against Hitler although this did not prevent her from engaging in a lively economic exchange with the Reich. Turkey's European neighbors, by contrast, were Bulgaria, which played a very modest role in the war as an Axis country; the Soviet Union, which absorbed its own refugees; and Greece, with which the Turks had a particularly troubled history. Nevertheless, the Turks received tens of thousands of fugitives, and Istanbul remained a major center of refugee activity. In each country a door to the Nazis' European prison opened on rare occasions to permit the passage of refugees. Each received fugitives, without upsetting a sometimes fragile relationship with the Third Reich.

Like the other Scandinavian countries, Sweden was reluctant to accept refugees from Nazism during the Depression years. Restrictionist sentiment was strong, strict immigration laws were enforced, and only in exceptional cases were fugitives permitted to remain in the country. By the spring of 1940, when the Germans occupied Denmark and Norway, there were only about 3,000 refugees from Hitler residing in Sweden although considerably more were in transit to other countries. Finnish refugees, however, were at the center of attention, and these numbered in the tens of thousands. From the vantage point of Stockholm, the great drama of 1939 was less the outbreak of war in Poland than the Winter War between Finland and the Soviet Union, which began nine weeks later. Swedes had enormous sympathy for the Finns and provided extensive private and governmental assistance for their refugees—many of them children evacuated for the duration of the conflict. Some of the refugees returned home when Finland reopened the war in 1941; the flow reversed itself in 1944, when the Soviets advanced against the Finns once again. Over the course of the war Sweden accepted a total of 131,000 Finnish refugees, about a third of them children.[86]

Clinging to her fragile neutrality, Sweden received new waves of refugees in 1940 as the conflict widened in the east and west. Fugitives from Norway and Denmark escaping Nazi occupation joined others from the Baltic countries fleeing the imposition of Soviet rule. Thereafter refugees continued to trickle across the thousand-mile frontier with Norway and the narrow stretch of water separating the Danish capital of Copenhagen from the Swedish port of Malmö. Thirty-two thousand Norwegian refugees had reached Sweden by the end of 1944, including about 800 Jews. Danish refugees began to cross the Sund to Sweden following a German crackdown on their country in September 1943. Sweden's important role in the rescue of 7,500 Danish Jews that autumn will be discussed shortly. To these one should add about 10,500 other Danish refugees who reached Sweden before the end of the war.[87]

In the last days of the fighting, moved by fears that the Nazi collapse would provoke a spectacular civilian bloodbath, Swedish officials engaged in a variety of rescue activities. One scheme that failed because of a German veto was the proposed passage westward of 250,000 Norwegians from the northern part of their country, menaced by the scorched-earth retreat of the Wehrmacht. More successfully, the Swedish embassy in Berlin became very active in 1944 and 1945, seeking the release of internees in German hands. The key figures in this complex drama were Count Folke Bernadotte, nephew of the Swedish king and vice-president of the Swedish Red Cross, and Felix Kersten, Himmler's personal physician, then living in Stockholm. With the blessing of the Swedish government, Bernadotte had several meetings with Himmler and other high-ranking Nazis in Berlin, resulting in the release of thousands of concentration camp inmates—mainly Scandinavians and including Odd Nansen, son of the polar explorer and League of Nations refugee advocate. In addition, the pastor of the Swedish Legation personally organized the escape of hundreds of Jews from Nazi deportations. Working with the American War Refugee Board, the Swedes sponsored the 1944 mission to Budapest of Raoul Wallenberg, a businessman turned diplomat who intervened with Nazi and Hungarian officials with a mixture of intrigue, bluff, and bribery to rescue thousands of Jews—usually by means of Swedish protective passports. In April and May 1945, Red Cross workers brought about 20,000 civilians to Sweden, 3,500 of whom were Jews. In all, it has been estimated, Sweden received nearly 237,000 refugees in the course of the war.[88]

The great majority of refugees who entered Turkey between 1939 and 1945 were Greek fugitives from the Balkan campaigns of 1940 and 1941 and others fleeing the subsequent occupation and the internal strife associated with the beginning of the Greek civil war. The Turkish Interior Ministry claimed that 67,000 refugees and internees entered the country during the war, of whom nearly 44,000 were Greeks. Many of the incoming refugees managed to continue their journey, crossing Syria and heading for British-controlled Palestine or Egypt. By mid-1943 there were about 5,000 Greeks in Egyptian camps, 4,600 in Cyprus, and others scattered about Africa and the eastern Mediterranean.[89] Turkey provided a particularly important escape route for Jewish refugees from various points in southeastern Europe. In late 1942 the Jewish Agency established a secret refugee and rescue service based in Istanbul, operating underground and in regular communication with Jewish officials in Palestine.

Generally speaking, the Turks were unfriendly toward the refugees. "In Turkish circles," wrote an American observer, "refugee problems always aroused misgivings and distaste." The Turks scrutinized Jewish fugitives carefully, insisting that the latter have the necessary Palestinian entry papers and Syrian transit documents before being admitted. Though not an Arab country, Turkey shared much of the opposition within the Moslem world to a Jewish national home in Palestine, and the Turks were eager not to be seen facilitating its growth. Moreover, Turkey attentively guarded her neutrality. Particularly in the last phases of the war, the Turks feared greatly that they would be drawn into the fighting. The Allies heavily bombed both Sofia and the Ploesti oil fields in Rumania, only a few hundred miles from Turkish soil; and the Russians swept into southeastern Europe, crumpling Turkey's neighbors. Franz von Papen, the German ambassador in Ankara, carefully manipulated Turkish fears of a Nazi attack, reinforcing the Turks' unwillingness to cause an open breach with the Reich by opening their doors to the Jews.[90]

One tragic incident highlighted the Turks' concerns that their country not become a conduit for illegal Jewish immigrants to Palestine. At the end of 1941 an ancient and derelict Bulgarian-owned ship, the *Struma,* reached the port of Istanbul, carrying Jewish refugees from Rumania and heading for Palestine without the necessary papers. Packed into this unseaworthy hulk, 769 passengers occupied space meant for a hundred. The Jews had paid exorbitantly for their passage and were robbed by the Rumanian officials just before leaving the port of Constanta. In urgent need of repairs, the ship remained offshore at Istanbul for two months while the Turks refused permission for its passengers to land, and the diplomats considered what should be done. Meanwhile, the refugees suffered terribly from hunger and overcrowding, and the Jewish Agency pleaded with the British on their behalf. Efforts to repair the ship's engines having failed, the Turks decided to send the ship back to the Black Sea, in obvious deference to the British desire to keep the Jews from Palestine. At this point, the British had second thoughts, becoming worried about the children on board. However, the Turks would not relent. They towed the ship toward the Black Sea to be set adrift, and the vessel promptly sank—likely the accidental victim of a Soviet torpedo. Everyone on board was lost, with the exception of a single survivor. Although the *Struma* affair echoed for several years in the British press and prompted an intense criticism of British policy, there were few attacks on the Turkish government and its role in the tragedy. The Allies were attempting to enlist the Turks in the war at the time, and the British Foreign Office wanted no attacks upon Turkish policy.[91]

Unlike the governments of several neutral states and a few allied to the British, the Turks declined a German offer to repatriate and thus protect from deportation about 3,000 Turkish Jews, some of them refugees, who were living in Nazi-occupied Europe. Ankara expressed interest in 631 of them and virtually abandoned the rest. Thanks to the initiative of an official at the Wilhelmstrasse, however, these Jews were saved, and the Turks eventually agreed to receive them.[92] By mid-1943, indeed, as the pendulum swung noticeably against the Nazis, the Turks appeared increasingly amenable to Allied pressure over refugees. Since his arrival in early 1942, the American ambassador to Turkey, Laurence Steinhardt, was reluctant to urge that the Turks receive refugees, especially Jews. A Jewish career diplomat, Steinhardt seems to have shared the restrictionist sentiment of the American State Department and its reluctance to see a confrontation with the British over Palestine. As the tide began to turn against Germany, however, Steinhardt seems to have become more active on behalf of refugees and more prepared to press their case with the Turks. In February 1944 the newly founded War Refugee Board sent the dynamic department-store executive Ira Hirschmann to Ankara to work with Steinhardt and speed evacuation through Turkey. Together they had some success in breaking the Turkish refugee bottleneck: during 1944, according to the World Jewish Congress, 14,164 refugees openly crossed Turkey on their way to Palestine.[93]

Palestine

The year 1939 at once dramatized the struggle of the Jewish refugees to enter Palestine and complicated Zionist opposition to British restrictions. As we have seen, the mandatory power increasingly saw Palestine during 1938 and 1939 through the prism of war preparation: in a word, the British knew the Jews would be on their side and appeased Arab opponents of Zionism to ensure their support in the struggle to come. Having set this course, the government determined to slow and eventually to arrest the growth of the Jewish National Home, particularly by restricting immigration. Unfortunately, this occurred just as doors everywhere else slammed in the faces of Jewish refugees. The White Paper of 1939, offering 75,000 Jewish immigrants in five years, remained the cornerstone of British policy during the war and infuriated refugee advocates. "It is a policy in which the Jewish people will not acquiesce," said David Ben-Gurion, speaking for the Jewish Agency in Palestine. Winston Churchill, one of the strongest opponents of the policy at Westminster, led a storm of criticism against the Chamberlain government: "This pledge of a home of refugees, of an asylum, was not made to the Jews of Palestine, but to the Jews outside Palestine, to that vast, unhappy mass of scattered, persecuted, wandering Jews whose intense, unchanging, unconquerable desire has been for a National Home. . . . That is the pledge that was given, and that is the pledge we are now asked to break. . . ."[94]

From the start, many Jews called for direct confrontation with the mandatory power in the form of illegal immigration, known to the Jews in Palestine as *Aliyah Bet*. Indeed, the Jewish Agency began this program even before the White Paper. Between mid-1938 and the outbreak of war, more than 17,000 Jews sought entry illegally, and most were captured and interned. The first shots fired in Poland only made the refugees more desperate and the British more determined to stop them. The Palestine Administration imposed wartime restrictions and added a new reason to keep the Jews out—the fear of enemy agents. Simultaneously, war forced the Zionists to reconsider how to oppose British restrictions. They faced a cruel dilemma. Britain was, after all, locked in combat with Nazi Germany, the very source of the Jewish crisis. In a famous attempt to define Jewish strategy, David Ben-Gurion called on Jews to fight the war as if there were no White Paper and also to fight the White Paper as if there were no war. Understandably, Ben-Gurion's call proved difficult to follow. Some Jews pressed urgently for illegal immigration, to batter down British restrictions with a great tide of seaborne refugees. One branch of the movement remained determined to act as provocatively as possible, to heighten pressure against the injustice of exclusion—even though the ships carrying illegal refugees stood little likelihood of successfully running the British blockade. Others preferred more discreet landings, to sneak greater number of refugees past the British. Still others developed severe misgivings about the whole process of illegal immigration, believing instead that the best chance for a change in British policy lay in military cooperation against Hitler through the formation of Jewish volunteer units in Europe and Palestine. This was the view of Chaim Weizmann, who signaled his disavowals of the illegal immigration to the British authorities in 1940 during the dark months of the Battle of Britain.[95]

Despite these theoretical disputes, illegal immigration continued: one after another, unseaworthy, vermin-ridden little ships appeared off the coast of Palestine, crammed with refugees. These vessels—"coffin boats," as Weizmann called them—were part of a black market in refugees, often run by Greek or Bulgarian gangsters, encouraged by the Nazis who were then trying to make Europe *Judenrein*. They were financed by the terrorized refugees themselves, together with their allies—the Jewish Agency and Zionist dissidents. Embarking at seaports on the Black Sea, the ships set forth for Palestine: their path took them through the

Dardanelles, the stormy waters of the Aegean, and the Mediterranean route to Haifa or Tel Aviv. Some of the boats were wrecked
in the Black Sea within a day or so of embarkation; one ship, the
Pancho, likely fell victim to an Italian mine. Almost invariably the
passengers who reached the Palestinian coast were twice survivors—first of the Nazis and second of a voyage of harrowing
danger and indescribable hardship. Everything imaginable happened to these vessels: the *Atlantic* reached its destination after
wandering aimlessly in the Mediterranean for three months, its
crew little more than pirates and its passengers threatened by typhoid fever; the *Salvador,* having sailed from the Bulgarian port
of Varna with 350 illegal immigrants, capsized in December 1940
in the Sea of Marmara leaving only seventy survivors; the *Rud-
nichar,* a Bulgarian riverboat, remarkably slipped through the
British blockade, landing all its passengers, 505 of whom were
promptly captured and interned. The Jewish Agency knew that
if the ships reached Palestine, the passengers would probably be
caught, imprisoned, and even deported. The hope was that the
resulting furor might move the mandatory power to reverse its
restrictions on immigration. As a political strategy, this tactic certainly failed: the authorities deducted new arrivals from the White
Paper quotas and remained adamant. Moreover, their knowledge that the Gestapo had encouraged such traffic only heightened their fears of the refugees' constituting a Fifth Column, intending to stir Arab opposition to the British.

In November 1940, as the British were about to deport 1,900
illegal immigrants to the island of Mauritius in the Indian Ocean,
the Jewish underground army, the Haganah, tried to disable the
deportation ship *Patria* with explosives, forcing its passengers'
disembarkation. Unfortunately, the charge was improperly set,
and the ship immediately sank, killing 267 people. This was the
culmination of a series of maritime tragedies provoked by the
refugee crisis. Although the *Patria* survivors were allowed to remain in Palestine, the cabinet determined that future consign-
ments of deportees would still be sent to Mauritius. Despite the
occasional challenge to government policy in Britain and despite
Prime Minister Churchill's personal sympathy with the Zionist
position, the British would not budge. On the contrary, military
men like General Archibald Wavell, the British commander in chief
in the Middle East, challenged the slightest relaxation of controls. Powerless to do more, the Jewish authorities in Palestine
became deeply embittered. During the first two years of the war,

what Arthur Koestler referred to as the "little death ships" rescued 10,672 Jewish refugees in seventeen voyages. But the cost was heavy: "one out of every four of the refugees transported in these seventeen ships," notes one writer, "drowned, died of exposure, violence, or some other hazard."[96] Not long after the *Patria* incident, both legal and illegal immigration practically came to a halt; the Nazis were firmly established in Yugoslavia and Greece, forbade all emigration, and were everywhere rounding up Jews for the Final Solution.

Embarrassed by having to arrest and intern the victims of these horrendous voyages, the British did their best to stem the refugee flow at its source. The Colonial Office was particularly obdurate. At the end of 1939, it suggested that the Turks be pressured into banning this traffic for their merchant marine; a few weeks later it proposed that Britain prevent Joint Distribution Committee aid from reaching a party of a thousand Jewish refugees blocked by ice from sailing down the Danube and stranded in Yugoslavia.[97] In mid-1942 the British finally relented, enabling individual Jewish refugees who reached Turkey or other neutral states to proceed to Palestine under existing quotas. But by this point there was practically no prospect for escape from Europe. Gradually, the mandatory administration released from detainment those Jewish refugees who had been caught. Yet there was no public announcement in favor of receiving refugees, and officials still tried to stop illegal immigration by all possible means. In March 1943 the government instructed its embassy in Turkey to issue Palestine entry certificates to Jewish refugees; this policy, however, remained secret, limiting its effectiveness. Long after the ghastly news of the Nazi mass murders was spread and only a few days after the solemn Allied declaration against the slaughter of European Jews, Colonial Secretary Oliver Stanley reflected the widespread reluctance to receive Jewish refugees:

> There was reason to believe that it was the policy of certain Axis countries, notably Rumania, to extrude Jews from their territories, as an alternative to the policy of extermination. This made it all the more necessary that the policy of His Majesty's Government to accept into Palestine only the limited number of Jewish children with a small number of accompanying women from Eastern Europe should be firmly adhered to.[98]

"Extrusion" remained a threat to the British as a long communication sent to the State Department in early 1943 made clear.

Change was agonizingly slow in coming, therefore, and far too late for countless Jewish refugees.[99]

Toward the end of the war, with the Nazis in retreat, British policy hardened toward refugees attempting to enter Palestine. Remarkably, the Palestine administration issued even fewer immigration certificates than those permitted under the White Paper: some 50,000 up to the end of March 1944, of which 20,000 were charged to illegal immigration. When the Russians broke into Rumania that year and a new wave of refugees to Palestine appeared possible, officials mobilized once again to stop the flow. When the fighting ended in 1945, the White Paper immigration quotas had still not been filled. The British counted over 58,000 Jewish refugees who came to Palestine between 1940 and 1945, a substantial proportion of whom were illegal entrants or "travelers" who remained beyond their allotted time. Their numbers were small indeed compared to the masses of refugees on the move elsewhere or the millions who were murdered. Yet this represented one-quarter of all Jewish refugees who escaped Europe. (One-half of the total managed to go to the United States, and the remaining quarter scattered about more than a dozen countries.[100]

Palestine also provided sanctuary for other refugees from Hitler who were swept into the Middle East by the tides of war. Some 9,000 Greeks ended up there, eventually moved to army camps near Gaza. Many of these had come via Syria, where they were interned in dreadful circumstances until the British ousted the Vichy French. Polish refugees also arrived, some of whom were part of General Anders' army, evacuated from Cyprus. A variety of others passed through the country on their way to Cairo, which became a veritable clearinghouse for East European fugitives.

Italian Sanctuaries for Jews

Italian soldiers and civilians provided a unique case in Nazi-dominated Europe of officially sanctioned refuge for Jews. The pattern first appeared in Croatia and Greece, parts of which came under Italian occupation following the campaigns of 1940 and 1941. According to an agreement between the Axis partners, the Italian army controlled a broad strip of land along the Dalmatian coast of the supposedly independent Croatian state. As early as the summer of 1941, the Italians learned of brutal persecutions of Jews and Serbs carried out by the fascistic government in Zagreb headed by Ante Pavelić. Almost immediately, refugees streamed out of areas controlled by Pavelić and his strong-arm Ustaše movement, heading for the coast. A year later, much the same happened in Greece. Italy had extensive occupation responsibility for that country, excepting only part of Macedonia, which was occupied by Bulgaria, and a German zone, which included other Macedonian territory, a narrow strip of Thrace along the Turkish frontier, some Aegean islands, and the port of Salonika. Here, too, there were refugees. Greeks attempting for a variety of reasons to escape the Germans found refuge in Italian regions. Some 85 percent of the Jews were in the German zone, and the great majority of these, about 56,000, lived in Salonika. When the Germans began roundups of Jews in that city in the summer of 1942, refugees struggled to reach protection in Italian-held territory. In both Croatia and Greece, Italian officers on the spot proved extremely resourceful in obstructing the efforts of their German allies to seize refugees and other Jews for deportation. Furious at the Italians' delay, evasion, and general recalcitrance, the Nazis dared not force the issue.[101]

Later that year the scenario was repeated in France, when Italy occupied a substantial part of Vichy territory east of the Rhône River in November 1942. In this case, the Italians clashed with Vichy almost immediately over the question of the Jews. Italian military officers forbade the enforcement of discriminatory legislation against both French and foreign Jews; more important, they refused to hand Jews over to the French police, then zealously obeying Nazi directives for the Final Solution and dispatching Jews to Auschwitz from across France. Pierre Laval, then head of the Vichy government under Marshal Philippe Pétain, insistently raised the point in Paris with Italian diplomatic authorities and the Gestapo chief Helmut Knochen. Confrontations rippled through the Italian zone, especially when a new series of deportations from France began in February 1943. French authorities attempted to round up foreign Jews for deportation, and the Italians blocked their efforts. Meanwhile, Jewish refugees from all over France tried to reach territory held by Fascist troops. To assist the new arrivals, the Italians requisitioned hotels along the Côte d'Azur, turning them over to Jewish groups to administer as refugee centers. Having 15,000 to 20,000 Jews before the war, the region harbored as many as 50,000 in mid-1943, according to a German report. In towns like Nice, the Italian protection was

conspicuous: carabinieri mounted guard over the local syn-
agogue, and Jews maintained an active, independent community organization. Grateful Jewish residents raised three million
francs for Italian victims of Anglo-American air raids. Elsewhere,
as in the Haute-Savoie, on the Swiss border, Jewish refugee networks organized, in at least one case with aid coming from an
American volunteer refugee service.[102]

A variety of factors account for the Italian sanctuary. On the
German side, quarrels among high-ranking Nazis probably explain their failure to press the issue at an early point: Foreign
Minister Joachim von Ribbentrop, extremely jealous of his pre-
rogatives in relations with Italy, fought off encroachment by subordinates concerned primarily with the Jewish question. Ribbentrop himself delayed raising the matter with the Italians until early
1943, when the Duce's regime was already beginning to totter.[103]
On the Italian side, Nazi persecutions revolted those in charge,
and the protection that the Fascist army gave to the victims
probably reflected widespread Italian opinion. Antisemitism had
certainly not been strong in recent Italian history, and the hatred
of Jews played no role in the growth of Fascism. There were few
Jews in Italy—less than 50,000 in 1938—and these constituted a
highly assimilated minority, with only a small proportion of foreigners. A few Jews had indeed played a role in the early phases
of Fascism, and Jews were fully accepted within the movement:
there were 10,125 Jewish members of the Fascist Party in 1938.
In 1938, however, largely for opportunistic reasons having to do
with the German alliance, Mussolini turned against the Jews.[104]
In July, Mussolini sponsored a racial manifesto embracing anti-
Jewish doctrine, and antisemitic legislation followed that autumn. Despite these moves, however, opposition to Jews failed
to strike deep roots in Italy, and most of the population remained impervious to these ideas. Far more powerful, in all
probability, particularly as the war progressed, was the Italian
dislike of the Germans and, in particular, of the latter's insuffer-
able claims of domination and superiority. Italian national self-
interest and the ordinary decency of Italians simply flew in the
face of Nazi-inspired persecutions of Jews.

As news of the Italian sanctuary spread, the Germans realized
that their Fascist ally was proving as unreliable in this respect as
in political and military matters. "The Italians are extremely lax
in their treatment of Jews," Goebbels wrote in his diary in De-

cember 1942. "This shows once again that Fascism does not really dare to get down to fundamentals, but is very superficial regarding most important problems."[105] A few weeks later, acting on the orders of RSHA boss Ernst Kaltenbrunner, Adolf Eichmann raised the matter of Italian protection of Greek Jews with the Wilhelmstrasse. Ribbentrop probably discussed the issue personally with the Duce at the end of February 1943—certainly to no avail. Although the Italians promised to take stern measures, it soon appeared that they had no intention of doing so. On the contrary, the Italians attempted to extend their protection to cover some Jews living in German-held Salonika and showed no sign of cooperation with the Germans in France or Croatia.

Unfortunately, the protective shield shattered everywhere in September 1943, when the Italians precipitately surrendered to the Allies and the Germans moved into Italian occupation zones. In anticipation of a Fascist collapse, Italian officers in Croatia concentrated a large number of refugees on the island of Arbe, off the Dalmatian coast. Between 15,000 and 20,000 were interned there, mainly Croats and Slovenes, and over 2,500 Jews. Before the Nazis arrived, many of the inmates managed to join Tito's partisans and thus escaped capture and deportation. Thousands of other refugees managed to reach Italy by a variety of routes. Jews in Greece and France were not so lucky. In the latter case, a prominent Italian Jew, Angelo Donati, planned a massive evacuation of Jewish refugees before the German occupation of Italian territory. Donati, a former director of the Banque France-Italie, secured four Italian ships that could convey between 20,000 and 30,000 Jewish refugees to North Africa. During August he conducted negotiations at the Vatican with British and American representatives and pursued separate discussions with the Italian government of Marshal Badoglio. According to Donati, the Italians agreed to admit those who would first be evacuated to North Africa and even began printing a special passport for their use. The sudden announcement of the armistice with Italy on 8 September intervened before preparations were completed, however, and the scheme collapsed. The Germans immediately entered the Italian zone, seized the Italian railway network, and rounded up all the Jews they could find. More than a thousand managed to escape to Italy at the last moment, but many were caught, placed in camps, and almost immediately deported to Auschwitz.[106]

RESCUE EFFORTS

During the last days of 1942, the Allied governments finally issued an official condemnation of Nazi genocide against the Jews.
Since the preceding summer, incontrovertible evidence of mass
murder inundated British and American officials; from various
quarters at once pressure mounted for a public pronouncement
on the matter. On 17 December such a statement was formally
released, made on behalf of Britain, the United States, the Soviet
Union, the French National Committee, and the governments in
exile of Belgium, Holland, Luxembourg, Norway, Czechoslovakia, Greece, Yugoslavia, and Poland. Immediately, the declaration was released to Nazi-occupied Europe in twenty-three languages. Numerous reports indicated, the statement began, that
the Nazis "are now carrying into effect Hitler's oft-repeated intention to exterminate the Jewish people in Europe." After a
graphic description of the slaughter and an estimate that the victims numbered "many hundreds of thousands of entirely innocent men, women and children," the declaration solemnly called
for the punishment of the perpetrators of these crimes. Along with
retribution, however, many who were moved by these words
demanded some sort of rescue. Something, they urged, had to
be done for the Jews who still survived. In practical terms, this
sentiment focused on the possibility of evacuating endangered
Jews to some place of refuge, perhaps to a neutral country like
Sweden.[107]

Appeals for some sort of rescue reached Allied capitals at the
very moment when the Nazis were beginning to lose the military initiative in the field. If there was a turning point in the European conflict, this was undoubtedly it: in October 1942 the battle of El Alamein in the Egyptian desert sent a great shock through
the Wehrmacht; the tremendous battle of Stalingrad, which ended
on 2 February 1943, shattered the entire German Sixth Army of
a half million men and kept the oil-rich Caucasus out of Nazi
hands. The following summer the British and Americans landed
on the coast of Sicily, soon to begin the fight up the Italian boot.
Meanwhile, the Soviets sent thousands of tanks against the Germans at Kursk-Orel, beginning a westward drive that never
stopped. By the end of the year the Russians reached the old
frontiers of Poland. These military moves provided the backdrop
before which rescue advocates issued their call for action. Allied
victories not only cleared territory, occasionally liberating civilian

refugees, but they also prompted neutrals and even some co-belligerents with the Reich to cooperate with evacuation schemes.

Despite the new opportunities, however, the results were meager. The Bermuda Conference on the refugee problem held in April 1943 in effect postponed serious efforts; the spectacular Danish rescue of several thousand Jews in October was an exceptional instance of rescue, made possible by unique circumstances of time and place. Only toward the end of the war did the Allies shift their priorities from the fighting itself, and only in early 1944 did serious refugee assistance begin, with the establishment of the American War Refugee Board.

New Prospects in 1943

In the months following the Allied declaration of December 1942, the plight of European Jews and the possibilities of rescue tugged at British and American opinion. Major Jewish organizations in the United States formed a Joint Emergency Committee on European Jewish Affairs to press for action. Nipping at the heels of this body and drawing on many unaffiliated Jews was a much more radical group, eventually known as the Emergency Committee to Save the Jewish People of Europe. Led by Peter Bergson, then a Zionist Revisionist activist and organizer of its underground army, the Irgun, the Emergency Committee helped keep the pot boiling with provocative demonstrations, full-page newspaper advertisements, and intensive congressional lobbying. On 1 March 1943 a huge rally in Madison Square Garden in New York announced a twelve-point program for rescue, calling for negotiations with Germany and her satellites for this purpose and resettlement of the refugees. From Geneva, the World Jewish Congress and the World Council of Churches issued a similar appeal, asking that neutral states grant temporary asylum to Jewish refugees and that Britain and the United States guarantee their reemigration after the war. In Britain, meanwhile, public opinion may have been even more aroused by news of the Holocaust than in the United States. In striking contrast to the phlegmatic commentary in the Foreign Office, angry calls for action sounded in Parliament, the newspapers, and church pulpits. Sir Neill Malcolm, former League of Nations high commissioner for refugees, added his voice to the chorus demanding the admission of more refugees to Britain and the colonies; Eleanor

Rathbone, an Independent M.P., bombarded Whitehall with a twelve-point program of her own, including a wide variety of concrete measures to facilitate the arrival of Jewish refugees.[108]

But caution quickly triumphed over imagination. In diplomatic correspondence with Washington in early 1943, the British responded to the mounting pressure at home by suggesting a private Allied meeting on the refugee question. London explained how complicated it felt the problem of refugees to be. There was a danger of "raising false hopes." Notably, the British worried that the Nazis or their satellites "may change over from the policy of extermination to one of extrusion, and aim as they did before the war at embarrassing other countries by flooding them with alien immigrants." Something, nevertheless, had to be done to assuage the aroused public feelings and to secure additional havens for refugees. Officials in Washington agreed with the British that it would be impolitic to put too much emphasis on Jews: "The refugee problem," they said, "should not be considered as being confined to persons of any particular race or faith. Nazi measures against minorities have caused the flight of persons of various races and faiths, as well as other persons because of their political beliefs." They concurred that a meeting would be useful. But on the concrete proposal they delayed.

When the British finally threatened to move on their own, the State Department launched what eventually became the Bermuda Conference of April 1943. The Americans first proposed Ottawa as a site for the meeting; but the Canadians, fearful that their own extremely restrictive refugee policy would become the object of critical examination, refused to go along.[109] Concerned in his own right that such a meeting might become a sounding board for appeals to open American gates, a State Department official, Breckinridge Long, a staunch opponent of accepting Jewish refugees, told Whitehall that the meeting was not to propose drastic solutions; the principal goal, it became plain, was to relieve the pressure of prorescue opinion. The Americans knew London to be extremely sensitive to the public relations issue and eager to appear as a champion of the Jews in Europe. In reality, the British were extremely sensitive to their declining influence on Washington and their increasing dependence on the United States. They feared being drawn in the wake of the Americans and exuded caution over the prospects of the conference. Foreign Secretary Anthony Eden was extremely nervous about having too many Jews thrust on the Allies, was defensive about

Jewish emigration to Palestine, and concerned about the availability of shipping space. Meeting with Roosevelt, Secretary of State Cordell Hull, and Under Secretary Sumner Welles at the end of March, Eden helped set the tone of the conference. In particular, he poured cold water on the suggestion that the Allies pursue the possibility of evacuating threatened Jewish refugees from southeastern Europe. Although London and Washington agreed to go forward with the Bermuda meeting, therefore, all parties approached it with their guard up, eager to see the gathering transpire without surprises and without any alteration of existing policy.

The Bermuda Conference largely achieved this aim. The Americans put no pressure on Britain over the White Paper quotas for Palestine; the British did not ask the United States to raise its own immigration ceilings. Jewish representatives and inquisitive reporters were kept at bay, and discussions were held in private. No report of the proceedings was ever published. The conference rejected suggestions for negotiations with the Germans or the dispatch of food and assistance to the Jews of Europe. The delegates determined flatly that "no shipping from the United Nations sources could be made available for the transport of refugees from Europe." Although fulsome in praise of what the British and Americans had already done for refugees, the conference report indicated no new prospects for resettlement. The conference did recommend the revival of the dormant Inter-Governmental Committee on Refugees (IGCR) with urgent new tasks. The IGCR ("this decorous and entirely ineffectual instrument" as Ira Hirschmann called it) was to try to obtain neutral ships to move refugees and to secure new places of refuge. The British and Americans were to negotiate the release of refugees held in Spain and consider their removal to North Africa. In an important declaration intended to encourage governments of neutral countries to accept more refugees, the participants announced that they would receive all their nationals at the end of the war and create conditions "as will enable all [refugees], of whatever nationality, to return to their homes."[110]

On both sides of the Atlantic, refugee advocates realized how meager were these results. Congressman Emmanuel Celler of New York denounced the meeting as "a diplomatic mockery of compassionate sentiments and a betrayal of human interests and ideals"; Rabbi Stephen Wise, head of the American Jewish Congress, pressed for a meeting with Roosevelt about the "inexplicable

absence of active measures to save those who can still be saved."
In London, overwhelmed by the news of the Warsaw Ghetto uprising and the failure of the conference, the Bundist representative on the Polish National Council, Shmuel Zygielboim, commited suicide, protesting the failure of the Allies to act. Eleanore
Rathbone, leader of the National Committee for Rescue from Nazi
Terror, told the House that the results were "pitiably little."[111]

The conference failed to accomplish much even in areas where
there appeared to be constructive agreement. On the matter of
encouraging neutrals' acceptance of refugees, for example, London and Washington each wanted an exception made. At the
suggestion of the American ambassador in Madrid, Carlton Hayes,
the State Department argued that pressure on Spain would be
counterproductive; similarly, the British urged their own exception, Turkey, fearing that Jews arriving there would head for
Palestine. Trouble also arose over the establishment of a temporary refuge in North Africa. On this issue, American fears of adverse reactions in the Moslem world to the concentration of Jewish refugees seem to have been decisive. Thus, American military
authorities objected to the establishment of even a small camp in
North Africa to drain off the accumulation of refugees in Spain.
Churchill forced the issue later that summer, and Roosevelt did
promise a camp for a few thousand refugees from Spain. More
delay followed, however, because of French objections. The results, in the end, were derisory—a few hundred refugees reached
North Africa by the end of June 1944, at which point the liberation of France was well under way.[112] Finally, the Inter-
Governmental Committee stirred into life in the summer of 1943
with Sir Herbert Emerson as director. Its staff and budget were
enlarged, and its membership broadened to include even the Soviet Union. But the IGCR was essentially a political organization
without a political mandate; it had no relief machinery of its own
and no real power to negotiate with neutral or enemy states. The
committee channeled most of its relief funds through the Joint
Distribution Committee and provided a secret conduit for sending relief assistance into Rumania. In every important respect, the
Allies kept the IGCR on a tight leash. As historian Bernard Wasserstein notes, "the Committee failed to acquire sufficient independent authority to play any significant role in the succor of
refugees from Nazi Europe; it remained a bureaucratic monument to the spirit of futility at Evian, where it had been born."[113]

For Peter Bergson and the Emergency Committee in the United

States the dramatic evacuation of almost 7,500 Jewish refugees from Denmark in October 1943 pointed to other possible rescue activities. The committee published a full-page advertisement in *The New York Times* with this message, headlined "It Can Be Done."[114] In reality, however, the Danish rescue was an extraordinary achievement, conceivable only in the particular circumstances of Danish geography and conditions of occupation.

Unlike the other defeated states of Western and Eastern Europe, the Danes were largely left alone by Hitler after their prompt surrender in the spring of 1940. In exchange for economic cooperation with the Reich, Denmark kept her king, her government, and even an army, maintaining a nominal degree of independence in internal affairs. For three and a half years, eight thousand or so local Jews lived in relative tranquillity. The Nazis remained unhappy about this, but feared provoking the democratically minded Danes over the issue. Stretched thinly across Europe and disinclined to launch an expensive policy of repression, the Germans kept their hands off Danish affairs. All this changed in the late summer of 1943, when the Nazis, under heavy air bombardment and suffering grave reverses in Russia, increased their economic exactions. Following strikes, riots, and some sabotage, the Germans proclaimed a state of emergency, sweeping aside the Danish government and proclaiming martial law. Nazi persecution of Jews and political dissidents came together with a sudden clamping down on Danish society. The Danes, who had formerly constituted the Germans' "model protectorate," now fought them on every front—including a proposed deportation of the country's Jews for the Final Solution.

In the famous rescue of Danish Jews, geography was paramount: the evacuees were overwhelmingly concentrated in Copenhagen, directly across a narrow stretch of water from the Swedish sanctuary. Timing was also critical. In the second half of 1943, persecutions of Jews were linked to the crushing of Danish independence because the two emerged at precisely the same moment. The autumn of 1943, moreover, was an obvious time for action. Danish resistance, like resistance everywhere in Europe, drew strength from the news of spectacular German misfortunes of the previous summer—the massive Allied bombing over the heartland of the Reich, the invasion of Sicily and the collapse of Mussolini, and the great tank battles in Russia.

Even with these circumstances, large-scale rescue was only possible because of the proximity of Sweden and the Swedish willingness to receive refugees. On more than one occasion, Stockholm had used its neutrality on behalf of victims of Nazi terror. Representing Dutch diplomatic interests in Berlin, the Swedish ambassador to Germany repeatedly inquired about the fate of Jews deported from the Netherlands. Swedish diplomats had intervened on behalf of Norwegian Jews on December 1942, declaring to the Germans their willingness to receive those about to be deported. Although the Nazis refused, the Swedes persisted. In addition, the Swedes knew that they did not act alone. As soon as he heard of the impending deportations, the Danish ambassador to Washington, Henrik Kauffmann, wrote to Secretary of State Cordell Hull, requesting help for Danish Jews and others persecuted by the Nazis; he also immediately cabled Stockholm, notifying the Swedes that Denmark would reimburse Sweden for all rescue costs. The State Department similarly encouraged Stockholm to receive the refugees. The Swedish government, therefore, did not hesitate in October 1943. As soon as the Swedes learned of the impending deportation, their representative in Berlin offered to accept into Sweden all the Jews in Denmark. When the Germans failed to reply, the government released its offer to the press. According to one historian, the Swedish ambassador warned the Nazi Foreign Office "of the indignation and the serious anti-German repercussions which a persecution of Jews in Denmark would provoke in Sweden."[115] Swedish officials encouraged the Danes involved in the rescue and cooperated in the logistical effort that brought the evacuees to Malmö. In Sweden, the Danes were permitted to organize relief committees up and down the coast. At the end of 1943, British representatives in Stockholm reported 11,600 Jewish refugees in Sweden, among them practically the entire Danish Jewish community.[116]

The months following the Bermuda Conference saw a clear shift away from the exclusionist forces of the State Department, led by Assistant Secretary Breckinridge Long, in favor of the advocates of rescue and assistance for refugees. Agitation on behalf of rescue, particularly with a focus on Palestine, increased within American Jewry. "You can't imagine the flood of correspondence that has poured in from all over the country," Long complained to Carlton Hayes in May 1943.[117] Protest mounted with the news of German reverses; and refugee advocates in the United States, notably the Bergson group, became more bold. At the same time, the various governments in exile, sensing now that the day of retribution was drawing near, pressed for declarations on war crimes and warnings about harm done to civilians. Among those moved by the accumulating evidence of the Final Solution was Treasury Secretary Henry Morgenthau, Jr., an assimilated Jew, whose father had called attention to the slaughter of Armenians in 1915, when he was American ambassador to Turkey. At the end of 1943, Morgenthau prepared a confidential report to Roosevelt, calling for rescue of Jews threatened by mass murder. Originally entitled "Report to the Secretary on the Acquiescence of this Government in the Murder of the Jews," the document contained a bitter attack on the dilatory State Department refugee policy. Aware of the depth of such feeling, which now extended significantly beyond the Jewish community, Roosevelt was ready to do something concrete. Following their meeting in mid-January 1944, the president authorized the Secretary of the Treasury to create the War Refugee Board (WRB) to pursue evacuation possibilities and provide relief and assistance. Operating unofficially under Morgenthau's department, conveniently outside the State Department, the WRB was given broad authority to plan rescue operations, evacuate and care for refugees, and negotiate with foreign governments. Roosevelt put John Pehle, thirty-four years old and an energetic Treasury Department official, in charge of the board. Ira Hirschmann, its first full representative, fresh from an important post at Bloomingdale's in New York, immediately sensed that this was an instrument with authority: "Here at last was not a gesture, but a powerful weapon forged by the people of the United States." Underlining the new priority, the director of the WRB was to report to the president "at frequent intervals concerning the steps taken for the rescue and relief of war refugees." Financed mainly through private agencies, the WRB, nevertheless, received a million dollars immediately from Roosevelt's emergency fund and considerably more within a few months.[118]

Rescue by Negotiations

Those who called for the evacuation of Jewish refugees in the last year or so of the war frequently had the most unrealistic perceptions of how many Jews remained to be saved. To some degree their efforts were predicated on the notion of huge masses of victims who could still be rescued. Despite extensive knowledge about the Final Solution, well-informed observers believed that there would be several million Jewish refugees once the Nazis were defeated. A 1943 publication sponsored by the American Jewish Committee seems to have assumed that Jewish deportees were somehow still alive, likely to reemerge after the war as displaced persons in Europe. The Joint Distribution Committee's *Digest* of October 1943, supposedly informed by the most accurate intelligence available, claimed that 1.25 million Jews had managed to escape the Nazis, half of whom fled into the Soviet Union; flight continued, the journal noted optimistically, venturing that at the end of the war there could be two or three million uprooted European Jews. Six months later Chaim Weizmann estimated that two million Jews might still survive when the fighting ceased. All these guesses turned out to be much too high: Russia received some 400,000 Jews, and nearly half of these perished. The machinery of destruction, we now know, murdered close to three million Jews by the beginning of 1943; the wheels continued to spin, eventually taking the lives of over five million people. The total number of Jewish survivors in Europe, according to Yehuda Bauer, was about a million.[119]

Understandably, rescue advocates clutched at straws extended from the enemy camp in the last year or so of the war, suggesting that thousands of Jews could be released from the Nazis' clutches through negotiation. There may, indeed, have been more substance in this possibility than in overly optimistic assumptions about postwar refugees. Certainly, a good deal of time and energy was poured into a handful of projects. By 1944 the WRB vigorously pursued several schemes, eventually provoking a split between the British and the Americans over the issue of refugees. Unfortunately, none of the projects bore fruit.

The first proposal that reached the West came from Rumania, the Nazis' least enthusiastic ally, a country that sent peace signals to the American State Department as early as beginning of 1943. In February of that year, *The New York Times* claimed that Bucharest was prepared to release between 60,000 and 70,000 Jews from camps in Rumanian-occupied territory at a cost of 20,000 lei (about one hundred dollars) a head. Gerhard Riegner of the World Jewish Congress in Geneva worked out a scheme whereby funds from Jewish organizations in the United States could be paid into a Swiss account that would be blocked until after the war. Rumanian-Jewish business in Rumania would raise money on the strength of this hard currency and would finance the Jewish emigration. The refugees would be dispatched to Palestine on two Rumanian ships flying the flag of the Vatican. Not long after, Jewish representatives fastened on a similar possibility of ransoming Jews from Bulgaria, another reluctant partner of the Axis. In both cases, the Jews who backed these ventures, notably the World Jewish Congress, were disappointed. State Department officials heard them out, and President Roosevelt apparently approved the idea, but delays intervened. The Americans were extremely cautious about the idea of transferring funds into the enemy camp. And the British were even more hesitant. Not only did they share the Americans' misgivings about payment, but they were also reluctant to encourage a process that would generate large numbers of Jewish refugees, thereby putting pressure on British immigration restrictions for Palestine. All concerned in London and Washington seem to have been happy to play the Jews along, perhaps expecting that the proposals would die a natural death. The more that final approval was postponed, the more the Rumanians and the Bulgarians had second thoughts. The Turkish government, which was first to receive the refugees, also expressed reservations. From the Allied standpoint the Bermuda Conference soon absorbed all their attention, and the venture ran into the sand.[120]

During the first half of 1944, with its days as an Axis partner numbered, Rumania appeared once again as a field for rescue strategists. The Russians had taken Kiev the previous November and reached the Dnieper in the early spring. Before long they would be at the gates of Bessarabia. Eager to extricate himself from the Nazis' hold, Antonescu was ready to listen to the Allies. American pressure focused on the thousands of Jews and anti-fascists who had been interned under appalling conditions in Transnistria, the huge province that the Rumanians had torn out of the Ukraine between the rivers Dniester and Bug. American envoys approached the Rumanians indirectly through Spain and Sweden. In Ankara, representing the War Refugee Board, Ira Hirschmann even conferred directly with the Rumanian ambassador. The American demand was to break up the camps in the region, permit victims to return to their homes, and allow several thousand refugees to leave the country. On the first point the Americans claimed success: the WRB announced in April 1944 that it had succeeded in removing more than 40,000 Jews from their internment into comparative safety. Emigration was another matter, however, requiring much more than the intimidation of the Rumanians. The British continued to drag their feet, apprehensive as always about their position in Palestine. Whether because of this opposition, technical mistakes, German objections, or Allied hesitation, a scheme for evacuating Jews from Constanta also collapsed. Under constant pressure from the Allies, the Rumanians did admit about 4,000 refugees from Nazi-occupied Hungary. Few of those endangered managed to leave the country, however, despite the apparent willingness in Bucharest to let them go.[121]

While negotiations over Rumania continued, officials in London and Washington also looked to Hungary, a much more difficult but potentially productive negotiating partner. Under Nazi occupation since March 1944, the Hungarians faced the prospect of seeing their country turned into a battleground. Air bombardment by the Allies foreshadowed massive destruction and loss of life. Meanwhile, the Red Army advanced from the east. Although subjected to persecution and discrimination, Hungarian Jewry had largely survived intact until the Germans invaded the country. Then, at the Nazis' orders, the Hungarians began a massive deportation of the country's 750,000 Jews. Hungarian gendarmes despatched victims at a rate of about 12,000 a day through the late spring and into the summer. In response, protest poured down on Regent Admiral Horthy. WRB agents in various neutral capitals accused Hungarian representatives of war crimes and urged a halt. Finally, on 6 July, after dispatching some 437,000 Hungarian Jews to Auschwitz, the regent stopped shipping Jews to Poland. On 15 October, however, Horthy was swept away, largely at the Nazis' instigation. In his place and under the protection of the SS emerged Ferenc Szálasi, head of the Hungarian Fascists, the Arrow Cross.

These events were accompanied by a series of quite spectacular overtures to the West about the possibility of rescuing many thousands of Jews. In the middle of May, while the Hungarian deportations were decimating the last surviving Jewish concentration in Nazi Europe, the British and Americans learned of an astounding offer, carried to Istanbul by two unlikely couriers—a member of the Budapest Jewish Refugee Committee named Joel Brand, and Bandi Grosz, a shady character suspected of having been involved in German counterintelligence. "As an alternative to the complete annihilation of all Jews remaining in Hungary, Rumania, Czechoslovakia, and Poland," London was told, "the Nazis are ready to evacuate 1,000,000 Jews from these countries to Spain and Portugal (though not, as they specifically stated, to Palestine)." In exchange, the Allies were to deliver to the Germans 10,000 trucks and various supplies. As a gesture of "good faith," the Germans were prepared, once the offer had been agreed on, to release between 5,000 and 10,000 Jews.[122] Several weeks later, immediately after the deportations from Hungary were suspended, Horthy himself advanced another suggestion: the Hungarians would release certain categories of Jews who had visas to other countries. The Germans, possibly linking this demarche with their own, had apparently agreed to permit the refugees to leave Nazi-held territory.

These messages carried enough verisimilitude to be taken seriously. British and American diplomatic channels buzzed with appraisals of these offers and suggestions as to how to reply. The Soviets, as might be expected, were completely unresponsive and objected to any contacts with the enemy. The British, for different reasons, also responded negatively. Churchill, through sympathetic to the Jews, would not countenance any deal whatever with Hitler. In addition, his government feared that the Nazis were using these efforts to split the Russians from the Western allies by leaving open the possibility that equipment sent to Germany would be turned against the Red Army. Beyond this, members of the war cabinet were clearly embarrassed at the prospect of large masses of Jewish refugees suddenly appearing. Where would all the Jews go? Shipping so many people could seriously hinder the war effort, and still the problem of resettlement remained. Sooner or later the worry about refugees fastened on Palestine—especially if large numbers were to be evacuated to Turkey. Officials in Washington shared many of these apprehensions, but were more prepared to string the Germans along, hoping to gain time. On hearing of the Brand mission, the War Refugee Board urged that discussions continue and pursued this line when Horthy's own offer reached Washington in July. Nothing ever came of these proposals in the end. An indirect result, however, was a rift between the British and Americans over refugee policy. Throughout the discussions over Hungarian Jewry, Whitehall fumed with resentment over the War Refugee Board, considered to be behind the United States' undue receptivity to rescue proposals. London believed the WRB's concerns were narrowly partisan and that the Americans were inordinately prone to such preoccupations. Washington, on the other hand, judged the British to be too inflexible, to eager to take public credit whenever something appeared to be achieved, and insufficiently disposed to rescue.

Too little is known about the German motivations for these proposals to assess whether the Allied differences of opinion, delay, or rejection of various offers made any real difference. It seems likely that the American willingness to talk led to other discussions over the possible release of Jews, notably to Switzerland and Sweden. In the last months and weeks of the Third Reich, some Nazi leaders were eager to establish contacts with the West and possibly wished to use approaches over the rescue of Jews as a wedge with which to open conversations for a negotiated peace. At the very least, some Nazi leaders may simply have been driven by fear and greed—believing that they could ransom the Jewish remnant for personal gain much as the Nazis of the late 1930s believed they could sell the release of German Jews. Whatever the Nazis' intentions, the outcome was not the sudden appearance of masses of refugees. The British continued to fear this prospect and did what they could to prevent it. As the war drew to its conclusion and as extremist Jewish groups renewed terrorist attacks on British targets, Palestine absorbed more of their attention and conditioned their views about Jewish refugees. In November, relations between the Jews and the British worsened drastically with the assassination in Cairo of the minister resident in the Middle East, Lord Moyne, killed by two members of the ultraright Stern Gang.

The Americans, meanwhile, mixed serious refugee and relief activity with tortuous negotiations, at one point even making direct contact with the Germans through Roswell McClelland, the WRB man in Switzerland and attaché to the United States embassy in Bern. On the frayed edges of the Nazi empire, the WRB did its work, picking loose some of Hitler's victims at the very last moment. Its agents leased boats in Sweden to carry refugees from the Balkans, helped smuggle fugitives from Hungary to Slovakia, and assisted the Red Cross and the World Jewish Congress in dispatching parcels to Bergen-Belsen and other camps. Working closely with the Joint Distribution Committee, the WRB conducted sensitive negotiations with neutral countries in an attempt to secure the passage of refugees from endangered areas. As we have seen, the War Refugee Board helped initiate Raoul Wallenberg's rescue mission to Hungary in cooperation with the Swedes. Through the efforts of Ira Hirschmann, the WRB succeeded in clearing a path for individual refugees through Turkey—no mean feat, given the Turks' obsessive fears of being sucked into the conflict at the last moment. By late 1944, Hirschmann saw new refugees crossing by ship and rail from Rumania and Bulgaria, now under the Russians. Fugitives arrived "bearing heart-rendering reports of suffering and privation."[123] Hauntingly reminiscent of their situation five years before, the first postwar refugees faced an old problem: Where could they go next?

5

The Postwar Era

The last year of the Second World War produced millions of new refugees in Europe because the Wehrmacht fought its losing battles so tenaciously, heaping an extra measure of chaos, destruction, and dislocation on an agonized continent. After the opening of a second front in 1944, the Allies needed a whole year to bring Nazism down—in stark contrast to the First World War, when the Second Reich collapsed like a house of cards shortly after Allied breakthroughs on the western front. Moving against Hitler from the East, the Russians began their advance from deep inside Soviet territory in the summer of 1943. They took almost twelve months to conquer all their pre-1914 territory, including Poland and the Baltic states. Thereafter they went south to the Danube Valley and north to Finland before renewing the drive on Germany in January 1945. Following the Allied landings in Normandy in June 1944, the Germans managed to stop the Anglo-Americans in September and nearly broke through their lines in the Battle of the Bulge three months later. The Allies crossed the Rhine in March 1945, but the Reich hung on until its final collapse in May.

This protracted fighting in 1944 and 1945 was by far the most destructive part of the war in Europe. Along with the colossal damage wrought by men and machines on the ground, the air war intensified dramatically. Preparing for their final assault, the British and Americans sent waves of bombers guarded by new long-range fighters over Germany in early 1944; within months German cities and strategic targets were being pounded around the clock. At the beginning of 1945, the Germans had lost more than 50,000 aircraft and were practically powerless before Allied bombardment. By the end of the war, hundreds of thousands of Germans had been killed, and millions were homeless, evacuated, and scattered across Central Europe.[1] Adding to the devastation, Hitler set out to destroy what remained of German society, continuing the conflict so long as possible as a kind of punishment for the German people who had not lived up to his expectations. In March 1945 he ordered a massive evacuation of civilians behind German lines and the wanton destruction of everything in the path of the enemy. "If the war is lost," he told Albert Speer, "the people will be lost also. It is not necessary to worry about what the German people will need for elemental survial. On the contrary, it is best for us to destroy even these things. For the nation has proved to be the weaker, and the future belongs solely to the stronger Eastern nation."[2]

The prolonged agony of Nazism produced its own deluge of refugees, adding to the forced displacement of the previous four years of war. Even following their liberation, European ports remained in ruins and almost entirely committed to supplying the war. Europe choked with refugees, but few ships were permitted to take them off the Continent until considerably after hostilities ceased. In Eastern Europe the Soviets, bled white in their struggle with Nazism, pushed westward with a vengeful, destructive fury that uprooted hundreds of thousands and eventually millions. Soviet suspicions and the imposition of exclusive Russian military and political control closed the region to the Western Allies, slowing recovery and preventing an early resettlement of refugees. In the West, the great task was to absorb millions of new fugitives from Eastern Europe, whose flight was triggered by the collapse of Lebensraum, the depradations of the Red Army, and the expulsion of ethnic Germans.

Europe had never seen so many refugees. According to one student of this problem, the total number of displaced Europeans during the entire course of Hitler's war was thirty million or more.[3] When the Allies crossed Rhine in 1945, they believed there were 350,000 displaced persons in their zones in Western Europe; shortly after the Nazis' surrendered, however, this total increased twentyfold. The Soviets similarly reported millions of refugees for a total of close to fourteen million in the autumn of 1945—nearly half of all those uprooted during the war.[4] In the months to come, great masses of displaced persons headed for what remained of their homes. Before the winter of 1945, most

of the eleven million Europeans requiring repatriation had returned to their countries—a phenomenal performance given the
collapse of normal communications. Moreover, against all expectations drawn from historical experience, medical technology
prevented the outbreak of epidemic diseases and the devastating
mortality that had so often accompanied such refugee floods in
the past. More survivors, of course, meant more refugees; and
with them science and technology dealt less successfully. Political ills were also less easily cured, and political conflict continued to drive Europeans from their homes even as millions of
others were being resettled. International efforts faltered, partly
through lack of experience and partly because the other political
and economic problems Europeans faced were much more absorbing than the fate of a diminishing number of refugees. Many
returnees were uprooted for a second time, and in some cases
former fugitives ousted their onetime persecutors. Other refugees remained in camps for years before a way was found to
provide them with homes. In the end, the job was done although not without a full measure of injustice to the victims of
war and its aftermath.

THE EUROPEAN PICTURE, 1945–47

Burrowed in the bureaucracies of the United States and Britain,
a handful of officials pondered refugee movements during the
war, attemping to foresee the immediate postwar situation. Before the Evian Conference in 1938, the Roosevelt administration
sponsored a President's Advisory Committee on Political Refugees to research the problem, and this body remained in place
throughout the conflict. In England, Churchill established a
Committee on Surpluses in late 1940 under the chief British economic adviser, Sir Frederick Leith-Ross, one of whose tasks was
to accumulate relief supplies for an eventual liberation of Europe. This committee yielded its work in 1941 to an Inter-Allied
Committee on Post-War Requirements, based in London, which
made projections of the anticipated numbers of refugees. At the
same time, independent investigation by Eugene Kulischer, Joseph Schechtman, and others addressed the same issue. All concerned spoke of staggering numbers of postwar refugees. Already in 1939, Roosevelt suggested that between ten million and
twenty million Europeans might have to leave Europe when the

fighting ceased. The Inter-Allied Committee reported in mid-1941 that there were twenty-one million displaced persons in Europe, of whom eight million had been brought to Germany as forced laborers and another eight million displaced in their own countries. Finally, on the eve of the D-Day landings in France, Allied military planners estimated that they would encounter over eleven million refugees and displaced persons in Central and Western Europe, not counting uprooted German civilians. By that time, detailed preparations were under way for dealing with a refugee crisis of unprecedented magnitude.[5]

The Allies and the Refugees, 1944–45

Throughout the last year of the war, all the liberating armies in Europe found masses of refugees under foot, whom few tried to differentiate. Once the Germans surrendered in May 1945, however, patterns began to emerge. At the end of September, the Western Allies cared for nearly seven million displaced persons; the Soviets claimed they took charge of an equal number. The largest group, in both cases, were Soviet citizens, over 7.2 million forced laborers and prisoners of war who had survived the ordeal of wartime Germany. Next came the French, with nearly two million, including 765,000 civilian workers; more than 1.6 million Poles, 700,000 Italians, 350,000 Czechs, over 300,000 Dutch, 300,000 Belgians, and many others. Among these was every possible kind of individual—Nazi collaborators and resistance sympathizers, hardened criminals and teenage innocents, entire family groups, clusters of political dissidents, shell-shocked wanderers, ex-Storm Troopers on the run, Communists, concentration camp guards, farm laborers, citizens of destroyed countries, and gangs of marauders. Every European nationality was present in both East and West. Some of the refugees remained shattered and bewildered by their experience, rooted to the soil where they were liberated. Others streamed in various directions, often without the slightest indication of what they would find at their destination. Endless processions of people trudged across the ruined Reich, sometimes with pathetic bundles of belongings, sometimes pushing handcarts, the ubiquitous refugees' conveyance, piled with household belongings. One refugee moving eastward through German territory cleared by the Soviets pictured the roads in 1944 and 1945 "like swollen mountain torrents in the spring, a Babel of people and languages, all former slaves of the Third Reich."[6] Awestruck by the spectacle, Western observers feared a complete breakdown of public order in Central Europe, which was already facing acute shortages, the collapse of established authority, and severe privation.

The soldiers brought a bureaucrat's zeal for classification to this storm-tossed mass of humanity. Allied military manuals first suggested a crude division between "refugees"—civilians in their own countries uprooted by war and eager to return home—and "displaced persons," commonly referred to as DPs—people outside their countries, many of whom had been deported by the Nazis and who required repatriation. This distinction had the advantage of separating civilians who could soon come under local jurisdictions from those who were clearly outsiders and who, it was hoped, would return as soon as possible. But it offered no clues as to the different needs and aspirations of the fugitives themselves. The Americans also referred to "kriegies"—U.S. slang for *Kriegsgefangenen* or prisoners of war—who had to be divided from political prisoners, evacuees, civilian expellees or escapees from the East, and the ghoulish-appearing slave laborers or concentration camp inmates. As time passed, categories evolved, and so did the survivors' ingenuity in assigning themselves, when possible, to the most advantageous classification.

Chaos and suffering lasted longest in Eastern Europe. Nowhere in Europe had wartime losses been greater. The Polish Office of War Damage later estimated over six million deaths on Polish soil—most due to the extermination of Jews and other civilians. The total cost of German looting was reckoned to be 20 to 25 billion dollars. The destruction of famous cities like Warsaw was awesome. In the middle of 1947, rubble still carpeted 80 percent of the Polish capital, and about 30,000 corpses still lay beneath the brick and concrete of what had once been the Warsaw ghetto. But liberation was as devastating as the Nazi occupation. Moving west, the Russians sometimes abandoned all restraint, destroying everything that remained in their path; their troops robbed, raped, burned, beat, and murdered their way through conquered territory. And after the soldiers were finished, reparation squads set about dismantling everything of economic value for transportation to the Soviet Union. Reconstruction, therefore, was painfully slow.[7]

Millions of refugees moved through the wreckage of Eastern Europe: Germans expelled by the Russians and various governments; provisional governments or angry mobs in various regions; hundreds of thousands of forced laborers released by the Nazi collapse; survivors of some 2.5 million Poles and Czechs returning from the Soviet Union, where they had spent the war; dazed former inmates of Nazi death camps, in addition to all those cast out of their homes by the fighting. Formerly German-populated areas suffered a particularly ferocious Russian advance, and some lost nearly all their inhabitants in the final phase of war. East Prussia, for example, was devastated by the Red Army in a cataclysmic campaign, possibly without parallel in European history. The diplomat George Kennan scouted the region afterward in an American plane, seeing entire areas where "to judge by all existing evidence, scarcely a man, woman, or child of the indigenous population was left alive after the initial passage of Soviet forces. . . ." To Kennan, "the sight was that of a totally ruined and deserted country: scarcely a sign of life from one end of it to the other."[8] In another case, German civilians fled west from the Silesian capital of Breslau during the terrible battles for the city in 1945; when its beleaguered garrison finally surrendered, all its native inhabitants were gone, to be replaced by Polish refugees, often from Polish territory incorporated into the Soviet Union. Thus, in a matter of days, German Breslau became Polish Wroclaw.

Unlike the situation in the West, starvation and disease persisted long after the arrival of "liberation" forces. Areas freed by the Allies in France, Italy, the Low Countries and Western Germany were at the end of a long chain of supply, ultimately drawing on the bounty of America; though sometimes slow in arriving, relief eventually came in great abundance. Soviet zones, on the other hand, largely had to fend for themselves. The Russians were hardly disposed to transfer food and supplies from the Soviet Union, itself in desperate straits in 1945, with its own civilian population weak, hungry, and overworked. Refugees from Eastern Europe who crossed into territory held by the Western Allies carried frightful tales of famine and typhus outbreaks long after the Red Army had arrived.[9]

Postliberation conflicts generated new refugees throughout Eastern Europe. We shall examine the most important case, that of the Volksdeutsche, in more detail shortly. But there were others. After Rumania withdrew from the Axis in August 1944, her armies turned against Hungary and with Soviet forces soon recaptured Hungarian-annexed portions of Transylvania. One result was the precipitate flight of Hungarians westward. Similarly, the imposition of Tito's exclusive control in Yugoslavia in 1945 forced thousands of dissidents out of the country, their political fate soon merging with that of large numbers of Croatians and various Nazi sympathizers, who had fled to what remained of the Reich a year before. In the moment of victory, Yugoslav partisans occupied formerly Italian territory in Venezia Guilia; to escape their clutches, tens of thousands of Italians, Slovenes, and Croats decamped for Italy.

Other refugees were by-products of population exchanges between the Soviet Union and her neighbors. Through an agreement with Poland, the Soviets received about two-thirds of the Lemko population of that country, declared by Moscow to be Ukrainians, allowing Poles living in the Ukraine to move in the opposite direction. The remaining Lemkos became refugees in an effort to escape being absorbed into the Soviet Union. To the south, Czechoslovakia ceded to the U.S.S.R. its eastern province of Carpatho-Ukraine (Carpathian Ruthenia), an area of about 8,800 square kilometers with a population of about 850,000. Along with the carefully defined "voluntary" exchange of population between the Ukrainian and Soviet Republic and the Czechoslovak state, a small number of fugitives illegally fled westward, often ending up in the American zone of Germany. Meanwhile, Czechoslovakia was busy expelling Hungarians from parts of Slovakia absorbed by Hungary following the Munich agreement in 1938. Finally, refugees accumulated from the civil war in Greece that erupted in 1944 between the British-supported royal Greek government and Communist movements spawned by the Resistance. To escape this fighting, tens of thousands of refugees joined those displaced by the war itself. More than 1,700 villages were destroyed, and hundreds of thousands were made homeless. In addition to Greeks and Macedonians, many thousands of Yugoslavs, Bulgarians, Rumanians, and Albanians roamed about Greece seeking shelter and assistance. In September 1949, by one account, there were still 700,000 refugees in Greece—one-tenth of the total population.[10]

Following the destruction of Nazism, refugees poured southward into Italy, which became one of the principal collection points for the displaced of Europe. Italy presented a wild profusion of refugees: former Nazis and their collaborators passed British and American checkpoints with little difficulty, adding to the odd assembly of Poles, Yugoslavs, Ukrainians, and Albanians who re-

TERRITORIAL CHANGES IN EASTERN EUROPE, 1947

After David Thomson, *Europe Since Napoleon* (New York, 1962).

303

fused repatriation. Jewish survivors of Hitler's death camps made their way down the Italian boot, usually hoping to sail for Palestine in ships organized by the Zionist underground. Tens of thousands of Italians were displaced by the war, including large numbers from Venezia Giulia and Dalmatia, lost to Italy with the collapse of Fascism. In addition, some 400,000 refugees from North Africa abandoned Italian colonies in Africa to "return" to a homeland many had never seen.[11]

Refugee problems were less severe in the West, not only because relief came sooner and in greater abundance but also because displaced persons moved about more easily than under the Soviets, returning quickly to their homes. Civilians began sorting themselves out as soon as the Germans were driven east. By early 1945 there was a flood of returnees. Two months before the end, German officials agreed to release to the Red Cross various inmates of concentration camps in the Reich. These reached Switzerland in early February and Sweden in mid-March. Immediately following the German surrender, the Allies diverted aircraft and military personnel to hasten the return of refugees. According to data provided by the Allied military command, 1,500 transport planes and bombers were busy with repatriation in May 1945, carrying up to 36,000 people daily in good weather.[12] Trainloads of former deportees and concentration camp victims arrived in Paris to be greeted by somber crowds. The American journalist Janet Flanner witnessed the first shipment of emaciated women from the camp of Ravensbrück, dressed in clothing "taken from the dead of all nationalities." People pressed flowers on the benumbed survivors: "As the lilacs fell from their inert hands, the flowers made a purple carpet on the platform and the perfume of the trampled flowers mixed with the stench of illness and dirt."[13] Ironically, concentrations of German refugees were among the last to clear. Just before the final collapse, the Nazis sent nearly 200,000 German evacuees to Denmark, still under Nazi control. Immediately after the war the Danes took charge of this great mass of displaced persons, amounting to 4 percent of the entire Danish population. Prevented by the postwar chaos from returning immediately, the unwelcome Germans sat for months in camps while the Danes provided them with food, shelter, schools, and other services.[14]

Repatriation to Western European countries proceeded quickly and often on an individual and unassisted basis. Severe problems remained, however. Much of France and the Low Countries had been turned into a battlefield in the 1944–45 period. In France more than half a million homes were destroyed, along with three-quarters of a million farms. Food was in short supply. Three-quarters of all French harbor installations and railway yards were out of action. Before relief supplies and equipment could be brought fully to bear, military needs had to be met, and preliminary reconstruction had to begin. Despite concerted efforts, this meant delay, often straining relations between the refugees and their liberators. Soviet representatives in Paris and Brussels complained bitterly about the treatment of their nationals who ended up in Western camps, urging less supervision and confinement by Western authorities. In early 1945 a group of Russian refugees went on a destructive rampage in Châlons, leading to sharp French protests up the chain of military command. Under such circumstances, repatriation soon became a major European political question.

Close to one-quarter of the entire population of the former German Reich was made up of refugees when the fighting ceased in 1945. Soon to receive the most gigantic population movement in European history, Germany was a scene of utter ruin, its cities turned into mountains of debris, reeking of unburied corpses, and its countryside pulverized by war and the last, spiteful convulsions of the Hitlerian state. Whatever its effects on German industry, Allied bombing devastated the urban landscape of the Third Reich. Measuring the results, the Allies counted five million Germans "dehoused," most of them in the last stages of the war, nearly 600,000 civilians died in the carnage, and 3.37 million dwellings were destroyed. Ten million German evacuees were living within the Reich; and others, on the rim of their former empire in places like Denmark or Poland.[15] Some communities were swamped, utterly incapable of sustaining the sudden torrent. Public order collapsed in the small town of Eutin in eastern Holstein, for example, after the influx of about 10,000 refugees from nearby Kiel and Hamburg. Throughout Germany, the Allies liberated millions of forced laborers and prisoners of war, many of whom roamed about the empty shells of German cities and the mutilated countryside of the Reich. Sometimes these managed to fend for themselves. An American military doctor came across a community of Ukrainians in the spring of 1945, camped near the town of Schwabach, baking their own bread, and tending a yard full of farm animals they had somehow acquired. Others, of course, were helpless. Soviet officers told the

Red Cross that there were 1.2 million German refugee children in their zone, of whom 60,000 were sick or undernourished—and this by the rough-hewn Soviet standards of the time.[16]

The most pathetic victims of Nazism emerged from close to 500 concentration camps operating in Germany in the spring of 1945. In all likelihood, the history of the Third Reich will forever be symbolized by these grotesque creations, filmed by Allied camera crews in the hour of liberation. In the annals of horror, they are only surpassed by the *Vernichtungslager* or extermination camps— centers devoted exclusively or in large part to mass murder, which the Nazis consigned to newly conquered territory in the East. The most important of the latter, Auschwitz-Birkenau, was liberated by the Russians in January 1945 as the camp was being dismantled. Most of the others were completely destroyed by the Nazis themselves and the record of their crimes crudely hidden long before 1945.

Within the Reich, it proved impossible for the Nazis to cover their tracks. The German camps were an integral part of the regime and evolved with it—from the first, hastily improvised centers in 1933, the "wild KZs," to the gigantic integrated system that collapsed so suddenly in 1945. The last weeks of their history were awesome in their descent even further into barbarism. Shortly before the Nazi collapse, many had inmate populations swollen many times their wartime size owing to the arrival of forced marches and sinister convoys from the East, bringing the living and the dead from areas threatened by the Red Army. As the liberating armies drew close, hunger and disease killed a large proportion of the inmate population that had managed, for so long, to survive overwork, torture, and execution. Of the 1.5 million who had passed through their gates, the camps contained only a few hundred thousand living persons at the very end, most of them only barely alive. In several cases, liberation occurred in the full glare of publicity. Just before the German collapse, General Eisenhower signaled to the Western media from Allied headquarters that reporters should come and see for themselves. During April and May 1945, graphic descriptions of the camps reached the West. The British and Americans were first to enter some of the largest German camps—Mauthausen, Dachau, Bergen-Belsen, Buchenwald, Dora-Nordhausen. The Russians, generally more secretive and security-conscious first arrived in Theresienstadt, Sachsenhausen, Ravensbrück, and Gross-Rosen.

What the soldiers found in these places haunted even hardened military men, presumably inured to the suffering wrought by war. Bergen-Belsen was one of the largest of these camps, liberated by the British on 15 April 1945. Built for eight thousand, it contained 40,000 skeletal prisoners when the British arrived, plus the corpses of another ten thousand. A medical officer first on the scene observed that "huts which should have contained at the most eighty to one hundred prisoners in some cases had as many as one thousand. Some huts had a lavatory, but this had long-since ceased to function and the authorities had made no provisions outside, so that conditions on the ground were appalling, especially when it is realized that starvation, diarrhea and dysentery were rife." Enormous piles of dead lay everywhere. A gallows stood at the center of the camp, and a huge open grave, half-filled. Surveying the human skeletons who still survived, a medical team decided that 25,000 needed immediate hospitalization, and of these 10,000 would probably die in a few weeks. In fact, 13,000 inmates perished.[17] To the Allied troops and to others who saw such sights, the victims were as if from another planet. Such were the 30,000 inmates of Theresienstadt, supposedly a model camp for Jews, exposed to the world and placed under International Red Cross supervision when the Germans suddenly left in early May 1945:

> Young though they were, [the prisoners] looked aged. They were too weak to walk or even to move. They were swarming with lice, covered with ulcers and running sores. . . . They wore the thin rags they had taken from corpses which still lay on the trucks. . . . The eyes of many were clouded by the enormity of their suffering that had passed the furthest limits of endurance: they were apathetic and indifferent to their fate. The eyes of some others were shining with a feverish brilliance, they greedily swallowed every scrap of food, and, their whole body trembling, they spoke of their insatiable hunger, of nights of torment in trucks, of their desperate fight for survival and of the agony and death of so many of their fellow sufferers.[18]

In these encounters with the liberated victims of concentration camps, Allied authorities had their first direct contact with the tragedy that befell European Jews. Nearly all the Jews liberated in Central Europe emerged from these camps. So degraded was their condition that communication was obstructed from the start. Richard Crossman, who visited the camps in the spring of 1945 as a member of the Anglo-American Committee of Inquiry on Palestine, noted how an abyss separated a place like Dachau from the rest of humanity. "Even the most sensitive and intelligent people who we met in Dachau seemed to accept it as the only reality, and to think of the outside world as a mirage. Similarly the incoming troops, after the first rush of indignation, seemed to slump back into accepting Dachau, not as 32,000 fellow human beings like themselves, but as a strange monstrosity to be treated on its own standards." However horrified, the liberators often failed to see the victims as fellow creatures: "One realized how little horror has to do with sympathy. Sympathy demands some common experience."[19]

Pitifully few Jews were enumerated in 1945, when the first estimates of displaced persons were made in Germany—between 50,000 and 100,000, or from 5 to 8 percent of the total number of displaced persons. More were to join their company after their emergence from hiding or their release from the Soviet Union and flight from Poland, Hungary, or Rumania. But only a small remnant of European Jewry survived. On every scale of suffering, the Jews were invariably at the uppermost registers, had been in captivity for the longest period, came from the worst camps, and had the most ghastly appearance. Allied military personnel, ill-trained and ill-equipped to deal with such trauma, frequently kept these tortured victims at arm's length, resented their needs, misunderstood their imprecations, and found them impossible to manage with customary military dispatch. Malcolm Proudfoot, who served in the Displaced Persons Branch of the Allied Military Command, both shared and described the outlook of so many of his colleagues. He found the Jews to be "understandably, in an unbalanced emotional condition." However comprehensible, he later wrote, "their mental state was frequently so abnormal and offensive that it required a real effort for even the most friendly non-Jews to keep from being goaded into discriminatory action." In a related observation, a Jewish investigator noted that "Jewish DPs are becoming an infernal nuisance, an annoying burden."[20] Before long, serious political trouble began to grow in this soil of tragic incomprehension.

Military Control in the West, 1944–45

Military projections of the European refugee situation during the latter half of the war suggested an enormous logistical task ahead, however they may have underestimated the attendant political complications. From the start it was assumed that the soldiers would be in charge. During the campaign in France in 1940, the British saw masses of refugees becoming tangled in military operations; crowds of French and Belgian civilians were driven across Allied communication lined by German aircraft, obstructing the movement of supplies and reinforcements. To underscore the priority of defeating the Nazis, Allied strategists assigned primary responsibility for refugees in 1944 to the troops in the field. Preparations began in earnest as part of Operation Overlord, the allied landings in France. Reports emanated from the Supreme Headquarters, Alllied Expeditionary Force (SHAEF), during the first half of 1944, coordinated by its G-5 office, the Refugee Displaced Persons and Welfare Branch. In June 1944, two days before D Day, this unit issued its first blueprint for dealing with the problem. G-5 expected 11,332,700 refugees and DPs in France, Belgium, the Netherlands, Denmark, Norway, and Germany (excluding German refugees); these civilians would speak at least twenty languages. The general assumption was that practically all the uprooted persons would return to their countries in about six months, with the soldiers ensuring an orderly passage and the provision of relief supplies. Soviet cooperation, essential for the smooth execution of the plan, was assumed. Eventually, it was further surmised, the care and relief of refugees could be turned over to the United Nations Relief and Rehabilitation Administration (UNRRA), which would resettle the few remaining refugees.[21]

Until the Western Allies broke into Germany in early 1945, refugee work proceeded smoothly. The military cared for, registered, and helped transport uprooted civilians in France, Belgium, and Italy. Many of these needed little help, and a large proportion returned home on their own. Civil affairs officers from the various armies cooperated with local authorities in liberated zones, speeding the process of repatriation. As they moved east, the Allies encountered large numbers of Soviet nationals and East Europeans and occasionally clashed with Russian repatriation officers sent to the scene. But these difficulties were insignificant compared to the storms that were later to break over such issues, and in any event the numbers were as yet small. Despite the sometimes desperate fighting, the soldiers seemed to have refugee matters well in hand. UNRRA acted as a junior partner under SHAEF direction, undertaking no operational tasks until January 1945. By that time there were 247,000 refugees to be cared for in the Low Countries and France.

In a few weeks, however, the Western Allies were swamped with nearly seven million displaced persons—not to mention the millions of uprooted Germans. Military authorities worked feverishly to get a grip on the situation. To the soldiers, the extraordinary effort involved was hardly their preferred line of work, and strains appeared from the start. Officers constantly made a case for other, more pressing priorities. To American commanders in Europe, for example, the overwhelming concern was to transfer three million men, together with their equipment, to the Pacific theater and the war against Japan. In the next six months more than two million soldiers were removed, at a rate of 400,000 per month. Meanwhile, seven million Germans who had surrendered to the Allies had to be discharged; 1.5 million German prisoners of war were to be returned from Allied countries; the Allies had to provide law and order throughout Central Europe, guard millions of tons of supplies and equipment, and reopen roads, highways, canals and ports.[22] Little wonder that the military was sometimes impatient with refugee issues, occasionally fumbled political problems, as we shall see, and resented the claims and demands of the rival agency, UNRRA.

Despite these obstacles, the military organization implemented the SHAEF plan energetically. On 5 May, three days before V-E Day, General Eisenhower broadcast a special message instructing all displaced persons to stay put and await orders from Allied personnel. In the next weeks SHAEF helped organize a massive repatriation to West European countries. Taking advantage of the summer months, occupation officials moved people using every possible manner of conveyance—passenger trains, open boxcars, barges, and army trucks. During the two months of May and June alone, 5.25 million persons were repatriated, at a rate of 80,000 a day. In July, when the Western allies pulled back from large areas of Germany according to an agreement with the Soviets, they transferred large numbers of nominally Soviet nationals to the Russians. In four and a half months more than 2.75 million were handed over, many of them involuntarily.[23]

SHAEF wound up its operations in mid-July 1945, leaving Western Germany and Austria under the military commands of British, French, and American zones. On refugee matters the soldiers retained supreme control. Each of these zones could now go its own way although each also agreed to continue working with UNRRA as before. Throughout the rest of 1945, as the inexperienced and sometimes inept UN organization dithered and fumbled, the military machinery was carried along by the momentum of the liberation period. Of seven million displaced persons under the control of the Western Allies, fewer than two million remained by the end of September.

But the more spectacular their success in helping to move such large numbers, the more impatient the soldiers became. During the extremely difficult winter of 1945–46, clashes between German civilians and displaced persons occurred, along with occasional instances of lawlessness and corruption in the camps. More often than not, the military blamed the DPs, increasingly seen as recalcitrant, uncooperative, and given to criminal activity. By this point, according to historian Leonard Dinnerstein, "the army wanted to be rid of the DPs and made no bones about it." General Lucius Clay, in charge of the American zone, determined to get on with the task of rebuilding the German economy. In February 1946 the Americans seemed on the point of closing down DP centers entirely, ending the remaining refugee problem, as it were, at the stroke of a pen. Only President Truman's intervention canceled the order at the last moment to avert political repercussions at home.[24]

Seen in retrospect, one of the great achievements of the military period was the control of many infectious diseases, especially typhus, that in earlier periods regularly decimated large groups of refugees. Historically, this may have been the most durable accomplishment of refugee workers after the Second World War. For centuries, typhus and other epidemics accompanied armies on the move, usually killing far more than died in the course of battle. In the Napoleonic period, to take a famous case, the emperor's *Grande Armée* set out for Russia with 265,000 men and lost 80,000 to typhus, dysentery, and other illnesses before serious fighting even began. After the Battle of Leipzig in 1813, an estimated two million people contracted typhus—about one-tenth the German population of the time—and about 200,000 perished from it. During the First World War, typhus killed about a quarter of the Serbian army in 1913, and about 300,000 people in all—one-tenth of the entire population. Although the disease was largely controlled on the western front between 1914 and 1918, typhus claimed about eight million victims in Central and Eastern Europe immediately after the war. The great influenza epidemic at that time killed around forty million persons world-wide, a heavier toll than the war itself.[25]

British Foreign Secretary Ernest Bevin had serious fears in 1945 that the huge numbers of European refugees would generate a fearful epidemic on the order of 1918–19, cutting deep into the working class and hindering postwar reconstruction. Drawing on the experience of the First World War, Bevin worried about influenza; others feared the impact of typhus, tuberculosis, or other former scourges associated with great masses of people living in conditions of severe privation.[26] These diseases, indeed, immediately appeared on the scene with the liberation of the camps in Germany and killed tens of thousands of people, most of them severely weakened by malnutrition. Alexander Donat described the ravages of typhus after liberation in a branch of the Natzweiler camp system in the spring of 1945: within days after the arrival of warm weather, 75 percent of the inmates were stricken.[27]

Unlike earlier periods, however, military refugee workers were well equipped to combat the postwar outbreaks. Against the principal threat, typhus, the great weapon was DDT, an insecticide developed for general use in the United States during the war. Blissfully unaware of eventual adverse effects, the Americans used it successfully in Naples in 1943 to kill typhus-bearing lice, bringing a full-scale epidemic to a dead halt. On the basis of this experience, refugee workers applied the chemical liberally to everything and everyone in sight in refugee centers. Everyone was "dusted"—including all congressmen, parliamentarians, journalists, and others visiting the camps. The Americans issued a little card to each displaced person after delousing; French officials soon blocked all westward movement across the Rhine without a DDT card. Marcus Smith, an American military doctor assigned to Dachau after the war and who had never seen a case of typhus before, found himself organizing a remarkably effective campaign using the new methods. All 32,000 inmates of the camp and nearby satellite installations were dusted in five days.[28] Other medical advances permitted army doctors and eventually UNRRA personnel to control smallpox, typhoid, tuberculosis, and venereal disease. Food, shelter, and fuel remained scarce; but field hospitals, military supplies, and dispensaries did their job well. The contrast with other refugee experiences suggests a historic achievement in this field. Indeed, given the conditions of life in Europe in 1944 and 1945, the health of the displaced persons was remarkably good, excepting, of course, those released from the concentration camps. According to Proudfoot, fewer than 1 percent of the DPs required medical care, and fewer still needed hospitalization.[29]

Forcible Repatriation to the Soviet Union

Immediately after the Nazi collapse, the Soviets and the Western Allies began complex discussions over the repatriation of more than two million Soviet nationals in the British and American zones. Throughout Germany, Russian and other Eastern European displaced persons formed the great majority of those liberated by the Allies. Some 5.5 million Soviet citizens were released from captivity in the former Reich. Over a million Soviet prisoners of war had survived the most cruel treatment imaginable and were clamoring to be released. With them were approximately two million forced laborers, a large proportion of whom had been torn loose from their homes and brutalized by the Nazis, but some of whom had actually volunteered for service in the Third Reich.

In addition, there was an extraordinary array of soldiers and civilians, holding every shade of political opinion, but almost all of whom detested the Soviet system. Since the beginning of the Russian advance in 1943, refugees and disaffected persons had gone westward to escape the Red Army and the reimposition of Soviet control. As the German defeat turned into a rout, their movement became a landslide, tearing away large masses of people from their homelands in the Caucasus, the Ukraine, Belorussia, and the Baltic regions. The Western Allies began to meet these people in 1944, along with many deserters from the Red Army and military units of various Soviet peoples who had fought against the Russians on the German side. Hundreds of thousands ended up in Western hands when the fighting ceased. Others clamored to reach the American zone, seriously inconveniencing the military administration, which blew up bridges to prevent a sudden inundation in May 1945.[30] A significant number gave allegiance to a onetime Red Army general Andrei Vlasov, who organized hundreds of thousands of men into an anti-Soviet force after having been captured by the Germans in mid-1942. Some were the remnants of forces explicitly committed to the counterrevolutionary cause in the Soviet Union, having assembled under German cover around a handful of aged exiles from the Bolshevik period. One group of Georgians, for example, was commanded by a brace of former princes, some of whom had been taxi drivers in Paris between the wars. The Cossack general Peter Krasnov, perhaps the most remarkable of these anti-Bolshevik leaders, was seventy-six years old when he surrendered to the British in 1945, his bearing and personal dignity reminiscent of the Russian Imperial Army of the tsarist period. Among the most bizarre in appearance were about 30,000 Cossacks who surfaced in Italy, where they were also taken in hand by the British. Having been mobilized by the Wehrmacht, these men wore German uniforms, but looked remarkably like their forebears who once terrorized Napoleon Bonaparte: squadrons of horsemen wore huge fur hats and knee-length riding boots and carried long curved swords; their baggage train included women, children, horse-drawn carts, farm animals, and camels—as if gathered for a film on the Russian campaign of 1812.[31]

The Eastern Europeans caused trouble for the British, French, and Americans from the moment the liberating armies landed in France in mid-1944. Having suffered more than most prisoners of the Nazis, they finished the war half-starved and in desperate need of medical attention. Notoriously undisciplined, they quickly acquired a reputation for boisterous, destructive behavior and drew the constant attention of hostile Soviet observers sent to monitor conditions in their camps. They were a continual threat to public order, particularly following drunken rampages. Malcolm Proudfoot notes that "they made a poor impression on the military government personnel, and particularly on the newcomers, many of whom were young, gullible and inexperienced. Controlling and caring for these disorderly East Europeans came to be regarded as an irksome chore." Seldom able to speak English or German, they frequently masked their Soviet identity to prevent their return home. Generally speaking, the Allies cooperated eagerly with Soviet representatives prior to a comprehensive agreement about repatriation. But the Russians insisted on a prompt transfer to Soviet jurisdiction and an early dispatch of the refugees to Russia. As Eisenhower wrote, "These displaced persons are a constant source of misunderstanding and controversial discussion with representatives of the Soviet Military Mission who make very exacting demands as to the care, administration, and conditions of their nationals. There is always the possibility of incidents which may cause frictions. . . . The only complete solution to this problem from all points of view is the early repatriation of these Russians."[32]

Western governments considered that the key to the problem was the agreement worked out with the Soviets at Yalta in February 1945. Concluded when the wartime goodwill of the Grand Alliance had reached its height, the accord reflected a Western disposition to accede to Stalin's insistent concerns rather than drive a hard bargain over matters considered of minor importance. As the hour of Nazi defeat drew near, the Anglo-American leaders were especially eager to conciliate Stalin, whose armies were within a hundred miles of Berlin. Still recovering from the Battle of the Bulge, Roosevelt wanted to convince the Russians of his good intentions and induce the Soviets to enter the war against Japan. Both British and American leaders hoped to reassure the Kremlin of their desire to continue cooperation among all three powers after Germany surrendered. Both Western allies feared that, if they delayed repatriation, the Russians would similarly slow the return of Allied prisoners of war then in Soviet hands.

Western representatives probably did not realize how many Russians dreaded being sent back to the Soviet Union. Consequently, when Stalin indicated that Soviet citizens should be returned as soon as possible, they immediately promised to send them back. Although the discussion included nothing about forcible repatriation, all parties seem to have taken this for granted. Soviet citizens were to be gathered together, housed separately, subjected to Soviet law, and handed over to Soviet authorities. In an important proviso, subsequently overlooked in the dispatch of many refugees, "Soviet citizens" were to be taken by the Western powers to be those who were Soviet nationals at the outbreak of war in 1939—thus excluding Balts, Poles, or Ukrainians from regions absorbed in 1939 and 1940. There were some differences of opinion between the British and Americans, with the latter being reluctant to use force against the Soviet refugees. But neither shrank from coercion or risked provoking Stalin's wrath. Both realized that the Soviets were extremely sensitive about two million of their conationals being in Western hands and displaying obvious disaffection from the Soviet system. The very last thing the Kremlin wanted after an exhausting war with Hitler was to see a huge émigré community emerge in the West, with many members already formed into military units, led by men who were sworn to destroy the Soviet system.

Acting on the strength of the Yalta decisions, SHAEF officers worked out an arrangement with the Russians in May for the dispatch of refugees, soon proceeding at a rate of tens of thousands per day. The French, anxious lest large numbers of Eastern Europeans ultimately enhance the population of Central Europe, negotiated their own transfer of Soviet refugees in June 1945. Following the lead of the wartime allies, the Swiss, Swedish, Belgian, and Dutch governments also facilitated the dispatch of Soviets to the Soviet Union. By early July, 1.5 million had been transferred to the Soviet zone. When the Western Allies pulled back from German territory delimited as part of the Soviet zone, they turned even more over to the Russians. By the end of September, when the transfers fell off dramatically, a total of 2,272,000 had been transferred.[33]

Western governments soon realized that the neat arrangements for the transfer of Soviet nationals involved ugly incidents and raised a host of sensitive political problems. The very determination of who was and who was not a Soviet national involved serious headaches. The Russians claimed refugees from areas recently annexed to the Soviet Union; about 120,000 natives of those regions denied they were Soviet citizens at all and refused to go home.[34] Russian repatriation officers protested loudly at this assault on their national integrity. Few of the Eastern Europeans involved had any doubt about the repression of nationalities in the Soviet state or of the solicitude of Stalin for those who had fought with Hitler against the Russians. Several Cossack officers killed themselves rather than be sent back. There were violent outbreaks when the British or Americans forced people to leave what they had assumed was political sanctuary. In the case of many Russian émigrés who had never been Soviet citizens, having lived continuously abroad since 1919, the British made the utterly unjustifiable determination to turn them over to the Russians, plainly exceeding what was required of them under the Yalta agreements. These aged counterrevolutionaries, part of the demonic folklore of the Soviet Union, some of them bearing Nansen passports, were dispatched along with all the others. Once in Soviet hands, large numbers were executed, among them old General Krasnov himself.[35]

Toward the end of 1945, with more than two million people transferred to the Soviets, Allied policy toward these refugees began to shift away from forcible repatriation. There were various reasons for the change. After more than six months of massive transfers, the serious economic burden of so many refugees was now lifted. The British and American prisoners had returned from Soviet-held territory by this point, thereby diminishing the leverage that Moscow could exercise on refugee issues. Protests against forcible repatriation were working their way through the military and civilian bureaucracies. Word reached the West, before long, of the brutal treatment that awaited the deportees, including those who had never held Soviet citizenship such as General Krasnov. Military men in Germany were increasingly reluctant to force the Russian refugees to return home against their will. Russian repatriation officers, meanwhile, provoked sharp reactions among Western officials, having shown themselves increasingly ruthless—actually kidnapping refugees, for example, in broad daylight in downtown Brussels. Lastly, the Grand Alliance was quickly losing its wartime cohesion, and the former allies were beginning to clash over the disposition of Eastern Europe. The Americans first announced a new course in December 1945, limiting forcible transfers to Soviet soldiers and known collaborators; the British were less eager to abandon their repatriation policy, but eventually adopted the American view. East-West relations continued to deteriorate in 1946 and 1947, relieving the Western Allies of their concerns about a one-sided breach of the Yalta agreement. About 500,000 Russians, according to one account, "escaped the sieve of the Yalta accord" either through their own efforts and subterfuge, bureaucratic fumbling, or the shift in repatriation policy.[36]

UNRRA

The first United Nations effort to assist refugees grew out of Allied preparations for the liberation of Europe. During 1942 British and American long-range economic planning included investigation of refugee relief operations. Activity began late that year in North Africa when troops of both countries encountered large numbers of refugees. The British operated a Middle East Relief and Refugee Administration with headquarters in Cairo and had cared for some 46,000 Balkan refugees by mid-1944. Working with the British, the Americans established their Office of Foreign Relief and Rehabilitation Operations, headed by a long-time refugee advocate, Herbert Lehman, former governor of New York. On the basis of this experience, anticipating that the gigantic

problems of postwar relief would require much more than infor-
mal Anglo-American cooperation, Washington pressed for a
broadly based international relief undertaking. During 1943, Assistant Secretary of State Dean Acheson conducted extensive negotiations with the Russians to build a framework for such an
organization. Discussions broadened to include other governments, and formal agreement was reached in the autumn, establishing the United Nations Relief and Rehabilitation Administration (UNRRA). Representatives of forty-four countries signed the
agreement. Signaling the importance of the United States in
pressing for this institution, the delegates assembled at the White
House for its launching in November 1943, heard an address by
Franklin Roosevelt, and repaired to Atlantic City for the first
meeting of the UNRRA council. Representatives there elected
Herbert Lehman as director general.

Looking back on the establishment of UNRRA, Dean Acheson
emphasized the emerging confrontation with the Soviets. The
American negotiator noted the Russians' extreme reluctance to
surrender their freedom of action to an international organization of any sort. Nothing was to be done, in their view, even in
the area of relief and rehabilitation, that might jeopardize Soviet
interests. The UNRRA discussions were one of the first occasions when Moscow put these concerns so insistently regarding
postwar reconstruction, and Acheson felt that the American representatives were thus "present at the creation of the pattern."
Specifically, the Russians insisted that all decisions taken by
UNRRA had to have its support, that nothing could be done
anywhere except through the relevant country's agencies, and that
to these ends an array of international safeguards had to be established. The Soviets believed that postwar aid should be closely
tied to political coloration, notably commitment to the struggle
against Nazism. "As far as the Russians were concerned," wrote
Acheson, "the organization existed to give prizes for fighting
Hitler." [37] An unspoken further assumption was that anti-Soviet
refugees, whatever other views they had, were not to become
charges of the United Nations. Recalcitrant Soviets were to be sent
home. American policymakers clearly did their best to conciliate
Stalin, hardly anticipating the East-West clashes of the later 1940s.
In deference to Russian wishes, SHAEF agreed that displaced
Soviet nationals would be segregated from other refugees and put
in special centers, to be run by officers sent from the U.S.S.R.

The inmates were to be subject to Soviet military discipline and repatriated to the Soviet Union "regardless of their individual wishes."[38]

Subsequent discussions defined the scope of UNRRA and set its refugee activity within the framework of relief and rehabilitation. As Acheson later recalled, all concerned were working in the dark, uncertain even what such terms as "rehabilitation" would mean in practice. Perhaps because of this, the role of UNRRA was strictly limited. Everyone agreed its operations would be temporary, providing supplies and services in cooperation with various public and private relief agencies and with the permission of local authorities. The general expectation was that refugees would seek repatriation as a matter of course, and so this sensitive issue, later to play so important a role in United Nations debate, did not trouble UNRRA at its inception. Enemy nationals were not to be eligible for assistance—a restriction that was later to cause significant difficulties. Finally, it was understood that UNRRA would operate in liberated territory with the consent of the military and later with the approval of the national authorities as they emerged.

UNRRA first encountered refugees in Italy, where the organization began limited operations in early 1944, providing relief and repatriation assistance. That summer, the agency took over refugee work previously undertaken by the British in the Middle East and North Africa, involving nearly 74,000 displaced persons by the end of the year. But this was merely a prelude to the massive European operations to follow. Following the Allied landings in Normandy, UNRRA began to cut its way through a thick tangle of administrative and jurisdictional obstacles to direct action. The Soviets did not permit UNRRA to operate in their zone, and all attention focused on Western and Central Europe. UNRRA carefully skirted the objections made by provisional governments and newly established authorities in liberated countries, all jealous of their own sovereignty in dealing with displaced civilians. Most importantly, UNRRA worked out its relationship with the military, finally achieved in the SHAEF-UNRRA agreement of 25 November 1944. UNRRA placed itself squarely under military patronage until an eventual "postmilitary period" and agreed to work within the SHAEF command; the UN body was to provide the technical staff, and the soldiers, in the first instance at least, would furnish the logistic and material support.

In mid-1945, UNRRA finally entered the field with several hundred field teams and began to take responsibility for assembly centers in Germany.[39]

On paper, UNRRA's achievement in the refugee field was staggering. Having begun massive operations in the second half of 1945, the agency helped oversee the repatriation of millions of displaced persons before the end of the year. By the beginning of 1946, three-quarters of the displaced persons in Europe had already been sent home, and the agency could address itself to the approximately one million remaining—most of them in Germany. Working under great pressure, UNRRA directors hurriedly thrust refugee workers into the field—sometimes too quickly, inadequately trained, and with unfortunate results. But the vast army of refugees shrank as fast as anyone could have expected. UNRRA employees cared for hundreds of thousands of displaced civilians, a particularly difficult task in view of the increasing proportion that resisted repatriation or was unable to return home. UNRRA put hundreds of doctors and nurses into refugee camps, coordinated the activities of voluntary agencies throughout Germany and Austria, and established an efficient Tracing Bureau to track down missing persons. Although its working language was English, UNRRA was a truly international organization, assembling a staff of 5,200 persons of thirty-three nationalities. American funds furnished nearly 75 percent of the UNRRA budget, and Britain provided nearly a quarter; the Soviet Union made only a token contribution—reflecting its distrust of the United Nations agency.[40]

To many who looked closely into UNRRA's field operations, however, the organization was a byword for waste, bungling, and incompetence. Heading its German operations was Lieutenant General Sir Frederick Morgan, a crusty British officer of real distinction, one of the planners of the Normandy invasion. Rather than being the organization's advocate, Morgan was one of its sharpest critics. From the moment he took charge, Morgan shared the soldiers' common appreciation of UNRRA as "that adventitious assembly of silver-tongued ineffectuals, professional do-gooders, crooks and crackpots."[41] Investigations indicated lavish expenditures and occasional involvement by UNRRA staff in black market trading and profiteering. To those who were used to a military chain of command, the celebrated international staff was both quarrelsome and incompetent. "Orders were often poorly translated, and half-understood," reports Proudfoot. "There was duplication, delay, and inefficiency." Top-heavy with administrators, UNRRA lacked men in the refugee camps, where they were needed.[42] Morgan thought that UNRRA's ranks were full of overpaid and insubordinate draft dodgers, cheats and shirkers. Among the French contingent were Sybaritic officers who had spent the war in captivity or house arrest; others were eager to avenge themselves on the local German population and evade responsibilities at home. No one, he complained, could be fired. And there were no proper regulations to enforce—"to the extent that it was at times not possible to persuade a typist to type a document of which the contents did not meet with the approval of the typist."[43]

Behind these cranky outbursts lay serious structural difficulties, many of which had little to do with UNRRA personnel. One basic problem was the perpetual subordination of UNRRA to the Allied armies. Because UNRRA funds were not to be used for housing, food, or clothing and because it was never intended to provide law and order, security, or transport, the organization remained throughout its existence subordinate to the military men, hardly well disposed to the agency's growth and development. Kept in a junior role, derided as a usurper by the armed services and diplomatic officers of Britain and the United States, UNRRA found its prestige drained and its capacity for independent action stripped away. In the rough-and-tumble world of postwar German and Austria, where effectiveness often depended on bold action, this was not a recipe for success. Morgan ruled his ramshackle bureaucracy from a picturesque castle in the Westphalian town of Arolsen—nearly 250 kilometers from the American army headquarters in Frankfurt and 650 kilometers from Munich, the main concentration point for displaced persons. Ira Hirschmann, sent as a troubleshooter from UNRRA headquarters, thought that the general might as well have set up shop in the middle of Montana. In the vacuum opened at an early stage by UNRRA's manifest lack of preparation for an enormous task, the military men took charge of a substantial amount of refugee activity. But the soldiers seemed equally ill-equipped to deal with displaced persons, particularly the steadily increasing proportion that could not or would not be repatriated. Gruff and impatient with their charges, military administrators often saw refugees as a bother to be overcome. Expressing an extreme though not uncommon view, American General George S. Patton said in September 1945 that the DPs should be treated as prisoners and put behind barbed wire. If not, he said, "they would not stay in the camps, would spread over the country like locusts, and would eventually have to be rounded up after quite a few of them had been shot and quite a few Germans murdered and pillaged."[44]

Horror stories about the maltreatment of Jewish DPs by the military spread in the United States immediately after the war. In August 1945, President Truman received a lengthy report on displaced persons by Earl G. Harrison, then dean of the University of Pennsylvania Law School, highlighting UNRRA's incapacities but even more sharply critical of the military. Harrison felt that the continued incarceration of these refugees, often under military guard, was scandalous. "As quickly as possible," he recommended, "the actual operation of the camps should be turned over to a civilian agency—UNRRA. That organization is aware of weaknesses in its present structure and is pressing to remedy them."[45] The report reached Truman's desk just as American authorities feared a serious breakdown of relief to war-torn Europe. When the gentlemanly and conservative Herbert Lehman resigned as head of UNRRA, therefore, Washington was eager to replace him with a person of energy and capacity. The new director general was precisely that—Fiorello La Guardia, the colorful former mayor of New York.

A legendary activist, La Guardia took charge of a rickety institution in March 1946, determined to set things right. His friend Ira Hirschmann recalled being summoned by the newly appointed international civil servant, affectionately called "the Boss": "I found La Guardia seated behind a huge desk mountainous with scattered papers and documents. He was literally steaming. His coat hung on the back of his chair; his broad frame seemed to hold his red suspenders taut; smoke poured from his corn-cob pipe like a small inferno."[46] La Guardia wanted to cut red tape, eliminate corruption, and clear the camps of their remaining refugees. In May he appointed the abrasive Hirschmann a special inspector general to survey UNRRA's operations abroad and look closely into General Morgan's administration. After extensive travel through Germany, Hirschmann denounced the role of the military, the incompetence of UNRRA's leadership, and its apparent coddling of former Nazis and their collaborators. La Guardia eventually fired Morgan, appointing instead a proven administrator—Myer Cohen, former assistant regional director of the Farm Security Administration. With only a few months of UNRRA's mandate remaining, Cohen and La Guardia strained mightily to empty the refugee centers. Against the wishes of Polish anti-Communists and both the British and American military, the director offered sixty days of rations to each Pole willing to accept repatriation. Against strong objections of the British, he championed Zionist demands that the Jewish DPs be allowed to enter Palestine. And in breach of all customary procedure for a United Nations relief official, he urged forty-eight countries associated with UNRRA to accept some of the refugees.

Even with La Guardia's bravado, however, there was a limit to what could be done. Most countries maintained tight restrictions on immigration. In the United States, isolationist opinion rejected the whole idea of American involvement in an international agency, especially one that dealt directly with the Soviets. Political voices now sounded against the forcible return of DPs, clashes occurred between various factions within refugee camps, and both London and Washington had second thoughts about dispatching people into Soviet hands. No doubt La Guardia's jolt accomplished a good deal. Between the autumn of 1945 and June 1947, when UNRRA ceased operations, more than a million displaced persons were repatriated. But nearly 650,000 UNRRA-supported refugees, virtually all of them Eastern Europeans, remained in the camps—bored, demoralized, disgruntled, feeding a voracious black market, and preyed on by black marketeers. Of these refugees, there were nearly 200,000 non-Jewish Poles, 150,000 Balts, and close to 110,000 Ukrainians. Nearly 170,000 Jews came under UNRRA protection, many of them clamoring to enter Palestine, just about to become a Jewish state.[47]

UNRRA's increasing preoccupation with Eastern European refugees put the organization on a collision course with the Soviet Union, unhappy that large masses of exiles in the agency's care refused to return to the Russian sphere of influence. The Soviets themselves were partly responsible for the delay in repatriating large numbers of Poles, the greatest mass of those refusing to return home. Through 1945, the Russians would not accept Polish repatriates in the Soviet zone of Germany, wishing to give the millions of their own nationals priority in the limited transport heading east. As a result, some 800,000 Poles spent the first winter after the war in Germany. During this time anti-Communist propaganda and developments in Poland encouraged many to remain. Representing the new Polish regime, Boleslaw Bierut castigated the West in 1946 for continuing to retain an alleged two million Poles. Backed by Moscow, Warsaw demanded immediate repatriation, by force if necessary. Quarrels between UNRRA officials and the Soviets erupted over the issue of propaganda in the refugee centers. To the Russians, Poles, and Yugoslavs, Western-style free expression in the camps was equivalent to supporting fascism. In February 1946, Andrei Vyshinsky denounced a systematic effort "to poison the souls and minds of these unfortunate men so as to thwart them in their natural aspirations and their legitimate desire to return to their native country."[48]

Caught between its desire to repatriate on the one hand and to protect the displaced persons from the bullying and coercion of Russian and Polish repatriation officers on the other, UNRRA was in an increasingly uncomfortable position. Its mandate precluded resettlement of refugees in new homes, the obvious next step for the hundreds of thousands still displaced. Gradually, its operations slipped into a dull routine of care and maintenance, the spirit of which was reflected in the dreary, mournful appearance of the refugee camps and centers that dotted Central Europe a year and a half after the war was over. In August 1946 the American representative to the UNRRA council was already calling for the end of the agency. In the United States, where most of the bills were paid, both Congress and the administration considered that internationally administered relief had been a failure. Truman plainly wanted American aid to pass through American hands, to serve American interests and responsibilities. In Europe, UNRRA encountered difficulties because of a severe food crisis; health conditions too began to get worse in 1947. Outside observers looked on the continuation of a refugee problem as an international scandal, and many, unfairly, blamed the hapless United Nations organization.[49]

REFUGEES ON THE MOVE

UNRRA-supported refugees were only a fraction of the uprooted humanity in Central Europe after the war, and their massive repatriation contrasted markedly with the sometimes aimless, unprotected wanderings of millions of others. Among the latter appeared increasingly large numbers of Volksdeutsche, the ethnic Germans, whom the Nazis once lionized as the standard-bearers of Aryan civilization in Eastern Europe. In a spasm of anger and revenge, the countries that had suffered from the Nazi occupation or involvement with Hitler now threw these Germans out. By the millions, the outcasts headed for a divided Germany, inundating the western zones with the greatest refugee flood in European history. Simultaneously, in a cruel irony, Jewish victims of Nazism moved in the same direction as the onetime master race. Fleeing the despair and even occasional pogroms that awaited them in Eastern Europe, the Jews also headed west. From their migration came a new refugee crisis, fatefully linked to the British impasse in Palestine.

Germans

Throughout the devastated former Reich, millions of Germans roamed about in the second half of 1945, directed and controlled by occupation officers, scantily cared for by local officials, and fed from soup kitchens run by the German Red Cross and various charities. Some were evacuees or civilians bombed out of their homes, some were former prisoners of war released by the Allies, and some were fugitives from the Soviets, who advanced by Anglo-American agreement in the autumn of 1945 to expand their zone of occupation. Eventually, millions of Germans expelled from Eastern Europe outnumbered even these refugees to form the largest single group of people displaced during the European war and its aftermath.

Foreshadowing their arrival, hundreds of thousands of Germans poured out of regions threatened by the Soviets in the last six months of the war. For nearly four years the Russians had fought on their own soil, seeing their country crushed and their neighbors annihilated. When they finally reached German territory in the autumn of 1944, after pushing more than a thousand miles, they turned on their tormentors in a campaign that nearly matched the Wehrmacht and the SS in destructive fury. Soviet divisions smashed their way through Silesia, East Prussia, Pomerania, and Brandenburg in a mammoth offensive involving almost three million men. For the first time encountering German civilians, the Red Army periodically ran amok, unleashing a wave of rape, plunder, and massacre. "Nothing in Germany is guiltless," said one Soviet propaganda leaflet, "neither the living nor the yet unborn. Follow the words of Comrade Stalin and crush forever the Fascist beast in its den. Break the racial pride of the German woman. Take her as your legitimate booty. Kill, you brave soldiers of the Victorious Soviet Army."[50] According to German sources, the Soviets massacred between 75,000 and 100,000 in territories east of the rivers Oder and Neisse alone. Looting and arson wrought considerably greater damage than bombardment and other acts of war. Meanwhile, the Nazis contributed their own share of mayhem. While Hitler refused to countenance retreat, what remained of his armies destroyed anything of value to make both town and country uninhabitable. Fleeing the advancing crest of the Russian wave, nearly five million Germans left their homes early in 1945, most in the dead of winter.[51]

Flight to the West was a harrowing experience. Russian tanks drove into crowds of refugees in East Prussia in October 1944, terrorizing the population that the local Gauleiter refused to evacuate. As the Red Army drew closer, 150,000 refugees fled to the supposedly impregnable fortress of Königsberg, where they were trapped until German troops briefly broke the siege, allowing some to escape. Thousands then crossed to Danzig on the perilous, icy wastes of the Frisches Haff, the narrow lagoon separating a strip of land on the Baltic from the inner coastline. Along the coast, seaborne evacuations continued until the very end of the war. Between two and three million soldiers and civilians, it has been estimated, were removed in this way. Thousands perished in the hurriedly assembled convoys whose ships were regularly attacked by submarines in the icy waters of the Baltic. To the south, the rout was at least as chaotic. In East Brandenberg the Russians advanced so quickly that they cut off nearly half the German refugees; close to 600,000 fell into Soviet hands trying to escape. One hundred thousand women, children, and old people left the Silesian capital of Breslau on foot in bitter cold weather just before the Russians encircled the city in February 1945. Virtually all these refugees left their homes intending to return. Few could have realized that their departure, in the carnage of war, would later be seen as part of a massive removal of Germans from Eastern Europe.[52]

Significantly, civilians continued westward after the Wehrmacht had fallen back, after the Soviets arrived, and even after the final Nazi collapse. Germans streamed out of eastern Central Europe, interrupted only by the Russian military authorities, who periodically rounded them up to serve in work battalions in the recently captured areas. Throughout Polish and Czechoslovak regions controlled by the Red Army, it was now the Germans' turn to be robbed, herded into camps, and frequently ejected from the regions where they had lived. Before long, similar episodes occurred in Hungary, Rumania, and Yugoslavia. Rather than diminishing with the end of battle, the current of persecutions and expulsions became even stronger, engulfing hundreds and eventually thousands of communities. When the Allies assembled at Potsdam for their tripartite conference of August 1945, the Americans, British, and Russians knew that a massive exodus was under way, involving millions of people of German descent. Stalin presented this as an inexorable force that it was pointless to resist. Accepting that some degree of expulsion was indeed inevitable, Churchill, nevertheless, pressed the Soviet leader strongly about the size of the evacuation. But clearly the British and American leaders did not consider the issue worth a major confrontation, especially as there were other, critical matters at stake— the future borders of Germany and Poland, for example. As a result, the Big Three agreed to formalize the expulsions. Article XIII of the Potsdam agreement mentioned the "transfer" to Germany of German populations still remaining in Czechoslovakia, Poland, and Hungary. In subsequent justification of their support for this measure, United States representatives spoke of the need for "orderly and humane" movements of people, in contrast to the brutalities that had been common to that point. The hope was that the Allied Control Council in Germany would study the matter, attempting to control the volume of refugee traffic, and achieve an equitable distribution of the expellees among the various occupation zones. The signatories requested Polish, Czech, and Hungarian officials to suspend further expulsions pending this investigation.[53]

The torrent continued, however, despite the unhappiness of the Western Allies and despite the letter of the Potsdam agreement. Western reporters relayed the pitiable state of the expellees, whose numbers grew astronomically and who seemed beyond anyone's capacity to reduce or assist adequately. Living conditions were near-catastrophic. All the refugees, as Churchill dryly put it, "bring their mouths with them." There were simply not enough provisions in Germany to go around; relief authorities hoped to provide a meager 1,500 calories per person daily, but they frequently failed to deliver even this. One British correspondent noted at the end of August 1945 that "there are eight million homeless nomads wandering about the areas of the provinces near Berlin." In the *New Leader*, Bertrand Russell described the transports of expellees from the east as reminiscent of Nazi deportation convoys during the war: "Many are dead when they reach Berlin; children who die along the way are thrown out of the window." Robert Murphy, the United States political adviser for Germany, cabled the State Department in October that ten refugees died daily in the Berlin railway station, suffering from starvation, illness, and exhaustion. It was clear that Potsdam failed to humanize the process of expulsion, which continued at a high rate until the winter. Many Germans now began to leave the affected region on their own, anticipating violent clashes with local mobs or fearing forced labor for Germans in eastern areas. By November 1945, between 2.5 and 3 million expellees had reached western zones, and 2 million were in Soviet-occupied Germany.[54]

To Stalin, the expulsions helped solidify a territorial settlement with Poland that he considered of fundamental importance. Moscow's great achievement at the Yalta and Potsdam meetings was to extend the Soviet frontier with Poland 150 miles west to the "Curzon Line," bringing an area with a Ukrainian and Belorussian majority within the Soviet Union. In compensation—and perhaps to weaken a future German state—the Poles were allowed to move their frontiers in the west, absorbing former German land up to the rivers Oder and Neisse. About 1.4 million Poles left the territory ceded to Russia, and many of these were settled in regions taken from the Reich. Throughout 1946 and 1947 the transfer continued. To the onetime German residents, it seemed plain that they left to make room for predatory newcomers. A doctor from East Prussia, Count Hans von Lehndorff, described the country town of Rosenberg, known to the Poles as Susz, where six thousand people once lived:

> Now the little town lies in ruins, the land, far and wide, has been devastated and the fine manor houses mostly reduced to rubble and ashes. Round the ruined core of the town the Poles have moved into any houses left standing, but only a minority appear to be really settled. The majority are in a state of perpetual unrest, and the train that runs once or twice over this section of the line is crammed with adventurers who come and go because they haven't yet found anywhere to stay, or are on the lookout for better openings.[55]

Along with their removal of the Germans, the new Polish administration sought to eradicate all vestiges of the former German presence. German names, institutions, administrative divisions were all to go the way of the Germans themselves. Finally, at the end of 1948, Warsaw declared the process finished. The former German territories east of the Oder-Neisse border were officially incorporated into the Polish state. Hardly any Germans remained.

Much the same occurred in former areas of German concentration in Czechoslovakia, Rumania, Yugoslavia, and Hungary. Expulsions from all but the last of these began even before the Potsdam conference and accelerated afterward with Soviet approval. The Russians may have favored the removal of Germans as a means to facilitate ideological unity and ease their ultimate domination of eastern Central Europe. But there is no doubt that the anti-German moves were popular in many regions, especially where people felt victimized by Nazi occupation and settlement policies. Against the protest of Eleanor Roosevelt in the United Nations General Assembly in February 1946, the Czechoslovak representative claimed "that we in Europe have a right to look at things in our own way. We have suffered more than many delegates in this room can imagine."[56] The Czechoslovak government-in-exile determined to rid its country of over two million Germans as early as November 1944; Czech officials turned on the Sudeten Germans immediately after liberation and proceeded to deport them to Germany with the full knowledge of the Potsdam participants. Rumania and Yugoslavia similarly moved on their own although they did not officially approach the Allies about the matter in August 1945. In the case of formerly Axis Hungary, where the new government was under enormous pressure from the Soviets, expulsion proceeded in various stages from the beginning of 1946. The Russians apparently pressed forcefully for the elimination of the German minority from Hungary, securing official Allied agreement at Potsdam. Fearing similar actions against Magyars living elsewhere, the Hungarian government wanted to limit the expulsions to known pro-Nazis; local Communists urged a much broader movement against the entire German-speaking minority, however, and eventually carried the day. Backed by the Russians, the Czechs supported this drastic solution. Prague had its own agenda for the consolidation of ethnic groups throughout the region: according to this scheme, tens of thousands of Magyars would be forced out of Slovakia and sent to Hungary, where they would take the places of the expelled Germans.[57]

No one foresaw the great avalanche of Germans that came crashing down on Germany in the postwar period. Expecting an orderly transfer of some three million Germans in 1945, the Allied Control Council could not hold back a much greater mass of expellees.[58] The first phase of flight and expulsion was the most violent; two million Germans are said to have died or been lost in the course of this colossal upheaval. Continuing for the rest of the decade, the forcible transfer of Germans eventually involved over twelve million people. According to German sources in 1950, 4 million of these officially designated refugees (*Vertriebene*) were in the Soviet zone, and over 8 million were in Western Germany. Subsequent calculations, often including fugitives from the Soviet zone, indicated that nearly ten million Germans were refugees in the Federal Republic in mid-1951—more than one-fifth of the entire population. A similar flow of Germans to Austria involved about 500,000 by 1952, about 7 percent of the Austrian population. Possibly half of all these refugees left their homes before the Nazi collapse or immediately after Germany's surrender; the rest were removed as a result of deliberate policy, to which the Western Allies formally agreed. The majority of the latter, about 3.5 million, came from Czechoslovakia; Poland was next with 1.5 million, and then Hungary with 500,000.[59] The uprooting of such an enormous mass of people left deep scars on the face of Europe, some of which took more than a decade to heal. For a time, relations between the expellees and other Germans were severely strained; opinion surveys of the German population in the postwar period showed high levels of resentment towards the newcomers and a corresponding sense among the refugees that they were poorly received.[60] In the extreme penury of these years, Germans frequently blamed each other for the scarcities of food, shelter, and basic amenities. In political terms, hostility and bitterness ran strong. The French strongly objected to the naturalization of so many Volksdeutsche, seeing in the enormous expansion of the West German population a significant strategic threat. Within Germany, a burning sense of the injustice of the expulsions persisted and prevented normal relations with Eastern European governments until the 1970s. Economic and social assimilation took considerable time and energy despite the efforts of the West German government, which established a special ministry for the purpose. On the other hand, the inexpensive skilled labor of the expellees is now seen to have helped fuel the *Wirtschaftswunder*, the remarkable economic recovery made by the devastated western zones of Germany after 1950. Aided by the Marshall Plan, the Federal Republic put the refugees to work, significantly outperforming all other European countries. To the surprise of most outsiders, West Germany proved capable of absorbing all eight million refugees, few of whom emigrated in the immediate postwar years.[61]

Jews

There were remarkably few Jews in the great ocean of refugees and displaced persons that appeared in mid-1945. As we have noted, the discovery of emaciated Jewish concentration camp inmates in places like Dachau, Buchenwald, and Bergen-Belsen horrified even the most hardened soldiers and military correspondents. But even more chilling, on reflection, was Hitler's evident success in carrying out a program of mass murder. Of the huge Jewish communities of Central and Eastern Europe, only fragments remained. In Poland, which once had a Jewish population of 3.3 million, there were 80,000 alive at the end of the war, for a loss of 97.5 percent; in Germany, where the proportion massacred was only slightly less, survivors numbered about 50,000 of a 1939 population of 221,000. In 1945 the Allies encountered no more than 100,000 Jews in Central Europe, between 5 and 8 percent of the DP population. More Jews than this had survived Hitler, of course. We now know that about one million remained in Europe outside the Soviet Union. Some of these were liberated outside the Reich, others did not declare themselves as Jews to the liberating armies, and still others were not enumerated as such by the refugee workers in charge.[62]

Ironically, the growing awareness of what the Jewish remnant had suffered did not prompt a mass outpouring of sympathy. On the contrary, the Jewish survivors were frequently thought a troublesome lot, as we have noted. From the start, military commanders in the western zones refused to accept that the Jews posed special problems or had particularly urgent needs requiring attention. The soldiers believed the Jews should be classified by their former nationality; could be housed with antisemitic and collaborationist elements from Poland, Hungary, the Ukraine, or the Baltic states; and would return eagerly to their former countries of origin. Military instructions and political considerations resisted any notion of Jewish particularity. In May 1945 the Displaced Persons Branch of SHAEF reiterated that "the problems of Jewish displaced persons are similar to those of other stateless and displaced persons persecuted by reasons of race, religion or political affiliation."[63] Reflecting American opinion, General Eisenhower insisted that there should be "no differentiation in treatment of displaced persons." The British were particularly worried that any implied recognition of a Jewish nationality or any acknowledgment of a Jewish claim to special consideration might lend weight to Jewish demands over Palestine. With thinly disguised irritation, British officials rejected all appeals directed only at Jews. "One must remember," Prime Minister Clement Attlee wrote to President Truman in November 1945, "that within these camps were people from almost every race in Europe and there appears to have been very little difference in the amount of torture and treatment they had to undergo [sic]. Now, if our officers had placed the Jews in a special racial category at the head of the queue, my strong view is that the effect of this would have been disastrous for the Jews. . . ."[64]

In reality, the experiences of these Jews made it impossible for them to think and behave like other displaced persons. A vast gulf of agony and humiliation yawned between them and the rest of humanity. Prematurely aged, weak, and suffering from extreme malnutrition, many presented a ghastly appearance. About 40 percent of the Jewish DPs liberated in Germany perished within a few weeks after the arrival of the Allies. Months after V-E Day survivors still wore their hideous striped concentration camp garb; others remained in the very camps where they were found by the Allied troops. Unlike other refugees and displaced persons who were frequently with family groups or neighbors, the Jews were usually alone, having lost most of their relatives and acquaintances. After the euphoria of liberation, many survivors sank into despondency, exacerbated by the lack of change in their situation. An American army intelligence agent, posing as a Jewish military police officer, reported that the survivors showed increasing "social and psychological isolation from the world around them." They were suspicious of all outsiders, even if Jewish. Disillusioned and irritable, they bitterly resented that their agonies had been ignored and their persecution had not received due recognition. "Sensitivity of Jewish DPs in the [American] zone has reached a neurotic point," he observed. "Any slightest remark or any official measure, be it one not even intended to apply to them, would be discussed on a single criterion: 'Is it or is it not antisemitic?' "[65] Above all, having been victimized so cruelly as Jews, they insistently advanced Jewish claims on their own behalf, beginning with the insistence that they be grouped together.

Military men sharply denied any hostility toward the Jews, and it seems plain that many were simply too impatient, ill-equipped, and temperamentally unsuited to deal with the difficulties Jews presented. Beyond this, however, there was also an undercurrent of anti-Jewish hostility in both the British and American armies that occasionally surfaced in repressive measures taken against Jewish DPs. The latter were commonly suspected of black market operations, theft, and indiscipline—as though these were peculiarly Jewish vices. Soldiers sometimes contrasted the Jews—demoralized, brutalized, and half-starved—with the healthier, well-scrubbed, or better-organized Balts or other groups. To be sure, not everyone in authority felt as General Patton, for whom "the Jews . . . are lower than animals." But clashes between Jewish survivors and both the British and American military occurred often enough to suggest a general pattern.[66]

During the spring of 1945 protests against these conditions mobilized in the United States. Earl Harrison's report, delivered to the president in August, was a bombshell and helped shift American policy in favor of Jewish refugees. Three months after V-E Day, Harrison indicated, little had changed for the Jewish displaced persons. They remained in deplorable conditions, often crowded into the most notorious camps, without rehabilitation and with frequently mediocre care. "As matters now stand," he concluded, "we appear to be treating the Jews as the Nazis treated them except that we do not exterminate them. They are in concentration camps in large numbers under our military guard instead of SS troops. One is led to wonder whether the German people, seeing this, are not supposing that we are following or at least condoning Nazi policy." The practical needs were obvious. But Harrison put a political gesture at the head of the list: "The first and plainest need of these people is a recognition of their actual status and by this I mean their status as Jews." It was desperately important to remove the Jews from the camps as soon as possible. Even at this point it was plain that many Jews, especially those from Poland or the Baltic states, had no desire to return home. To the consternation of the British, the Harrison Commission announced that the Jews overwhelmingly preferred to go to Palestine. That issue, the report said, "must be faced." "To anyone who has visited the concentration camps and who has talked with the despairing survivors, it is nothing short of calamitous to contemplate that the gates of Palestine should be closed." This was undoubtedly the strongest recommendation Harrison made and one that Truman soon adopted as his own.[67]

The Harrison report raised the delicate question the British were anxious to avoid: What were the Allies to do with the Jews remaining in the camps? By considering Palestine an obvious answer to the Jewish refugee question, Truman signaled a break with London. At Potsdam, the American president urged Churchill to allow the Jews to enter Palestine; accepting Harrison's recommendation, he went on to advise the new government of Clement Attlee that it issue 100,000 Palestine immigration certificates to Jewish refugees. This, Truman assumed, would remove the embarrassment that the Jews presented in Europe. And it would do so, he may also have reflected, without any confrontation with Congress over American immigration restrictions. Responding to this unwelcome advice, the British proposed an Anglo-American Committee of Inquiry on Palestine, hoping to engage the United States in seeking a more palatable solution. Among the tasks of that committee was to ascertain what the Jewish displaced persons themselves wanted and what was the practical possibility of sending them to Palestine.

Completing its work in April 1946, the Anglo-American committee reported what Harrison discovered a year before. The camps were a nightmare. Richard Crossman, of the British delegation, observed that "an appalling demoralization had set in which could not be halted by good food or accommodation or educational facilities." Repatriation, the obvious solution, was utterly impractical. The Jews in Central Europe had little desire to return to their countries of origin, often seen as a graveyard for the Jews. With more authority than the Harrison inquiry, the committee reported that the Jewish DPs wanted desperately to go to Palestine. Of 138,320 Jewish DPs enumerated in the British, American, and French zones (the Soviets did not permit entry to the committee), 118,570 declared Palestine their preferred destination. Many of these would have doubtless chosen a quiet life in America, but this was hardly an option at the time. With the doors to the United States still closed, Palestine was widely considered their only hope.[68] Urging that all countries accept some of the Jews who could not return to their homes, the committee also adopted Truman's earlier recommendation: 100,000 Jews should be allowed to enter Palestine.

It is often said that if the British had acceded to this demand, they would have solved the Jewish refugee problem in Europe and defused a major Zionist argument on behalf of the Jewish state. The Jewish Agency, we are told, trembled lest the British decide to allow 100,000 Jews to immigrate. Refusing to budge, however, London soon saw its own position deteriorate—not only in Palestine itself, where violent Jewish pressure against the British mandate rose to a climax, but also in Europe, where the pool of Jewish refugees began an extraordinary growth that evoked much sympathy for the Zionist cause. This last was the great surprise for UNRRA workers and the military men in charge: as the overall number of DPs continued to diminish, the Jewish total steeply climbed. And it was no small irony that the one country whose Jewish population was growing was Germany, which had just unleashed the most destructive genocidal attack in Jewish history. In December 1945, UNRRA listed only 18,361 Jews receiving assistance in the various zones of Germany; by June 1946, it counted 97,333; and a year later, 167,531. The combined total for Germany, Austria, and Italy was now close to 250,000.[69] Unexpectedly, Jewish survivors of the Holocaust crowded into Central Europe, many of them hoping to reach Italy and then sail for Palestine.

One force behind these events was the catastrophic situation that many Jewish refugees encountered when they returned to Eastern Europe. Often it was only on repatriation that these Jews fully grasped the results of Nazi genocide: practically everyone they knew had disappeared, and everything they owned was destroyed or in other hands. Moreover, the stunned and bewildered Jews encountered massive hostility, particularly in the Polish areas, where many hoped to resettle. In this sense, little seemed to have changed from the prewar era. Reporting from Warsaw in December 1945, British Ambassador Victor Cavendish-Bentinck reported that "the Poles appear to me to be as antisemitic as they were twenty-five years ago."[70] Bitter quarrels erupted over claims for the restitution of apartments or other property that had been plundered, stolen, or confiscated. Nationalist thugs accused Jews of being in league with the Communists. Before long there were serious outbreaks of violence. Anti-Jewish riots began well before the European war ended, likely launched by right-wing extremists. Time proved that these were not isolated incidents. During the next two years a series of pogroms rumbled across Poland, Hungary, and Slovakia. The worst occurred in the

summer of 1946 in the Polish town of Kielce, about 200 kilometers south of Warsaw. Of the Jewish community of 250, a mob
killed more than 40 and seriously injured many more—apparently with the help of local army units. According to one student
of these events, "1,500 Polish Jews were murdered or died in pogroms between the end of the war and the summer of 1947." More
Jews were killed in this period than in the entire decade before
the Second World War. Writing in November 1946, Zorach Warhaftig declared that "the war is not yet over for European Jewry."
Incredibly, a murderous antisemitism still continued.[71]

Fleeing these outrages, Jewish returnees headed west. With
them went also a proportion of the Jews who had somehow survived in Poland during the war and also thousands of Jews who
had escaped eastward during the Nazi occupation and had managed to find refuge in the Soviet Union. In the spring of 1946 the
Russians released about 150,000 of these Jews, and many of them
continued west after a brief taste of life in postwar Poland. A fair
proportion would have probably agreed with the refugee who
explained his motives to Richard Crossman: "A Jew can no longer
remain in the city where he used to be happy with the other
members of his family, and where he can now see, in the face of
every man he looks upon, a possible murderer of his family."[72]
The Polish government encouraged the exodus, readily granting
documents to the Jews valid only for exit from the country. Im-
pelled by similar forces, Jews from Hungary, Slovakia, and Galician regions incorporated into the U.S.S.R. took to the roads,
part of the broad migration to areas occupied by the Western Allies.

Within West Germany, Jews also moved, seeking, when possible, to enter the American zone, where conditions for them were
somewhat better than elsewhere. Following the Harrison report
and new presidential instructions to Eisenhower about DP camps,
the Americans began to group Jews together and make some efforts at improving their situation. In practice, these American
moves probably facilitated the tasks of the Zionist underground,
which was now engaged in a vast, often clandestine effort to move
the Jews westward, to focus refugee pressure on Palestine, and
actually to smuggle some of the refugees to the Holy Land on
ships from southern France and Italy. This gigantic effort, known
by the Hebrew code word *Brichah* (flight), accelerated powerfully
after the Kielce pogrom in July 1946.

Having begun spontaneously in 1945, the migration of Jews

received support from the Jewish Agency in Palestine and the Joint Distribution Committee. From various quarters at once, the leaders of *Brichah* mobilized to channel the mass movement of refugees, directing them along political paths leading to a Jewish state. Veterans of the Jewish Brigade that had fought in Italy, Holocaust survivors, and emissaries from Palestine all lent a hand. Occasionally, organizers also received help from sympathetic American military personnel. By 1946 an elaborate network of agents and couriers was in place, moving tens of thousands of people through frontiers and checkpoints by rail, using newly purchased trucks, and sometimes on foot. Evading hostile occupation officials, making ample use of forged documents, local guides, and caches of food and clothing, *Brichah* resumed the course of illegal immigration to Palestine started before the war.[73]

To Allied representatives, particularly the British, the "infiltration" of Jewish refugees into their zones threatened the orderly processes of repatriation and subverted military authority over displaced persons. As early as January 1946, Lieutenant General Morgan, then head of UNRRA operations in Germany, publicly denounced the migration as a well-organized Zionist plot. At a press conference in Frankfurt, Morgan claimed that Jews arriving in Berlin were "well-dressed, well-fed, rosy-cheeked, and had plenty of money." Morgan disparaged the talk about pogroms in Poland and alluded to the dark political purposes behind the exodus. The goal of the operation, he wrote in his memoirs, was "the promotion and sustenance of armed aggression by the forces of Zionism against the British garrison stationed in Palestine. . . . I could not exclude from my mind the thought that the ultimate aim of all this endeavour was death to the British."[74] Expressing other fears, American intelligence officers worried that many of the refugees, especially those who had served in the Red Army, could be hostile to the United States or even Soviet agents.

Tensions mounted as the flow of refugees increased. British troops did their best to stem the tide, and General Morgan, despite being chastised for his outspokenness at UNRRA headquarters, remained in place, opposing aid to the "infiltrees." In the summer of 1946, thousands of Jewish inmates still in Bergen-Belsen demonstrated against British attempts to turn away Polish-Jewish refugees. The British tried to seal their own zone against the newcomers, putting additional pressure on the Americans. The camps in Austria overflowed with Jewish refugees, and UNRRA aid facilities threatened to collapse under the weight of new arrivals. Despite efforts to hold back the flood, the Jews poured into Italy, seen as the springboard for Jewish emigration to Palestine. In early 1947 some 12,000 Jewish refugees were grouped into self-governing colonies of about 200 persons called *hachsharot*—the Hebrew word for preparation.[75]

The British failed to reduce the pressure accumulating in the refugee camps and centers. Realizing there was considerable sympathy for the Jewish remnant, especially in the United States, Foreign Secretary Ernest Bevin hoped that the refugees could be resettled anywhere outside Palestine. Unfortunately, the prospects were few. On the instructions of President Truman, the Americans admitted displaced persons on a preferential basis after December 1945, and about 12,000 Jews thereby managed to enter the United States.[76] But there were no other opportunities, and nothing could offset the arrival of ever more refugees from Eastern Europe. In the spring of 1946 the Foreign Office even sought Soviet assistance, hoping that Moscow would allow some of the Jewish refugees to settle in Birobidzhan, the Soviet-sponsored homeland for Jews in Soviet Central Asia. The reply from the deputy minister of foreign affairs, Vladimir Dekanazov, was both brutal and disappointing: "We've got enough. You've got the whole empire to put them in."[77] Calling attention to the stalemate, the Zionists began sending shiploads of refugees toward Palestine from secret locations in Italy. Few of these vessels managed successfully to run the British blockade, and almost all were intercepted by British naval patrols. Once captured, the refugees were taken to Cyprus for internment. Twenty-six thousand Jewish DPs were gathered there in 1948. Meanwhile, over 250,000 Jews remained uprooted in Central Europe.

Dramatizing the impasse in mid-1947, the Zionists sent a battered Chesapeake ferry, renamed the *Exodus*, from the French port of Sète crowded with 4,500 passengers. Closing in on the vessel off the coast of Palestine, the British boarded the *Exodus* after a furious struggle. For several hours the Jews relayed an account of the battle by radio to thousands of listeners in Palestine, and from there to the outside world. In London, Foreign Secretary Bevin decided to make an example of this ship by sending the refugees back to Europe. Predictably, this move backfired. Returned to France in British vessels, the Jews refused to disembark despite the stifling heat and cramped quarters. In the full glare of publicity, the refugees denounced the British and highlighted their own determination to settle in the Holy Land. Finally, the passengers were taken back to Germany, where many had to be forcibly carried ashore at Hamburg. Far from having chastened the Zionists, the *Exodus* affair drew more sympathetic attention to the Jewish refugee problem in Europe than ever before. The British were treated to three months of unfavorable press at an extremely sensitive moment in negotiations over the future of Palestine. The Jewish refugees had proved thereby that they were not quite so impotent as they had first appeared.

In the last act of this drama, the British decided to submit the Palestine question to the United Nations in February 1947. Later that autumn, following the recommendation of its own investigative committee, the General Assembly voted to partition Palestine, creating a Jewish and an Arab state. Fighting between Jews and Arabs broke out immediately, to flare into full-scale war the next year.

To the very last, even as the British administration in Palestine was withering away, the British held firm on the question of Jewish immigration, attempting to keep the rate roughly at the 1939 level—1,500 a month. With the independence of the new state of Israel only a few months away, London allowed an additional trickle of Jews to leave the camps for Palestine. Three hundred a month were permitted in an operation code-named Grand National—after a famous English horserace and supposed, by at least one observer, to allude to the unusual numbers of obstacles still set across the course.[78] The final accounting testifies to the force of British restrictions: from 1 January 1946 to 15 May 1948, when Palestine passed from British control, 48,451 European Jews entered Palestine; 30,000 of these were illegal entrants, subsequently recorded by the mandatory administration.[79] Immediately following the formal establishment of the Jewish state, its government flung open the doors to Jews. Everywhere Jewish inmates streamed out of the camps. First to come were 25,000 formerly illegal immigrants interned by the British on Cyprus; through the remainder of 1948 new arrivals averaged over 13,500 per month, as war raged between Israel and her neighbors, prompting a mass exodus of Palestinian Arabs. Immigration rose to 20,000 per month in 1949. During the first year and a half of independence, 340,000 arrived; mass migration continued until the end of 1951, doubling the country's Jewish population in five years and putting an end to the Jewish refugee problem in Europe.[80]

THE IRO AND THE "LAST MILLION"

The International Refugee Organization (IRO) emerged after sometimes heated United Nations debate over the "last million" refugees in Europe—those who could not be repatriated by UN-RRA and to whom the Western nations felt a continuing obligation. Realizing that hundreds of thousands of displaced persons would remain behind after the work of UNRRA was completed, British representatives to the Inter-Governmental Committee on Refugees and the United Nations urged additional UN involvement in this field toward the end of 1945. Deeply preoccupied with the Palestinian mandate, London may well have been seeking to diffuse responsibility for unsettled refugees, thereby weakening the Zionist case against British immigration restrictions in Palestine. In any event, the proposal received powerful American support. The United States UN delegate, Eleanor Roosevelt, favored international action for broad humanitarian reasons in the interests of world peace. The Soviets, Yugoslavs, and eventually other Eastern European governments were extremely suspicious of the idea, but were nevertheless drawn into extensive discussions over the proposed organization. Outlines of the new agency worked their way through the complex circuitry of the United Nations after the first session of the General Assembly in London in January 1946. Referred to the General Assembly's Third Committee, which dealt with social, humanitarian, and cultural questions, the plans later proceeded to the Economic and Social Council. The latter established a Special Committee on Refugees and Displaced Persons to work out details. In the second half of 1946, the process reversed itself, finishing at the General Assembly, which approved the IRO constitution and its budget on 15 December. An entire year was thereby spent debating the issue, which turned out to be much more controversial than most suspected when the matter was first raised.

The IRO debates divided East and West much as they were divided over Iran, German reparations, and many other questions in the immediate postwar period. Although international relations remained fluid in 1946, one can see in this confrontation the delineation of two power blocs, led by the Americans and the Soviet Union. From the standpoint of the Russians, the refugee problem was nonexistent: the real difficulty was the continued presence of large numbers of what Andrei Vyshinsky called "quislings, traitors and war criminals" in the DP camps, refusing to return home after liberation.[81] Supported by Soviet delegates from Belorussia and the Ukraine, Moscow's representatives charged that Western propagandists were manipulating refugees from Eastern Europe and permitting the mobilization of anti-Soviet groups in refugee centers. Representatives from Poland and Yugoslavia agreed and were sometimes even more vituperative in their denunciations of émigré elements. These delegates demanded the prompt return of their own nationals and believed that this should be the sole function of an international refugee agency. Under no circumstances did they want their governments to contribute to the maintenance and resettlement of hostile émigrés in Europe. Western representatives reacted cautiously toward the Soviets and were eager to avoid a break with Moscow over this issue. They agreed that the new refugee organization would emphasize repatriation, and they supported various measures to encourage the return of Eastern European nationals during the preparatory period before the formal establishment of the IRO. But they refused to continue the forcible repatriation previously undertaken, and they resisted Soviet demands to control free expression in the camps. An American negotiator, Ernest Penrose, later emphasized the efforts of Britain and the United States to compromise with the Russians, to respond to their understandable concerns. He found the Soviets extraordinarily persistent, returning to specific points again and again. The British and American delegates avoided the harsh language that later characterized cold war discourse, meeting privately with Soviet representatives to reach agreement on particularly sensitive issues.[82]

East and West disagreed sharply over which groups were genuine refugees entitled to international assistance. Western countries sought to include large numbers of dissident and anti-Communist elements; Eastern countries tried to exclude those whom they believed were deadly political enemies. Both sides agreed that the international community had no obligation to the many millions of Germans who remained unsettled in Europe in 1946. Beyond this, the IRO reflected a broad compromise between East and West, designating broad categories of persons to be assisted rather than offering an abstract definition to be used for all cases. The organization's constitution considered as "refugees" the uprooted victims of the Nazi and Fascist regimes (including those expelled before the war), Spanish Republicans, and other prewar exiles; among these were also persons unable or unwilling to return home "as a result of events subsequent to the outbreak of the Second World War." "Displaced persons" were deportees and forced laborers who were similarly victims of the Axis powers. Specifically excluded from the mandate of the agency were "war criminals, quislings, and traitors" and other wartime collaborators. While stressing that the organization was not to assist hostile activities against countries of former nationality, the constitution set strict conditions under which individuals could refuse repatriation. In this, the drafters of the document broke new ground and eventually paved the way for a broader notion of refugees. Refusal to return home could be based on "persecution, or fear, based on reasonable grounds, of persecution because of race, religion, nationality or political opinion, providing that these opinions are not in conflict with the principles of the United Nations." Even more broadly, the IRO would permit "objections of a political nature judged by the organization to be ,valid." In thus considering the individual's reasonable expectations of persecution to play a role in refugee decisions, the IRO foreshadowed the definition of refugees adopted in 1950 by the United Nations High Commission on Refugees.[83]

Conceived in the protracted discussions of 1946, the IRO needed another two years to come formally into existence, awaiting the ratification of its constitution by at least fifteen governments. In the interim, a Preparatory Commission of the IRO (PCIRO) set up shop in Geneva and took charge of an estimated 1.5 million refugees worldwide (excluding those in the Far East). Intended to be temporary (its assignment was assumed to require no more than three years), the IRO was to encourage an early return of refugees to the countries of origin and, failing that, to assist resettlement of those who remained. Notably, the United Nations kept the new organization at arm's length, reflecting the sharp divisions provoked by its establishment. The IRO was termed a nonpermanent specialized agency of the UN—permitted to operate outside UN supervision, with its own budget and membership. Of the thirty governments that supported the IRO agreement, only eighteen ratified its constitution in the end, and these countries alone became part of the agency. Despite considerable efforts to propitiate the Eastern European states, none of the latter ever joined the organization or supported its work. Throughout the life of the IRO, which did repatriate some 54,000 refugees to countries under Soviet control, Communist officials consistently accused it of protecting traitors and serving as an instrument of American policies. From another quarter, Americans suspicious of the Soviets denounced the degree of compromise with the Russians already undertaken, pointing out that the U.S.S.R. and its satellites did not join the organization anyway.[84]

In practice, the IRO became the instrument of the Western powers, chiefly the United States, which contributed over half of its operating funds. Sharply critical of the previous UNRRA operation, American policymakers determined to control the workings of the new organization—justified, they felt, by the huge American share of the IRO budget. Three successive executive secretaries who ran the administrative operations were Americans, and while they directed an international staff from forty nations, there was no mistaking their style. A Frenchman who entered the IRO headquarters in Geneva described the "bouffée d'air américain" that greeted him: offices and corridors "in democratic uniformity, simple wooden furniture, banks of telephones, cleanliness and austere comfort, along with an atmosphere of good humor and cordiality." Unlike UNRRA, which worked constantly at the end of a tight leash held by the military, the IRO had its own resources and could negotiate agreements with zonal authorities as well as with various governments. To facilitate the movement of refugees, what has been called "the greatest organized transatlantic exodus in history," the IRO acquired its own fleet of forty vessels—mostly rented from the United States. To activate concern over refugees and engage governments to receive them, the IRO worked closely with sixty volunteer relief agencies from various countries. In the achievement of its task, the IRO spent some 450 million dollars, an extremely large sum for the time, and did so while maintaining a high reputation for efficiency and integrity.[85]

On 1 July 1947 the IRO's Preparatory Commission took charge of 643,000 UNRRA refugees in Europe, in addition to nearly 70,000 formerly in the care of the Inter-Governmental Committee on Refugees and the Allied zonal authorities. All of these were still living in camps and refugee centers in Germany, Austria, and Italy. The largest group were the Poles—nearly 300,000, including 125,000 Jews. There were, in addition, 150,000 Balts, 107,000 Ukrainians, 30,000 Yugoslavs, and miscellaneous other groups, including some as yet unsettled Nansen refugees of prewar vintage. Outside the camps and centers were another 500,000 refugees within the new mandate, living mainly in Germany but also in Austria, Italy, and elsewhere.[86] Complicating the work of the new organization, there was a constant influx of new refugees claiming to fall within the designations set forth in the IRO's constitution. These included Jews swept along in the *Brichah* migrations, people fleeing the establishment of Communist regimes in Eastern Europe, and various other non-German displaced persons who had not been deemed refugees by UNRRA.

After a few months of operation, it became clear to IRO officials that there was little prospect for repatriation, given the hostility of most Eastern European refugees to the Soviet regime and its domination of much of Eastern Europe. Fifty thousand were repatriated to Eastern European countries in the first twelve months, but only a few thousand followed in subsequent years. The Soviets energetically attempted to encourage return, sending teams of aggressive repatriation officers throughout Central Europe; the results were meager, however, prompting bitter Russian protests against allegedly unfair restrictions on their salesmen and the dissemination of anti-Soviet propaganda. Meanwhile, dramatic events in Europe confirmed the most pessimistic assessment of Stalinist expansionism: in February 1948 came the Communist "coup" in Prague, followed in a few months by the establishment of full Communist control in Hungary and the imposition of the Berlin blockade. In response to each of these events, the numbers of repatriates declined precipitously. Poland, Yugoslavia, and Czechoslovakia then closed the IRO liaison offices formerly operating in each of their capitals.

IRO workers hoped to resettle refugees in Western Europe, identifying ways in which new immigrants could assist postwar reconstruction. The first response was caution: in the countries ravaged by war and demobilizing large numbers of men, it took time for needs to become clear and for priorities to be set. Given the enormous size of their task, IRO officials began to go further afield—to North and South America, as well as to Australia and New Zealand. By 1948, the tempo of resettlement quickened. IRO representatives attempted to link the economic needs of various countries with particular skills among the DP population. The organization, in Proudfoot's term, functioned as "an international employment agency" for two dozen different states. In all, the IRO resettled 1,039,150 persons between mid-1947 and the end of 1951. Over three-quarters of these went to four countries: the United States, which cleared the way to large-scale immigration with the Displaced Persons Act of 1948, received 329,000;

Australia, which negotiated a compulsory labor scheme in 1948, took 182,000; Israel, which financed and organized Jewish settlement without UN help, given the organization's reluctance to do so during the Arab-Israeli war, brought 132,000; and Canada, which mainly preferred manual laborers, settled 123,000. Of the European states, which received a total of 170,000, the United Kingdom allowed entry to 86,000; then came France with 38,000 and Belgium with 22,000.[87]

Given the IRO's focus on finding employment for refugees, its sequel was understandable enough: by mid-1949, after the bulk of migration was finished, nearly 175,000 people remained in the full care of the organization, unlikely candidates for emigration elsewhere. This was the "hard core" of unsettled refugees, people whose occupation, health, age, or some other condition made their removal extremely difficult. In the middle of the following year, the German and Austrain governments assumed responsibility for the bulk of these displaced persons as the work of the IRO wound down. Despite the humanitarian gestures of a few governments, most of the "hard core" were to remain in Central Europe—having "passed through the sieves of nations," as one refugee worker put it.[88] From Eastern Europe, between 1,000 and 1,500 refugees flowed monthly into Turkey, Greece, Trieste, Italy, and Austria.[89]

In keeping with the limitations of its mandate, the IRO began to restrict its commitment to new refugees in mid-1949. That year, the United Nations General Assembly formally launched a UN High Commission for Refugees, intended to deal with a continuing global refugee problem. By this point, the Americans were pouring capital into Europe through the Marshall Plan and assumed that economic development would soon permit the Continent to absorb refugees on its own. Despite widespread support for continuing the agency, the United States, on which its future depended, insisted on termination. Much work remained to be done. Millions of German refugees remained unsettled, along with the "hard core," along with many thousands of others who had not qualified for IRO assistance. Having performed admirably in a crisis situation, the IRO was not empowered to search for durable solutions to refugee problems. International refugee work returned to the United Nations, which first took up the task in 1943.

EPILOGUE

Contemporary Europe

FOR Europe, the era of refugees may be finished. The attention of refugee agencies has passed to the Third World, where in recent years millions of desperate wanderers pose the most serious challenge to the international order. Within the Old Continent many continue to seek asylum, and thousands of refugees remain excluded from the national community. But these are relatively few in number. Other outsiders, millions of foreign workers, have now become the principal expression of European homelessness. How has this change occurred? Prosperity is part of the answer. Economic recovery in Western Europe during the 1950s provided favorable conditions for integrating millions of uprooted people. Similarly, economic growth encouraged overseas countries to help drain the refugee camps. At the same time, refugee flows have been staunched by the stabilization of Europe into two power blocs dominated by the Soviet Union and the United States. The Soviets' blockade of Berlin, lasting eleven months until May 1949, consummated the division of Germany, leading to a consolidation of the two-bloc system by the mid-1950s. This postwar settlement produced hundreds of thousands of refugees at its inception, and continues to generate fugitives fleeing political repression in Soviet-dominated countries. Occasionally, spasms of revolt in Eastern Europe send out new waves of exiles. However, the absence of armed conflict between European countries and the firm control by the Soviets and their allies over their societies have prevented the great inundations of times past.

REFUGEES AND THE ORIGINS OF THE COLD WAR

Since the end of the 1940s, the confrontation between the Soviet bloc and the Western powers has been the main source of Europe's refugees. Following the historic path of exiles from east to west, hundreds of thousands of people from the Soviet Union or Soviet-dominated countries have left their homes, seeking asylum in Western Europe or elsewhere. Given both the repressive conditions in the regimes from which they came and notably the circumstances of their departure, Western authorities have usually referred to them as refugees. Unlike most exiles in the Nazi era, however, they have not been expelled from their former domicile; generally, they have emigrated illegally or in flagrant contravention of government policy, and they see themselves as unable to return because of the likelihood of persecution at home. Invariably, these fugitives became themselves a source of political conflict. By the late 1940s the primary focus of refugee controversy was no longer relief for huge masses of people; rather, East and West quarreled over how refugees affected the international order and whether solemn international agreements were being respected.

As Communist elements imposed an ever firmer control in Poland, Hungary, Yugoslavia, Rumania, and Bulgaria, a wide variety of fugitives headed west—often young people, some seeking to avoid military service, middle-class opponents of the left, and political opposition figures. Others, like the communities of Eastern Orthodox Christians fleeing Yugoslavia, were religious exiles, anticipating the installation of a Marxist regime. Most of the refugees went to Germany, avoiding Soviet controls by moving through Czechoslovakia and ending up in the American zone. Significant numbers of non-Jewish Poles followed the path of *Brichah* in the immediate postwar years, moving through East Germany and often seeking the American-occupied zones of Central Europe. Anti-Tito elements in Yugoslavia streamed out of their country through Trieste, making their way to Italy and elsewhere in Western Europe. Through the Balkans, corridor for many hundreds of thousands of fugitives for close to a century, refugees were on the move once again. Rumanians and Bulgarians, feeling the drift of their own governments to the Left, went into Greece, particularly after the peace settlements of early 1947 with Italy, Rumania, Hungary, and Bulgaria sanctioned the emerging Communist regimes in southeastern Europe.[1]

During this period, the Soviets concentrated their energies on their European frontier, deeply aware of their own international weaknesses and vulnerability in the face of the West. Having achieved victory at such enormous cost and fearing the predatory designs of the Americans, Stalin seems to have decided on a defensive posture, but one that required the consolidation of wartime territorial gains and a broad band of friendly states on his western borders. Before long, this led to serious confrontations. Western observers were affronted when, in one country after another, the Soviets manipulated postwar coalition governments to remove non-Communist members. Soviet fears and suspicions grew following the proclamation of the Truman doctrine and the announcement of the Marshall Plan in 1947. In a series of brutal countermeasures, Stalin sponsored the systematic elimination of non-Communist opposition elements, a coup in Czechoslovakia in February 1948, and the blockade of Berlin in June. By the summer, Stalin's victory in Eastern Europe was complete. "People's Democracies," a thin theoretical screen for full Communist control, ruled throughout the western marchlands of the Soviet Union outside of Finland.

Soviet concerns about refugees from Eastern Europe emerged sharply in the years 1947–48, when Communist parties throughout the region dominated the postwar governments. Up to that point, the Russians had certainly been difficult negotiating partners on refugee matters and had insisted that Western countries return those who refused repatriation. But in private conversation Soviet representatives freely acknowledged that many of the displaced persons rejected the Communist system and that it was unrealistic to force them all to go home.[2] As cold war positions hardened, however, the Soviets lashed out at the operations of the IRO, clamped down on those attempting to flee, and demanded the return of those who had managed to do so. The charge against the IRO was that it allowed Western governments to use refugee centers as "slave camps," recruiting cheap labor to fuel reconstruction projects. According to the Soviet version, Western authorities terrorized and propagandized uprooted refugees seeking simply to return to the U.S.S.R. More plausibly, Moscow charged that the West recruited spies and anti-Soviet intelligence experts from the large numbers of pro-Nazi East Europeans who had fought against the Soviet Union during the war.[3] Responding to these efforts, the Soviets and their satellites grew more restive on the refugee issue. On hearing of 9,000 Ukrainians detained in Italy as "surrendered enemy personnel," the Russians demanded custody of the fugitives; when Baltic refugees attempted to escape to Finland, Moscow obliged the Finns not to accept the refugees. Other Eastern European countries followed suit. In early 1948, Poland clamped down: Polish authorities refused to issue passports without a promise of a visa from a country of resettlement, virtually halting the emigration process. The Bulgarians reacted with extraordinary harshness against anyone attempting to leave the country without permission: escapees were declared traitors, punishable by death; their relatives were similarly subject to imprisonment and heavy fines. Mobilizing entire populations now for the rebuilding process, the Soviets and Eastern Europeans were hardly in a mood to see large numbers of defectors pursue a comfortable life in the West.[4]

The most bitter confrontation, signaling a hardening of views on both sides, occurred during 1947 and 1948 over thousands of unaccompanied children whom the Soviets and the Poles insisted be sent back to their countries of origin. The Nazis originally brought many of these young people west in the infamous *Lebensborn* program, designed to recover "Aryan blood" from Eastern Europe. German racial planners sought to replenish the German stock by literally kidnapping children deemed to have desirable "racial characteristics"—typically tall, blond, blue-eyed youngsters, seen as fit human material to join the master race. Many thousands were uprooted in this way.[5] After the war, many were recovered and quickly repatriated. But thousands more could not easily be identified, having been adopted by German families or lost in the confusion of the Nazi collapse and postwar chaos. Others were orphaned during the conflict and were not immediately reclaimed by Eastern representatives. During 1948 the Soviets and their allies insistently demanded that the children be returned—whether or not their parents were alive, irrespective of the children's wishes, and regardless of the suitability of the homes that might have opened to them in the West. Spokesmen from the Soviet Union, Belorussia, and Poland demanded that "the kidnappings which occurred under Hitler's regime should be stamped out."[6]

Western representatives did not concert together on the issue before 1948 and offered no difficulties in the relatively few, straightforward cases where the children's nationality was known and the parents were alive. During the UNRRA period about 5,000 were repatriated. The Allies also sent home hundreds of Polish children who finished the war in India and East Africa, having been evacuated via Siberia. But when the Eastern European parents were not alive or when the nationality could not easily be determined, Western representatives raised serious difficulties over repatriation. Military authorities in the British and American zones similarly resisted turning over to the Soviets those Baltic or Ukrainian children who came from areas outside the 1939 boundaries of the U.S.S.R. Conflicting opinions on the matter rained down on the IRO officials, who referred the issue to the United Nations. After heated debate, the Economic and Social Council reached its decision: the children were to be united with their parents, but when this could not be done, "the best interests of the child" was to be the determining factor. Against the wishes of the Eastern Europeans, this opened the door to a whole series of quasi-judicial hearings held under the auspices of the Allied authorities, with principles and procedures utterly foreign to the Soviets and their satellites. Close to 2,000 children were finally repatriated to Eastern Europe, mainly Poland, and twice that number resettled, the overwhelming majority being dispersed among three countries—the United States, Israel, and Canada. The resolution of the issue left a great residue of resentment in the Eastern camp, particularly in Poland. The Poles claimed that 80,000 Polish children had been taken, of whom only a fraction were ultimately returned.[7]

Much of the refugee movement during the early years of the cold war occurred under the immediate threat of war. To Europeans traumatized by six years of destruction, the predictions of a new outbreak that periodically swept the Continent seemed utterly plausible. Emigré groups sometimes heightened these apprehensions for their own purposes. In the summer of 1946, Ira Hirschmann, then visiting a refugee camp in Wiesbaden, found a newspaper published by Polish anti-Communists warning the DPs not to go home: the Warsaw government, according to this view, would soon fight the Soviets, and the Poles would return afterward as conquerors. Surveys of refugees and displaced persons in the early 1950s revealed that even in the Eastern bloc countries many felt war was imminent. The Communists' own "peace" campaign against the Americans at the time of Korea convinced some that that liberation of Eastern Europe could not be long delayed.[8] Occasionally inspired by such predictions, flight was intended to remove people from a war zone and to seek protection in the more powerful West.

Others fled Communist countries in search of a better life, tracing the traditional path of European migration from east to west. Throughout Eastern Europe, the backbreaking efforts to rebuild shattered economies were slowed by the subordination of these societies to the central priority of Soviet reconstruction. After 1948 Soviet-style economic plans were adopted throughout the region, deliberately emphasizing heavy industry at the expense of consumer goods and imposing collectivized agriculture on an unwilling peasantry. Meanwhile, economic recovery in Western Europe, aided by the Marshall Plan, provided one of the most remarkable examples of successful postwar development, leaving the East far behind. The resulting disparity contributed significantly to the flow of refugees. Those interviewed in the West complained of a grim, Spartan life under Communism, attributed largely to the Soviet exactations.[9]

Harsh economic conditions were accompanied by political repression, invariably referred to by émigrés seeking asylum in the West. Particularly after the Soviets' dispute with Yugoslavia and the latter's expulsion from the Cominform in June 1948, Stalin was anxious to see that no other Communist-bloc countries followed this independent path. The result was imposed conformity, which permeated Eastern European societies. The late 1940s and early 1950s saw numerous show trials; purges; officially sponsored antisemitism; and accusations of espionage, disloyalty, and subversion. Within Eastern Europe, there was practically no escape. Some Communist dissidents managed to enter Yugoslavia, which offered the attractions of a planned economy minus Soviet control. Between 1948 and 1951 the Yugoslavs received some 8,000 of these schismatics, about half of whom soon decamped for unknown destinations.[10]

With the establishment of openly Communist governments in 1948, flight to the West became increasingly hazardous. Elaborate efforts were made to seal the borders to escapees; frontier police emerged as elite formations, entrusted with the crucial task of preventing flight. "Satellite border exit controls have been and are being constantly strengthened," an American refugee adviser reported in 1953. Trees were cut down along frontier regions; barbed wire, strung across open fields; land mines, implanted, along with signal rockets and detection devices. Field workers developed a new vocabulary—"escapee," "defector," "asylum-seeker"—to distinguish between various kinds of fugitives evading new restrictions on migration to the West.[11] Before long, police action cut the movement of refugees from Eastern Europe to very small numbers. Following the Communist take-over in Prague in February 1948, about 5,000 people left Czech-oslovakia for the American zone of Germany, and about 10,000 entered Austria. In all, some 50,000 Czechs quit their country be-fore these borders, too, were closed. Masses of refugees were able to leave East Germany because of the peculiar situation of Berlin, deep within the Soviet zone yet partly occupied by the Western Allies. Hundreds of thousands slipped into the West via this route, including many from other satellite countries who managed first to enter East Germany. When formal procedures for granting asylum were established in West Germany in 1949, 42,000 refu-gees were enumerated for the last three months of that year; during 1950 the total was 197,000; in 1951, 165,000; in 1952, 182,000; and in the turbulent year of 1953, when serious riots flared in East Berlin, 331,000.[12]

The one exception to this pattern of reduced emigration from Communist-dominated Europe occurred along the frontier be-tween Bulgaria and Turkey, across which refugees had washed for at least a half century. The authorities in Sofia, clumsily struggling to collectivize agriculture on the Stalinist model, faced serious difficulties integrating the substantial Turkish-speaking minority into the political, economic, and cultural life of the new socialist order. Charging in August 1950 that the Turks were un-willing to cooperate with an emigration program designed to re-duce this minority, Bulgaria suddenly demanded that Turkey re-ceive 250,000 emigrants. When Ankara failed to respond satisfactorily, the Bulgarians began expelling huge masses of un-wanted people, mostly Turkish-speaking but including thou-sands of Gypsies who were neither Turks nor Moslems. Tension between the two countries mounted until an agreement was worked out at the end of the year, largely to the Bulgarians' sat-isfaction. During the next six months, over 140,000 were vir-tually forced to emigrate from Bulgaria, most of them penniless and close to half either too young or too old to support them-selves.[13]

The West did not spontaneously open its doors to fugitives from Soviet-dominated Europe, most of whom failed to get assistance from the IRO or help from countries of first asylum. "Nothing is done for them when they arrive," wrote a reporter for *The New York Times* in December 1950, as refugees straggled into Western Europe. Nine months later the same newspaper carried an article headlined "West Bitter Haven For Red Refugees." According to the latter, escapees who reached the American zone of Germany stood a reasonable chance of being jailed for illegal crossing of a frontier. By 1952, nearly 200,000 anti-Communist refugees jammed into camps and centers in Berlin and West Germany, sometimes living in appalling conditions. To supplement the rudimentary assistance then offered in refugee camps, the American government acted on its own to resettle the refugees, doubtless hoping to facilitate the growing number of defections from the Soviet bloc. The United States may also have been eager to reduce the congregation of refugees in frontier regions adjacent to Eastern European borders, fearing the buildup of political tensions in the area. With the authority of the Mutual Security Act of 1951, President Truman allocated 4.3 million dollars for a United States Escapee Program (USEP) to assist refugees entering Western Europe and to make easier their integration or emigration elsewhere. Acting in concert with voluntary agencies, USEP established its headquarters in Geneva and built links with various refugee agencies. By 1955, it was reported, this program assisted the emigration of 10,000 refugees from West Germany. As cold war tensions increased, American public opinion became more sympathetic to the admission of escapees from Eastern Europe. Congress facilitated immigration with the Refugee Act of 1953, and in the next decade about 190,000 refugees from behind the Iron Curtain were resettled in the United States.[14]

SHAPING A UN AGENCY: UNHCR

Tightly bound by its mandate to resettle "the last million" refugees in Europe, the International Refugee Organization (IRO) began work as massive waves of refugees appeared suddenly throughout the world—a cruel reminder that solutions to refugee problems had constantly to be sought anew and were by no means limited to Europe. The partition of the Indian subcontinent in 1947 precipitated the slaughter of close to a million persons in religious riots between Hindus and Muslims, and launched millions of fugitives in the direction of India or Pakistan. Hundreds of thousands of Arabs fled Palestine during the Israeli War of Independence, to be dispersed among neighboring countries, and similar numbers of Jews were forced out of Arab lands, settling in Israel. The Korean War, a few years later, saw millions of

noncombatants streaming into the southern half of the peninsula. Meanwhile, on a much smaller scale but highly visible to
the Western media, escapees continued to sneak through the
barbed wire of Eastern Europe, joining the "hard core" of unsettled European refugees, many of them still languishing in camps
left over from the Second World War. Near the end of the IRO's
term, according to one careful estimate, there were about 15 million unsettled refugees throughout the globe who desperately
needed attention. Of these, 700,000 were residual IRO refugees
and fugitives from Communist-dominated countries, mostly living in Europe.[15]

As the millions of refugees moved to and fro, the UN Economic and Social Council pondered the prospective termination
of the IRO and heard suggestions for future United Nations refugee activity beginning in 1949. East and West snarled angrily at
each other over the issue, but without serious effect: the Soviets
and Poles boycotted the meetings of the council to protest the
presence of Nationalist China, and the deliberations proceeded
in their absence. At other points, the Soviet bloc, hostile to what
emerged from the discussions, protested loudly but ineffectually
along lines similar to their objections to the IRO. However, the
practical divisions were among the Western countries: all favored international action on behalf of refugees, but they differed sharply as to the scope and nature of refugee activity.
Western countries concurred on the need to restrict UN activity
to refugees for whom no state would assume responsibility—thus
excluding, for example, some 700,000 persons made homeless by
the civil war in Greece and millions still uprooted on the Indian
subcontinent. Furthermore, Arab and Korean refugees were assigned to their own, separate agencies, heavily supported by the
Americans—the United Nations Relief and Works Agency for
Palestine Refugees (UNRWA) and the United Nations Korean
Reconstruction Agency (UNKRA). Generally speaking, the United
States argued for a strictly limited role for the new body, fearing
to be drawn into unending obligations to the world's refugees.
Many governments from countries of overseas settlement took
the same view. West European representatives, on the other hand,
were often sympathetic to a more active organization that could
lift some of the burdens still posed by several hundred thousand
refugees on the Continent.[16]

What emerged at the end of 1950 was a compromise—an agency
closely circumscribed in the kinds of services it could offer, but

broad in its definition of refugees and the possibilities of future action worldwide. Organized as part of the UN Secretariat, the United Nations High Commission for Refugees (UNHCR) was to be an international spokesman for refugees and their legal rights while leaving to governments and voluntary agencies the task of maintaining and settling them. Granted a small budget by the UN—only 300,000 dollars in the first year—the High Commission could not expend a penny for direct assistance and could not ask governments for additional funds or make a general appeal without the approval of the General Assembly. At the same time, its scope included a wide range of fugitives—not only those made refugees before the statute came into force but also any other alien outside his home country who fears persecution in his country of origin because of threats based on race, religion, nationality, or political opinion. Beyond this, there were a few practical limitations: war criminals would not be helped; the UNHCR was not concerned with refugees already given the rights and obligations of nationals in the countries where they lived— effectively excluding German expellees, Chinese, Indians, and Pakistanis; Arab and Korean refugees were also excluded because they received aid from other UN agencies.[17]

Although carefully hedged with diplomatic language, the UNHCR definition represented a victory for those seeking a general designation of refugees that could be used in a variety of cases rather than the listing of national groups and specific categories of refugees such as had been done with the League of Nations High Commission, UNRRA, and the IRO. At the heart of the new definition was the notion of a "well-founded fear of persecution." Clearly, this was attuned to the European situation, with an accumulating mass of escapees from Iron Curtain countries who had never been expelled by the governments under which they had lived. In addition, the open-ended nature of the definition, being capable of application to future refugees, was a startling innovation. However, though they were prepared to empower a relatively inexpensive agency to "protect" such persons, many states were reluctant to assume obligations to future refugees or those outside the European continent. As a result, the 1951 Convention on the Status of Refugees, which was intended to prescribe international obligations to refugees and which also emerged from the UNHCR debates, defined the latter more narrowly than the UNHCR statute. The convention was explicitly limited to persons who became refugees as a result of events occurring before 1 January 1951; moreover, states becoming parties to the convention could further restrict its scope to events occurring in Europe.[18]

Following its establishment in 1951, the UNHCR had only a minimal impact on the situation of refugees in Europe. Shackled by a tiny budget and a staff of only ninety-nine persons, the UN organization surveyed 1.25 million refugees that came within its mandate—350,000 of whom were still unsettled in Europe, including 175,000 still in camps. Meanwhile, between 20,000 and 60,000 refugees flowed annually into Germany, Austria, Italy, Greece, and Turkey from various Communist countries.[19] The Dutch diplomat G. J. van Heuven Goedhart, high commissioner until his death in 1956, struggled against the limitations imposed on his office, particularly the severe financial constraints. Goedhart concentrated mainly on Europe, attempting to clear the remaining IRO refugees and to use his influence to resettle the more recently arrived fugitives from Eastern Europe. But there was never enough money even to provide proper legal and administrative services. The UNHCR obtained permission to raise an emergency fund, but was largely unsuccessful; like Nansen's High Commission, it became an international mendicant, constantly faced with the UN delegates' and the governments' flagging interest. After the High Commission was awarded the 1954 Nobel Peace Prize, Goedhart sold for 14,000 dollars a gold bar purchased by the Nansen Office with the proceeds of its own 1938 prize in order to meet emergency needs. Not only could the high commissioner not advance the "permanent solutions" so hopefully alluded to in the UNHCR statute, but he was also unable to organize any long-range projects on a significant scale, concentrating rather on encouraging integration of the refugees in the countries where they first arrived.[20]

Overseas, opportunities for immigration diminished in the early 1950s, and in parallel fashion governments and public opinion lost interest in the refugee problems of Europe and elsewhere. The 1951 convention relating to the status of refugees, a landmark international agreement that enshrined specific rights for refugees, did not come into force until 1954, when Australia's ratification provided the minimum six signatories to the document. Meanwhile, the dramatic economic recovery in Germany blotted up millions of refugees dumped into Central Europe after the Second World War. Goedhart seems to have spent much of his time as high commissioner simply convincing people that there remained a refugee problem. The United States quite obviously lacked enthusiasm for the agency, and the Eastern bloc countries wanted it disbanded—arguing that it simply perpetuated the refugee problem in Europe, which should be resolved by repatriation.

COLD WAR REFUGEES

After the consolidation of Communist dictatorships behind the Iron Curtain, the flow of Eastern Europeans to the West declined significantly. In part, this was due to the increasingly effective systems of control adopted by the various regimes—not only along their borders but also within, in terms of the repression of dissidents, the dissemination of propaganda about harsh conditions in the capitalist world, and the cultivation of an official anti-Western ethos. Moreover, political and economic conditions improved significantly in the decade after 1956, bringing a much more relaxed Communist order, fewer shortages, and an increased standard of living. Although an important disparity between East and West remained, the differences gradually narrowed. As this process continued, serious questions were raised as to whether those who did emerge from Communist Eastern Europe should be considered refugees. Although such people certainly lived under authoritarian rule, many were to some degree economic migrants interested in achieving a better material life, driven by the age-old motivation of Europeans on the move from east to west. Much turned on the notion of "persecution" referred to in the UN definition. Political scientist Aristide Zolberg distinguishes between those who have suffered from "nefarious political routine," probably the most common experience of governance in the world, and others "who are singled out targets for the wilful exercise of extraordinary malevolence on the part of some agent."[21] The implication is that only the latter are truly refugees.

In practice, however, exits often involve great danger, running a gauntlet of bureaucratic and police harassment, and frequently risking life itself. A significant proportion of Soviet and Eastern European émigrés have suffered such ordeals and are widely deemed refugees as a result. Others have fallen within the designation because of the political storms that have periodically shaken Eastern European regimes. Sometimes those caught up in these events have simply seized the opportunity to make a break for the West, still moved by economic motives; but frequently, too, these refugees are the product of the familiar cycle of opposition, protest, rebellion, and repression that has produced countless political exiles in the past.

Throughout the Communist world the death of Stalin in 1953 opened the door to one such upheaval. With the dreaded tyrant finally gone, creative energies long buried by layers of bureaucracy came into the open. The effervescence was first felt in Germany, where serious riots erupted in East Berlin in 1953, suppressed with the aid of Soviet troops. During 1956, this process accelerated rapidly in Poland and Hungary, two satellites with a strong anti-Russian tradition. In Poland, the independent course set by Wladyslaw Gomulka was contained within the framework of Communist orthodoxy, evolving peacefully into a compromise with the Soviets; in Hungary, however, the challenge to Stalinism begun in October produced an open break with the Kremlin and prompted fearsome repression. Swept into office by a popular uprising, Prine Minister Imre Nagy presided over a genuine grass-roots revolution that pushed toward a Western-style democracy. Unwilling to yield control of Hungary, the Soviets struck back at the revolutionaries: on 4 November a powerful Russian military force rolled through the country and smashed the Budapest-led revolt. The cost was heavy, including the lives of at least 3,000 Hungarians. Nagy fled to the Yugoslav embassy, and the Stalinists returned to power in the Hungarian capital. As a punitive measure, thousands of Hungarians were then deported to the Soviet Union.

The flight of Hungarians from their country in late 1956 has been termed "the largest spontaneous movement of a civilian population in Europe since the Spanish Civil War," and there are few comparable outpourings of population anywhere in modern times.[22] Refugees began to appear on the frontier in great numbers on 28 October, following the first, bloody clashes in Budapest between the insurgents and Soviet forces. According to some, the earliest "vintage" of refugees, those who left before the Soviets arrived en masse on 4 November, were for the most part nonpolitical and included Communists associated with the crumbling Stalinist order who feared popular retribution once the revolutionaries took control. Then, with the Russian tanks ensconced in the capital, more politically committed and generally youthful émigrés left the smoldering ruins of the uprising. Finally, along with the latter and continuing to the end of 1956, a great tidal wave of mainly middle-class Hungarians poured out of the country, moved by an urge to settle abroad. Through the extremely cold autumn the refugees came on foot, proceeding mainly to the Austrain border, which was lightly guarded by the Hugarians and the Soviets. They surged across at numerous frontier points, through the day and night, continuing on into the winter. By mid-December, when the exodus was winding down, 200,000 people had fled—2 percent of the entire Hungarian population.[23]

The great majority of these refugees, about 180,000, came to neighboring Austria, fully independent and officially neutral since the previous year, and finally free of Soviet occupation troops. By an extraordinary coincidence, Hungarian officials in Budapest had ordered the barbed wire and land mines along the Austrian frontier removed a mere three weeks before the uprising; and, together with the absence of most border guards, this cleared the path to the West. On 28 October the Austrian government decided to grant asylum to all refugees and mobilized to assist them once they arrived. The Austrian response was remarkable: small and poor and with a population of only seven million, the republic was still caring for 150,000 German expellees and 30,000 additional East European refugees when the Hungarians arrived. Austrian church groups, Red Cross societies, and civic leaders greeted the refugees with an extraordinary wave of generosity. One veteran observer of the Austrian scene attributed this response not only to the popular revulsion against the Soviets' repression but also to a historical sense of solidarity between Austrians and Hungarians, a feeling of a historical community, or *Schicksalsgemeinschaft*, that had for so long bound together the peoples of the Danubian basin.[24]

Quite the opposite tradition existed in Yugoslavia, where for centuries popular antipathies had existed between Serbs and Magyars. Nevertheless, in a remarkable display of independence, Tito's government also granted asylum to nearly 20,000 Hungarian refugees. Carefully balancing between the East and the West, the Yugoslav ruler regretted the brutal use of force against a Communist neighbor, yet disapproved at the same time of the Hungarians' radical break with Communism. The Russians, on the other hand, having conferred extensively with Belgrade as the crisis developed in the autumn, snubbed Tito at the end of November after the die was cast: in flagrant disregard of an agreement made between the Yugoslavs and the Soviet-sponsored Hungarian government of János Kádár, the Russians simply abducted Imre Nagy and his associates, who were then under Yugoslav diplomatic protection.[25]

Having languished somewhat as an internationally accredited refugee agency, the UNHCR stepped quickly into the breach as a new refugee crisis loomed. Asked by the UN General Assembly to organize emergency assistance, the high commissioner coordinated operations in Vienna, bringing to bear a wide variety of national, international, and voluntary agencies. In short order, everything fell into place: local authorities immediately offered assistance, vast quantities of supplies began to arrive, and immigration countries announced their willingness to receive the refugees. In retrospect, the UNHCR contribution was seen as critical. According to refugee worker Elfan Rees, the High Commission saved the day. Had the organization "not been in existence and available to give a hand, the chaos would have been fantastic. . . . That order did emerge, however, from this chaos; that each agency was set to the task it was best equipped for; that border reception, emergency feeding and clothing, temporary housing, and ultimate movement were planned and coordinated is irrefutable testimony to the importance of an office of a United Nations High Commissioner for Refugees."[26]

In less than six months the refugees were granted permanent asylum in thirty-five countries—a remarkably speedy operation, in notable contrast with the lingering refugee problems of the Second World War. About 18,000 Hungarians were repatriated after the UNHCR worked out a detailed arrangement with the new authorities in Budapest. Settlement elsewhere was organized by the Inter-Governmental Committee for European Migration, which drew on massive aid from the United States. Half of the 154,000 who left Austria went overseas, and half remained in Europe. More than two-thirds of those who settled in Europe were dispersed among the United Kingdom, France, Germany, and Switzerland; abroad, the overwhelming majority went to the United States (35,185), Canada (22,575), and Australia (9,458).[27]

In addition to Hungary, the main source of refugees during the cold war years was East Germany, from which some 3.5 million migrants flowed to the German Federal Republic in the decade after 1951.[28] Most of these poured through Berlin, which remained under four-power wartime occupation arrangements that provided for intercity access. In this way, Berlin became the great hole in the Iron Curtain. Over the years, the passage of Germans from east to west proved not only an embarrassment to the Soviets and their Eastern European clients, but it was also extremely costly; the emigrants were overwhelmingly among the best-educated, most thoroughly trained in German society, constituting a loss the regime could ill afford. Moving to stem the tide in the summer of 1961, as tensions between the superpowers mounted over the Berlin issue, East German police sealed the exits from their half of the city. Within a matter of hours the Berlin Wall snaked its way across the city. Watchtowers, barbed wire, armed guards, and other prison camp appurtenances now scarred the very heart of the former German capital, an ugly monument to the fears and failures of the Communist side. For a few years, Germans continued escape attempts in the city, and shootings by border guards were a familiar occurrence. But the exodus was reduced to a small trickle. Gradually, the refugee camps in West Germany cleared, and the few arrivals in the Federal Republic could be handled with dispatch. Through a move of extraordinary brutality, the German refugee issue seemed at last to be resolved.[29]

Despite the periodic improvements in East-West relations under Stalin's successors, the Communist world continues to send refugees into Europe and deny to others the possibility of leaving. In 1963 the high commissioner for refugees reported that about 10,000 flowed yearly from the Warsaw Pact countries, along with a new influx of Cuban refugees into Spain. A government-sponsored campaign of antisemitism in Poland, begun after the outbreak of the Six-Day War in the Middle East, attacked the tiny Jewish community in that country, diverting attention from serious challenges to those in power. Clothed in "anti-Zionist" rhetoric by Wladyslaw Gomulka, this campaign involved the expulsion of most of the 30,000 Jews remaining in Poland.[30] In August 1968 the Soviets and some Warsaw Pact allies invaded Czechoslovakia, putting an end to the flurry of experimentation begun under Prime Minister Alexander Dubček known as the "Prague Spring." At the time about 80,000 Czechs either fled their country or remained abroad awaiting a resolution of the crisis. The majority of these became refugees, to be settled in Western Europe or overseas.[31]

In the last decade and a half, an important movement of Soviet Jews to the West, involving over 260,000 people, has been the subject of much speculation. Beginning in a significant way in 1971, when Moscow allowed 15,000 Jews to leave the Soviet Union for Israel, the flow increased to more than 30,000 annually in 1972 and 1973. Then, after a brief decline in the mid-1970s, Jewish emigration soared to over 50,000 in 1979. Recently, this emigration has been drastically cut, with only 2,700 departures in 1983.[32] Students of Soviet policy still puzzle over the shifts reflected in this extraordinary emigration. Coinciding with the departure of over 85,000 Volga Germans, plus lesser numbers of Armenians and other minorities, the emigration of the Jews seems likely related to Moscow's effort to defuse internal dissent and to appeal to Western opinion for various trade and other concessions.[33]

In 1980, the status of Jewish émigrés leaving the Soviet Union became an issue of political importance. That year, Israeli Prime Minister Menachem Begin and the head of the Jewish Agency argued that because these Jews left Russia legally, were not homeless, and had no obstacle before their becoming Israelis, they should not be deemed refugees. This objection, however, seemed far more an effort to ensure that the émigrés ended up in Israel than a reflection on the circumstances of their departure from the Soviet Union. An increasing proportion, known as "dropouts," elected to go elsewhere after they reached Vienna; by 1980, when this proportion reached 70 percent, some Israelis were extremely eager to redirect them to the Middle East.[34]

Surveys of the geographic origins of Soviet Jews who left, the anti-Jewish pressures in Soviet society, and the specific harassments of Jewish activists all suggest a complex variety of motivations for Jewish emigration. Some have emigrated for what might be termed "ethnic" reasons, insisting on a right to live a Jewish life. Such emigration accounts in part for the large share of departures from non-Slavic parts of the Soviet Union or from regions absorbed into the country as a result of the Second World War. Jewish culture survived longest in these regions, and Jewish identities were better preserved. Some émigrés have a political orientation, finding the Soviet system intolerable; and others have left for social or economic reasons, seeking a better life in the West. Furthermore, as Zvi Gitelman has pointed out, "the principal cause of emigration is prior emigration": once underway in the early 1970s, the emigration movement became a social pattern providing one option in a society where such choices are not always present in great profusion. Whatever the motives, however, there seems little doubt that after applying to leave the U.S.S.R., most applicants had a "well-founded fear of persecution." Prospective emigrants are sharply attacked in the Soviet media, are likely to lose their employment and become isolated in Soviet society. The growing preference for a destination other than Israel after the Arab-Israeli War of 1973 does not mean that the circumstances of departure have become any easier. Quite the contrary seems to have been the case as the Soviet authorities have moved to terminate the process of Jewish emigration in the early 1980s. Under these circumstances, as departure becomes extraordinarily difficult, with mounting persecution of those seeking to leave, the few who do succeed look persuasively like refugees.[35]

Most recently, the deteriorating economic and political conditions in Poland provoked a huge exodus from that country, including many who sought asylum as refugees. During 1981 the flow of emigrants reached record levels, even before the imposition of martial law in December of that year. Four times as many Poles asked for refugee status in Austria in the first seven months of 1981 as in the whole of 1979. According to one UNHCR official, as many as 500,000 Poles may have been outside their country as the crisis became evident: many sought to remain in the West, and 200,000 were still unsettled in Western Europe a year later. Although recognizing the refugee status of a minority among the Poles, international opinion has tended to consider them economic migrants: European governments often urged the Poles to return home as soon as the crisis eased. But there is little doubt about the pressure these new arrivals brought to bear on the facilities in the various countries of reception and on the UNHCR. Austria bore a particularly heavy burden, receiving more Poles than any other country, followed by the German Federal Republic.[36]

DIMINISHED IMPORTANCE OF EUROPE

It took a decade and a half to resolve the major European refugee problems left by the collapse of the Third Reich—to empty the camps, resettle millions of uprooted civilians, and assimilate and restore victims of the most colossal upheaval the Continent has ever seen. To be sure, some European refugees remained, their suffering continued, and both national and international agencies faced crises that spilled additional exiles into receiving countries. Moreover, new refugee pressures emerged as fugitives from crumbling colonial empires were driven out by liberation movements, and tens of thousands of Third World refugees entered Europe. But in June 1959, beginning "World Refugee Year" proclaimed by the UN General Assembly, public attention addressed a worldwide, rather than a strictly European, problem. During the next decade the catastrophes of the Third World claimed the attention of the world's refugee agencies. For Europe, the urgency was gone: in its place, one hopes, is a treasury of experience in dealing with refugee issues that the world can peruse with profit.

Settling Old Business

During the early 1950s, before the significant leap forward of European prosperity, many considered the Continent overpopulated, showing a chronic imbalance between resources and numbers of people. As the International Refugee Organization wound up its operations, this assessment prompted much concern to maintain the momentum of overseas migration from Europe that that agency helped to stimulate. On American initiative, but with substantial European support, the Western countries established an international organization commited to assist the flow of Europeans overseas. The Inter-Government Committee for European Migration (ICEM), as it became known after its foundation in 1952, was also supposed to facilitate the resettlement of refugees through the provision of transport facilities and negotiation with countries of asylum. Working closely with the UNHCR and other refugee agencies, the ICEM continued the efforts of the IRO to clear the refugee books of European countries, especially those of first asylum. Taking charge of much of the staff and equipment of the IRO, the ICEM assumed responsibility for the movement of refugees during the Hungarian crisis of 1956. Not burdened with sovereign diplomatic responsibilities of its own and independent of the United Nations, the ICEM could afford to adopt a loose definition of refugees, extending its range of humanitarian aid to Indochinese, Soviet, and East European émigrés. Assisting tens of thousands of migrants annually, the ICEM significantly eased the refugee pressure on emigration countries like Austria and Italy, which also received large numbers of fugitives from Eastern Europe.[37]

Another push in the direction of resettlement came from the UNHCR, which as we have seen was strictly limited in its ability to finance solutions to refugee problems. Chafing under these restrictions in the mid-1950s, the high commissioner successfully launched the United Nations Refugee Fund (UNREF) to help countries integrate as yet unsettled or unassimilated refugees. Set in motion at the beginning of 1955, UNREF drew on millions of dollars of private and government donations. Intended to carry the burden of internationally assisted resettlement for a fixed period, the program expired at the end of 1958, moving some 50,000 refugees, mostly from Germany and Austria.[38]

Along with material aid and resettlement, the search for durable solutions to refugee problems has involved extending the range of definition of refugees, widening the sphere of protection states which were willing to accord those who sought asylum, and increasing the numbers of states that accepted international agreements on refugees. During the postwar decades, European countries contributed to this international effort, which resolved some outstanding legal issues. Most important, this involved the ratification of the 1951 convention on refugees and accession to the Protocol of 1967, which significantly amended it. As we have seen, the 1951 convention related only to persons who had become refugees as a result of "events occurring before 1 January 1951." Moreover, some signatories opted for this to mean events occurring only on the European continent. During the Hungarian Revolution in 1956, the countries that received refugees extended the benefit of the convention to the new fugitives, a move permitted by the convention itself. In the decade after it first appeared, all the Western European states except Spain ratified the convention; of the East European countries, only Yugoslavia accepted the agreement. In 1967 a new protocol removed the limitations of time and geography on the termination of refugees: the final formulation, accepted by more than ninety countries, deems a refugee to be an alien, not protected by his own government, who fears persecution in his country of origin because of threats to specific aspects of his human rights. In relatively short order, all the European states that had ratified the 1951 convention signed the new document, significantly improving the legal situation for refugees. Spain, too, has joined the list.[39] In addition to these fundamental charters, a battery of rules protecting individuals who may be refugees was built into the Treaty of Rome in 1957, a foundation document for the European Common Market. Similarly, the 1951 European Convention of Human Rights, a child of the largely consultative Council of Europe established in 1949, significantly protected aliens in European countries of refuge from being forcibly returned to states where they were subject to persecution. The drift of decisions by the European Commission on Human Rights since 1961, though not interfering with the right of any state to refuse admission, clearly limits the ability of council members to deport refugees.[40]

New Refugees

The new refugees of contemporary Europe come mainly from the underdeveloped or developing countries, once part of European colonial empires and now transforming themselves into modern nation-states. Since the beginnings of decolonization in the 1950s, the former imperial masters have had an important part in this drama and in the crisis produced by some ten million homeless and uprooted wanderers in the world today: Europeans have helped to resettle tens of thousands, participate in international efforts on behalf of others, and chip away at the problems in the light of their own, considerable experience.

As decolonization proceeded, tens of thousands of settlers from the former imperial powers were forced to leave. Most returned to the countries to which they had always given allegiance. With them came large groups of natives and other elements previously associated with the European domination or known to have opposed the nationalists who assumed control. After four years of fighting, the Dutch cleared out of Indonesia in 1949, opening their doors to about 250,000 people. Half of these so-called "repatriates" had never set foot in the Netherlands before. In a similar situation, the Belgians brought some 90,000 Belgian subjects "home" to Belgium when they began leaving the Congo ten years later amid a collapse of local authority. France received tens of thousands of uprooted civilians from Indochina in 1954, Morocco in 1956, Tunisia in 1957, and the Republic of Guinea in 1958, when the latter broke away from the new French Community. British subjects from Uganda, Kenya, and other former possessions, including many Asian subjects of the British Crown, entered the United Kingdom in the early 1960s.

Algeria produced the largest number of postcolonial refugees—well over a million, including those who left before the withdrawal of Europeans became a panicky flight. Most of these settled in France, but about 50,000 went to Spain; 12,000, to Canada; and 10,000 Sephardic Jews, to Israel. With the French of European background also went 180,000 so-called *harkis*, with their dependents—Muslim auxiliaries attached to the French police and gendarmerie and who fought against the nationalists. During the first five months of 1962, with the end of European rule in sight, the refugee tide brought 75,000 to France. In June, when even the diehard OAS signed a truce with the Algerian nationalists, the withdrawal became a rout; through that summer 600,000 fugitives arrived from Algeria by sea and air. French authorities were entirely unprepared for this onslaught: their working assumption had been that 400,000 would "return," over a period of five years. As a result, the government faced extraordinary difficulties in integrating the new arrivals—particularly the poorest among them who required substantial state support.[41]

The most recent postcolonial refugees involve Italy and Portugal, two of the weakest former colonial powers. Few Italians left their Libyan possession in the immediate postwar period despite the latter's preparation for independence in 1949. A United Nations General Assembly resolution that year forbade the Italian minority to export money or property until 1962 and subjected them to a heavy duty for alien registration. Over the next decade the Italian residents became a deeply pessimistic, disgruntled, and sometimes bitter minority—subject to vexatious political and economic controls. Their numbers dwindled, and many ended up on Italy's doorstep. In the autumn of 1969, Colonel Qaddafi's coup d'état triggered a confiscation of Italian and Jewish properties, leading in turn to a substantial exodus.[42] Most of the long-term European residents took this occasion to leave the country. Shortly afterward the end of the Salazar era in Portugal triggered vast changes in that country and its African territories. Having held off independence until the mid-1970s, Portugal had to assume the greatest postcolonial influx proportionately of any European state. Flooding into the country in 1975 and 1976, an estimated 600,000 *"retornados"* raised the local population by nearly 7 percent. At the same time, several thousand Portuguese liberals, draft evaders, and other opponents of the previous regime were able to leave their European refuge and return home. Small and poor and with troubles of its own, Portugal appealed for interational assistance. The UNHCR came to

Lisbon's aid with about a million dollars annually, and the ICEM helped move the returnees overseas.[43]

Ex-colonial refugees normally had explicit claims on the former colonial powers, which the latter were prepared to honor. In each of the cases just mentioned, the European governments set the fugitives on a proper legal footing in their new legal homes, with full rights of citizenship and occasionally significant financial help. Nevertheless, the cost has been heavy. Particularly in France and Portugal, the integration of "returnees" has proved long and difficult. As a result, governments have been extremely wary of increasing their burden through refugee immigration from Third World countries. London, for example, faced in 1968 with a threatened huge influx of Asians from Kenya owing to the latter's Africanization program, took away the right of free entry into the United Kingdom of all those with British passports "but who lacked substantive personal connection with the country." Henceforth such people ran into obstacles similar to those set before other Asian or African immigrants. Britain agreed to accept some 28,000 Ugandan Asians following a similar crisis in 1972, but this was clearly an exception, as the home secretary made clear at the time.

During the 1960s, Europeans watched the center of refugee attention move away from Europe to countries in Africa and Asia that had previously drawn attention only because of the plight of Europeans living there. Appropriately, the new UN high commissioner for refugees in 1965, succeeding the Swiss diplomat Felix Schnyder, was Prince Sadruddin Aga Khan, a man of cosmopolitan culture and background with particularly strong links to the Third World. Coinciding with this shift, UNHCR expenditures for Europe dropped dramatically, surpassed in the middle of the decade by those for Africa and Asia. The organization's overall budget climbed astronomically all the while, reaching half a billion dollars in 1980.[44] Faced with the standing appeals on behalf of destitute and sometimes starving refugees from the poorest parts of the globe, European governments have preferred to receive a relatively small immigration from areas where they have traditionally had political influence and cultural presence. Tens of thousands of Cuban refugees, for example, fleeing Castro's revolution, took advantage of the air links maintained between Havana and Madrid. Between 1961 and 1977, 140,000 Cubans followed this route, with 10 percent of them finding asylum in

Spain. Similarly, the French opened their doors to Tunisian Jews, expelled during the Arab-Israeli war in 1967, and to more than 83,000 Indochinese refugees between 1975 and the end of 1981.

Apart from these cases, however, the Continent's response to the world's present refugee crisis is governed by the deeply conservative impulses we have studied in this book. Throughout Western Europe the present drift is toward restrictions—increasing limitations on the right of asylum, cutting back on the numbers of refugees from Third World countries who are received. West Europeans still feel their first obligation to East European fugitives, thousands of whom manage to penetrate the Iron Curtain annually. Even these arrivals, however, are seen as a heavy burden. Over the last decade, applications for asylum have increased tenfold in Europe; there is every indication now that governments feel strained to the limit and will not accept more refugees. Burdened with flagging economies and millions of unemployed, most Western European governments have already adopted a more restrictive approach. In contrast to the expansionist views of the decade before 1973, when immigration and the importation of foreign workers were encouraged, outsiders are now being sent home. Finally, with the exeption of Yugoslavia, the eastern bloc governments make no contribution to the UNHCR, the UN Fund for Palestinian Refugees, or the various international agencies involved with emigration. Deeply offended by the use that has been made of the latter by escapees to the West, the Soviet Union and her allies stand as aloof on refugee issues today as they did in the 1950s.

European countries have received relatively few of the refugees from Third World upheavals—the emergence of Bangladesh; the wars in Biafra, Ethiopia, and Sudan; the effective partition of Cyprus after the Turkish invasion of 1974; and the Vietnamese "boat people": only France and Switzerland rank among the top ten countries in the world ordered by the ratio of refugees to population from 1975 to 1980 (standing eighth and tenth, respectively).[45] Like refugee workers on the scene, European governments emphasize the need to assist refugees where they are, encouraging eventual repatriation or resettlement to countries that might want the labor or skills of those who are uprooted. Meanwhile, Europe's own refugee problems have largely been solved. The World Refugee Survey for 1983 estimated over 7.8 million "refugees in need" throughout the globe; only 30,700 of these were in Europe, mostly people passing through Austria on their way to permanent resettlement. Since the downturn in West European economies in the early 1970s, with resulting high levels of unemployment, governments have been increasingly reluctant to see greater imigration. Indeed, most are preoccupied with reducing the pools of foreign workers allowed to grow to approximately ten million during the previous decade. Unemployment and hard times have also stimulated a wave of racism and xenophobia that renders most politicians even more cautious toward prospective new arrivals. Similarly, given the unpredictable nature of refugee flows, making sudden tidal waves entirely possible, authorities are reluctant to assume abstract obligations to refugees in the future.

All of this is woefully familiar to those who survey the history of refugee movements in Europe. So also are the despair and suffering that afflict so many homeless people outside the European continent today. What is extraordinary, however, is the apparent end of a European refugee problem, which has bedeviled political leaders since the First World War. Of course, no one can say that a new tide of refugees may not one day engulf the Continent, and few believe that existing legal structures or international agencies could themselves prevent such an inundation. If this history suggests any lesson, it is that ultimately the flow of refugees can only be controlled by favorable economic circumstances and the stabilization of the international order. Until the achievement of these elusive goals worldwide, the consciences of Europeans will forever be tested by refugees, wherever they appear.

Notes

INTRODUCTION

1. François Berge, "Editorial," in *Personnes déplacées* (Paris, 1948), 5.
2. Dorothy Thompson, *Refugees: Anarchy or Organization?* (New York, 1938), 5.
3. Arieh Tartakower and Kurt R. Grossman, *The Jewish Refugee* (New York, 1944), 1.
4. Malcolm J. Proudfoot, *European Refugees, 1939–52: A Study in Forced Population Movement* (New York [1956]), 21; Joseph B. Schechtman, *Postwar Population Transfers in Europe, 1945–1955* (Philadelphia, 1962), 363.
5. Hannah Arendt, *The Origins of Totalitarianism* (Cleveland, 1958), 277.
6. Jacques Vernant, *The Refugee in the Post-War World* (London, 1953), 52.
7. Aristide R. Zolberg, "Contemporary Transnational Migrations in Historical Perspective: Patterns and Dilemmas," in Mary M. Kirtz (ed.), *U.S. Immigration and Refugee Policy. Global and Domestic Issues* (Lexington, Ky., 1983), 21.
8. Martin Van Creveld, *Supplying War: Logistics from Wallenstein to Patton* (Cambridge, 1977), chap. 1; Geoffrey Best, *Humanity in Warfare: The Modern History of the International Law of Armed Conflicts* (London, 1980), 181.
9. *Trübners deutsches Wörterbuch* (Berlin, 1940), II, 399.
10. Jane Kramer, *Unsettling Europe* (New York, 1981), xiii.

CHAPTER 1

1. Bernard Porter, *The Refugee Question in Mid-Victorian Politics* (Cambridge, 1979), 26.
2. R. F. Leslie, *Polish Politics and the Revolution of November 1830* (London, 1956), 259.
3. Georges Dupeux, "France: l'immigration en France de la fin du XVIIIᵉ siècle à nos jours," in Commission internationale d'histoire des mouvements sociaux et des structures sociales, *Les Migrations internationales de la fin du XVIIIᵉ siècle à nos jours* (Paris, 1980), 162.

4. Carl Ludwig, *La Politique pratiquée par la Suisse à l'égard des réfugiés au cours des années 1933 à 1955* [Annexe au rapport du Conseil fédéral à l'Assemblée fédérale sur la politique pratiquée par la Suisse à l'égard des réfugiés au cours des années 1933 à nos jours] (Bern, 1957), 6.

5. David McLellan, *Karl Marx: His Life and Thought* (London, 1973), 225; Oscar J. Hammen, *The Red '48ers: Karl Marx and Friedrich Engels* (New York, 1969), 402; Beatrix Mesmer, "Die politischen Flüchtlingen im 19. Jahrhundert," in André Mercier, ed., *Der Flüchtling in der Weltgeschichte: Ein ungelöstes Problem der Menschheit* (Bern, 1974), 209–39; Alte Grahl-Madsen, *The Status of Refuges in International Law* (2 vols., Leyden, 1966), II, 8–11.

6. Alexander Herzen, *My Past and Thoughts*, trans. Constance Garnett (6 vols., London, 1924–27), III, 59–60; H. Nobholz et al., *Geschichte der Schweiz* (Zurich, 1938), II, 486–87.

7. Porter, *Refugee Question*, 1.

8. Ibid., 120, 157.

9. Herzen, *Past and Thoughts*, V, 2.

10. J. Langhard, *Die politische Polizei des Schweizerischen Eidgenossenschaft* (Bern, 1909), 158–59; Herzen, *Past and Thoughts*, III, 2.

11. Herzen, *Past and Thoughts*, III, 120–21.

12. Geoffrey Best, *Humanity in Warfare: The Modern History of the International Law of Armed Conflicts* (London, 1980), 190; *Geschichte der preussischen Invasion und Okkupation in Böhmen im Jahre 1866. Gesammelte Beilage der Zeitschichrifte "Politik"* (Prague, 1867), 305.

13. Karl Obermann, "Allemagne: les grands mouvements de l'émigration allemande vers les Etats-Unis d'Amérique au XIXᵉ siècle," in Commission internationale, *Migrations internationales*, 419; Dupeux, "France," ibid., 163.

14. Celina Bobinska and Adam Galos, "Poland: Land of Mass Emigration (XIXth and XXth Centuries)," ibid., 483–85; Benjamin P. Murdzek, *Emigration in Polish Social-Political Thought, 1870–1914* (Boulder, Colo., 1977), 41.

15. Jacques Rougerie, *Procès des Communards* (Paris, 1964), 17–20.

16. Otto Kirchheimer, *Political Justice: The Use of Legal Procedure for Political Ends* (Princeton, 1961), 374 ff.; Betty Garfinkels, *Belgique, terre d'accueil: problème du réfugié, 1933–40* (Brussels, 1974), 23.

17. Georges Haupt, "Emigration et diffusion des idées socialistes: l'exemple d'Anna Kulischoff," *Pluriel*, 14 (1978), 8–9; David Saunders, "Vladimir Burtsev and the Russian Revolutionary Emigration (1888–1905)," *European Studies Review*, 13 (1983), 39–62.

18. Jacob Lestchinsky, *Jewish Migration for the Past Hundred Years* (New York, 1944), 10.

19. Hans Rogger, "Tsarist Policy on Jewish Emigration," *Soviet Jewish Affairs*, 3 (1973), 28; idem, "Russian Ministers and the Jewish Question, 1881–1917," *California Slavic Studies*, VII (1975), 15–76; Joseph Kissman, "The Immigration of Rumanian Jews up to 1914," *YIVO Annual of Jewish Social Science*, 2–3 (1947/1948), 178; Heinz-Dietrich Löwe, *Antisemitismus und reaktionäre Utopie: Russischer Konservatismus im Kampf gegen den Wandel von Staat und Gesellschaft, 1890–1917* (Hamburg, 1978).

20. Mary Antin, *The Promised Land* (Boston, 1912), 8; Joel S. Geffen, "Whither: To Palestine or to America, in the Pages of the Russian Hebrew Press *Ha-Melitz* and *Ha-Yom* (1880–1890)," *American Jewish Historical Quarterly*, 59 (1969), 182.

21. Michael R. Marrus, *The Politics of Assimilation: A Study of the French Jewish*

Community at the Time of the Dreyfus Affair (Oxford, 1971), 154; Pierre van Passen, *Days of Our Years* (London, 1939), 22; John A. Garrard, *The English and Immigration, 1880–1910* (London, 1971), 162.

22. Rogger, "Tsarist Policy," 32.

23. Arthur Ruppin, *The Jew in the Modern World* (London 1934), 114, and passim.

24. Lloyd P. Gartner, *The Jewish Immigrant in England, 1870–1914* (London, 1960), 29.

25. Jonathan D. Sarna, "The Myth of No Return: Jewish Return Migration to Eastern Europe, 1881–1914," *American Jewish History*, 71 (1981), 256–68; Jacob Lestchinsky "Jewish Migrations, 1840–1956," in Louis Finkelstein, ed., *The Jews* (New York, 1960), II, 1565.

26. Raphael Mahler, "The Economic Background of Jewish Emigration from Galicia to the United States, *YIVO Annual of Jewish Social Science*, 7 (1952), 255–67.

27. André Chouraqui, *L'Alliance israélite universelle et la renaissance juive contemporaine (1860–1960)* (Paris, 1965), 87–100; Howard Morley Sachar, *The Course of Modern Jewish History* (New York, 1963), 258; R. W. Seton-Watson, *A History of the Roumanians from the Roman Times to the Completion of Unity* (Cambridge, 1934), 348–51.

28. Kissman, "Immigration of Rumanian Jews," 160.

29. Ibid., 177–78.

30. *Days of Our Years* (London, 1939), 23.

31. *The New Exodus: A Study of Israel in Russia* (London, 1892), 268.

32. Jack Wertheimer, " 'The Unwanted Element'—East European Jews in Imperial Germany," *Leo Baeck Institute Year Book*, XXVI 23, 32 n. 23; S. Adler-Rudel, *Ostjuden in Deutschland, 1880–1940: Zugleich eine Geschichte der Organisationen, die sie Betreuten* (Tübingen, 1959), 163–64; Mark Wischintzer, *To Dwell in Safety: The Story of Jewish Migration Since 1880* (Philadelphia, 1948), 113; Steven E. Aschheim, *Brothers and Strangers: The East European Jew in German and Jewish Consciousness, 1800–1923* (Madison, Wis., 1982).

33. Bernard Gainer, *The Alien Invasion: The Origins of the Alien Act of 1905* (London, 1972), 209; Garrard, *English and Immigration*, 104–05, 109.

34. *New Exodus*, 274.

35. V. D. Lipman, *A Century of Social Service* (London, 1959), 94–96.

36. Stanford J. Shaw and Ezel Kural Shaw, *History of the Ottoman Empire and Modern Turkey*, Vol. II, *Reform, Revolution, and Republic: The Rise of Modern Turkey, 1808–1975* (Cambridge, 1977), 189; Robert J. Donia, *Islam Under the Double Eagle: The Muslims of Bosnia and Hercegovina, 1878–1914* (Boulder, Colo., 1981), 31, 73–75.

37. André Wurfbain, *L'Echange greco-bulgare des minorités ethniques* (Lausanne, 1930), 100; Dimitri Kossev, Virginia Paskaleva, and Stephan Dojnov, "Bulgarie: les migrations bulgares de la fin du XVIIIe siècle à la seconde guerre mondiale," in Commission internationale, *Migrations internatinales*, 560–61; Robert Lee Wolff, *The Balkans in Our Time* (New York, 1978), 87; L. S. Stavrianos, *The Balkans Since 1453* (New York, 1958), 519–20.

38. Carnegie Endowment for International Peace, *Report of the International Commission to Inquire into the Causes and Conduct of the Balkan Wars* (Washington, D.C., 1914), 151; Stephen P. Ladas, *The Exchange of Minorities: Bulgaria, Greece and Turkey* (New York, 1932), 15; A. A. Pallis, "Racial Migrations in the Balkans

during the Years 1912–1924," *Geographical Journal*, 66 (1925), 325–31; Wurfbain, *Exchange greco-bulgare*, 30.

39. C. A. Macartney, *Refugees: The Work of the League* (London, n.d.), 113; Carnegie Endowment, *Report*, 257.

40. Joseph P. Schechtman, *European Population Transfers, 1939–1945* (Oxford, 1946), 12.

41. Wurfbain, *L'Echange greco-bulgare*, 31; Ladas, *Exchange of Minorities*, 21; Dimitri Pentzopoulos, *The Balkan Exchange of Minorities and Its Impact upon Greece* (The Hague, 1962), 54–56; Carnegie Endowment, *Report*, 257–59.

42. Ralph H. Major, *War and Disease* (London, n.d.), 142.

43. Dragolioub Yovanovitch, *Les Effets économiques et sociaux de la guerre en Serbie* (Paris, 1930), 46.

44. Ibid., 41.

45. Bureau international du travail, *Les réfugiés et les conditions du travail en Bulgarie* (Geneva, 1926), 6 and passim.

46. Aristide R. Zolberg, "Contemporary Transnational Migrations in Historical Perspective: Patterns and Dilemmas," in Mary M. Kritz, ed., *U.S. Immigration and Refugee Policy. Global and Domestic Issues* (Lexington, Ky., 1983), 18–19.

47. Arnold Toynbee, "Greece," in Nevill Forbes et al., *The Balkans* (Oxford, 1915), 248.

CHAPTER 2

1. Geoffrey Best, *Humanity in Warfare: The Modern History of the International Law of Armed Conflicts* (London, 1980), 224.

2. Madeleine de Bryas, *Les Peuples en marche: les migrations politiques et économiques en Europe depuis la guerre mondiale* (Paris, 1926), 56.

3. Daniel W. Graf, "Military Rule Behind the Russian Front, 1914–1917: The Political Ramifications," *Jahrbücher für Geschichte Osteuropas*, 22 (1974), 396.

4. Fred C. Koch, *The Volga Germans in Russia and the Americas from 1763 to the Present* (University Park, Penn., 1977), 244.

5. Sir John Hope Simpson, *The Refugee Problem: Report of a Survey* (London, 1939), 64; Eugene M. Kulischer, *Europe on the Move: War and Population Changes, 1917–47* (New York, 1948), 32–35.

6. Graf, "Military Rule," 402.

7. Violetta Thurstan, *The People Who Run. Being the Tragedy of the Refugees in Russia* (London, 1916), 42.

8. Isaac Deutscher, *The Prophet Armed. Trotsky: 1879–1921* (New York, 1965), 466.

9. Joyce NanKivell Loch, *A Fringe of Blue: An Autobiography* (London, 1968), 70–74.

10. Bryas, *Peuples en marche*, 20–21; Kulischer, *Europe on the Move*, 130–31.

11. Bryas, *Peuples en marche*, 21; Fred Kupferman, *Au Pays des Soviets: le voyage français en Union soviétique, 1917–1939* (Paris, 1979), 54.

12. *New York Times Book Review*, 30 December 1979, 7.

13. Martin Gilbert, *Winston S. Churchill*, Vol. IV, *The Stricken World, 1916–1922*, (Boston, 1975), 365; Sir Arthur E. Shipley, "The Collapse of a Nation," *English Review*, 38 (1924), 837–40; Fridtjof Nansen, *Russia and Peace* (London, 1923), 150–51.

14. Martin Gilbert, *Sir Horace Rumbold: Portrait of a Diplomat, 1869–1941* (London, 1973), 231–33, 267; P. J. Noel-Baker, *The League of Nations at Work* (London, 1927), 118–19; Fu-yung Hsu, *La Protection des réfugiés par la Société des nations* (Lyon, 1935).

15. Bryas, *Peuples en marche*, 78.

16. Hans von Rimscha, *Der russische Bürgerkrieg und die russische Emigration, 1917–21*, (Jena, 1924), 80–87.

17. Simpson, *Refugee Problem*, 80–87.

18. S. W. Baron, *The Russian Jew Under Tsars and Soviets* (New York, 1964), 191.

19. Maurice Paléologue, *La Russie des tsars pendant la grande guerre* (3 vols., Paris, 1921–22), I, 335.

20. Arieh Tartakower and Kurt R. Grossmann, *The Jewish Refugee* (New York, 1944), 19.

21. Ibid., 14–15; Zosa Szajkowski, "East European Jewish Workers in Germany during World War I," *Salo Wittmayer Baron Jubilee Volume* (3 vols., New York, 1974), II, 906; Pawel Korzec, *Juifs en Pologne: la question juive pendant l'entre-deux-guerres* (Paris, 1980), 55; Robert C. Williams, *Culture in Exile: Russian Emigrés in Germany, 1881–1941* (Ithaca, N.Y., 1972), 56.

22. Zvi Y. Gitelman, *Jewish Nationality and Soviet Politics: The Jewish Sections of the CPSU, 1917–1930* (Princeton, 1972), 162–163. Estimates of the numbers of Jewish casualties vary widely. See Tartakower and Grossman, *Jewish Refugee*, 23; S. Ettinger, in Ben-Sasson, ed., *A History of the Jewish People* (London, 1976), 954; Norman Davies, "Great Britain and the Polish Jews, 1918–1920," *Journal of Contemporary History*, 8 (1973), 119–42; Baron, *Russian Jew*, 220–221.

23. Ezra Mendelsohn, *The Jews of East Central Europe Between the World Wars* (Bloomington, Ind., 1983), 40.

24. Thurstan, *People Who Run*, 153.

25. "Dr. Nansen's Message," *The Jewish Guardian*, 14 October 1921.

26. Käthe Liepman, "Statistical Notes on Refugees in Rumania," Archives of the Royal Institute for International Affairs (henceforth RIIA) (1937–38); Tartakower and Grossmann, *Jewish Refugee*, 25.

27. *Bulletin of the Executive Committee of the Jewish World Relief Conference*, 20 September 1921.

28. Arthur Ruppin, *The Jews in the Modern World* (London, 1934), 47.

29. Bryas, *Peuples en marche*, 70; Tartakower and Grossmann, *Jewish Refugee*, 25.

30. Korzec, *Juifs en Pologne*, 106–07; League of Nations Archives (henceforth LNA) 1919–1927, 45/32992/145091; George Earley and Roberta Earley, "Bavarian Prelude, 1923: A Model for Nazism?" *Wiener Library Bulletin*, XX, N.S. 43/44 (1977), 56–58.

31. Lucien Wolf, *Russo-Jewish Refugees in Eastern Europe* (London, 1921), 15.

32. Herman Dicker, *Piety and Perseverence: Jews from the Carpathian Mountains* (New York, 1981), 34–37.

33. Wolf, *Russo-Jewish Refugees*, 17.

34. Harold Nicolson, *Peacemaking 1919* (London 1964), 31–32.

35. See Benjamin Akzin et al. *La Nationalité dans la science sociale et dans le droit contemporain* (Paris, 1933).

36. Joseph Schechtman, *European Population Transfers, 1939–1945* (New York, 1946), 6.

37. Quoted in Hannah Arendt, *The Origins of Totalitarianism* (Cleveland, 1962), 277, n. 22.

38. Ibid., 289; Marc Vichniac, "Le statut international des apatrides," Académie de droit international, *Recueil des cours*, 43 (1933), 145–67; Simpson, *Refugee Problem*, 229, n. 1.

39. Bryas *Peuples en marche*, 44 ff.; Kulischer, *Europe on the Move*, 166–74.

40. Bryas, *Peuples en marche*, 151.

41. Franz Werfel, *The Forty Days*, trans. Geoffrey Dunlop (London, n.d.), 103–04. See Gérard Chaliand and Yves Ternon, *Le Génocide des Arméniens* (Brussels, 1980), 41–42; Norman Ravitch, "The Armenian Catastrophe," *Encounter* (December 1981), 70–71.

42. James L. Barton, *Story of Near East Relief (1915–1930): An Interpretation*, (New York, 1930), 82, 110.

43. Richard G. Hovannisian, *The Republic of Armenia*, Vol. I, *The First Year, 1918–1919* (Berkeley, Calif., 1971), 126; Firuz Kazemadeh, *The Struggle for Transcaucasia (1917–1921)* (Oxford, 1951), 211; Barton, *Near East Relief*, 122; Anaïs Ter-Minassian, "Migrations des Arméniens en Russie et en Union Soviétique aux XIXe et XXe siècles," in Commission internationale d'histoire des mouvements sociaux et structures sociales, *Les Migrations internationales de la fin du XVIIe siècle à nos jours* (Paris, 1980), 223, Société des Nations, *Plan d'établissement des réfugiés arméniens: exposé général et documents principaux* (Geneva, 1927), 74.

44. Hovannisian, *Republic of Armenia*, I, 130, 135; idem, Vol. II, *From Versailles to London 1919–1920*, (Berkeley, Calif., 1982), 57.

45. Ibid., II, 402; James Barros, *Office without Power: Secretary-General Sir Eric Drummond, 1919–1933* (Oxford, 1979), 106–8.

46. Simpson, *Refugee Problem*, 44; Frederick A. Norwood, *Strangers and Exiles: A History of Religious Refugees* (2 vols., Nashville, Tenn., 1965), II, 267.

47. Barton, *Near East Relief*, 126–36 and passim.

48. Gilbert, *Churchill*, IV, 295.

49. Benjamin M. Weissman, *Herbert Hoover and Famine Relief to Soviet Russia, 1921–1923* (Stanford, Calif., 1974), 28.

50. Ibid., 199.

51. E. E. Reynolds, *Nansen* (London, 1932), 211; Liv Nansen Høyer, *Nansen: A Family Portrait*, trans. Maurice Michael (London, 1957), 228; Noel-Baker, *League of Nations*, 114–17.

52. F. P. Walters, *A History of the League of Nations* (2 vols., London, 1952), I, 188–89; Norman Bentwich, "The League of Nations and Refugees," *British Yearbook of International Law*, 16 (1935), 114.

53. Simpson, *Refugee Problem*, 202.

54. "Russian Refugees and the Soviet," *The Times* (London), 27 August 1923.

55. See the correspondence in LNA 1919–1927: 45/16187/12319 and 45/18950/12900; Høyer, *Nansen*, 242.

56. Paul M. Hayes, *Quisling: The Career and Political Ideas of Vidkun Quisling, 1887–1945* (Newton Abbot, England, 1971), 26–30.

57. Nansen, *Russia and Peace*, passim; Nansen memorandum, March 1922, LNA 1919–1927 47/19846/19846; "Dr. Nansen's Message," *The Jewish Guardian*, 14 October 1921; F[ridtjof] N[ansen], "Refugees and the Exchange of Populations," *Encyclopaedia Britannica*, 13th ed. (London, 1926).

58. Sir Norman Angell, *Freer Migration and Western Security* (Liverpool, 1951), 17.

59. Paul Fussell, *Abroad: British Literary Travelling Between the Wars* (Oxford, 1980), 26–31.

60. Hsu, *La Protection des réfugiés*, 72–73; Lawrence Preuss, "La dénationalisation imposée pour des motifs politiques," *Revue internationale française du droit des gens*, 4 (1937), 18.

61. Simpson, *Refugee Problem*, 106.

62. Hsu, *La Protection des réfugiés*, 49.

63. Gilbert, *Churchill*, IV, 854.

64. Henry Morgenthau, *I Was Sent to Athens* (New York, 1929), 180–81.

65. League of Nations, *Greek Refugee Settlement* (Geneva, 1926), 12–13.

66. Barton, *Near East Relief*, 152; C. A. Macartney, *Refugees: The Work of the League* (London [1930]), 84–85.

67. Stephen P. Ladas, *The Exchange of Minorities. Bulgaria, Greece and Turkey* (New York, 1932).

68. Loch, *Fringe of Blue*, 122.

69. Nansen's note of 18 November 1922, LNA 1919–1927: 48/24722/24357; Macartney, *Refugees*, 86.

70. Dimitri Pentzopoulos, *The Balkan Exchange of Minorities and Its Impact upon Greece* (The Hague, 1962), 67.

71. Angelos Hadzopoulos, *Die Flüchtlingsfrage in Griechenland* (Athens, 1927), passim; André Wurfbain, *L'Echange greco-bulgare des minorités ethniques* (Lausanne, 1930), 118; Violet R. Markham, "Greece and the Refugees from Asia Minor," *Fortnightly Review*, 1 February 1925, 179.

72. Hadzopoulos, *Flüchtlingsfrage in Griechenland*, 32; Bryas, *Peuples en marche*, 40, n. 2; League of Nations, *Greek Refugee Settlement*, 4, 93; Simpson, *Refugee Problem*, 14–17.

73. League of Nations, *Greek Refugee Settlement*, 92; Bryas, *Peuples en marche*, 92–93; Stojan Kiselinovski, *Gruckata Kolonizacija vo Egejska Makedonija, 1913–1940* (Skopje, 1981), 174–176.

74. Morgenthau, *Athens*, 180–81.

75. Dorothy Thompson, *Refugees: Anarchy or Organization?* (New York, 1938), 23–24.

76. Simpson, *Refugee Problem*, 11.

77. C. A. Macartney, *National States and National Minorities* (London, 1934), 448.

78. Wallace McClure, *World Prosperity as Sought Through the Economic Work of the League of Nations* (New York, 1933), 322–23; Schechtman, *European Population Transfers*, 22.

79. Wurfbain, *L'Echange greco-bulgare*, 81; Ladas, *Exchange of Minorities*, 722–723.

80. Cf. James Barros, *The League of Nations and the Great Powers: the Greek-Bulgarian Incident, 1925* (Oxford, 1970).

81. Nansen, "Refugees," 322.

82. John George Stoessinger, *The Refugee and the World Community* (Minneapolis, 1956), 15–16.

83. Nansen memorandum, 22 August 1924, LNA 1919–1927: 45/38178/12319.

84. Walters, *A History of the League*, I, 189.

85. Hsu, *La protection des réfugiés*, 37.

86. Georges Mauco, *Les Etrangers en France: leur rôle dans l'activité économique* (Paris, 1932), 33–34; Gary S. Cross, *Immigrant Workers and Industrial France: The Making of a New Laboring Class* (Philadelphia, 1983), chap. 3.

87. Jean Pluyette, *La Sélection de l'immigration en France et la doctrine des races* (Paris, 1930), 2; Cross, *Immigrant Workers*, 126.

88. Stoessinger, *The Refugee and the World Community*, 212; Nansen, "Refugees," 321; H. Stuart Hughes, *Contemporary Europe: A History* (Englewood Cliffs, N.J., 1961), 137.

89. Sir John Hope Simpson, *Refugees: A Review of the Situation Since September 1938* (London, 1939), 12.

90. *Palestine Royal Commission Report, 1937,* Command paper 5479 (London, 1937), 13, 46, 62; Noah Lucas, *The Modern History of Israel* (London, 1974), 109–10, 335; Norman Davies, *God's Playground: A History of Poland,* Vol. II, *1795 to the Present* (Oxford, 1981), 404–09; Howard Morley Sachar, *A History of Israel: From the Rise of Zionism to Our Time* (New York, 1976), 144, 154–55; Ezra Mendelsohn, *Jews of East Central Europe*, 36–43; Korzec, *Juifs en Pologne*, 148–50.

91. Chaim Weizmann, *Trial and Error* (New York, 1966), 277.

92. Yehuda Bauer, *My Brother's Keeper: A History of the American Jewish Joint Distribution Committee, 1929–1939* (Philadelphia, 1974), 62.

93. Ibid., 61, 63, 103.

94. Simpson, *Refugee Problem*, 36.

95. Nansen, *Armenia and the Near East* (London, 1928), 320.

96. Ibid., 26, 323.

97. Société des Nations, *Plan d'établissement des réfugiés arméniens.*

CHAPTER 3

1. Charles F. Delzell, ed., *Mediterranean Fascism, 1919–1945* (New York, 1970), 104.

2. Philip V. Cannistrano and Gianfausto Rosoli, *Emigrazione, chiesa e fascismo: Lo scioglimento dell'Opera Bonomelli (1922–1928)* (Rome, 1979).

3. Aldo Garosci, *Storia del fuorusciti* (Bari, 1953), 12.

4. Charles F. Delzell, *Mussolini's Enemies: The Italian Anti-Fascist Resistance* (Princeton, 1961), 51; Alan Cassels, *Mussolini's Early Diplomacy* (Princeton, 1970), 370.

5. Delzell, *Mussolini's Enemies*, 44, n. 2.

6. Sir John Hope Simpson, *The Refugee Problem: Report of a Survey* (London, 1939), 117–25; Armando Zanetti, "Italian Refugees," Archives of the Royal Institute for International Affairs (henceforth RIIA) (1938), 10.

7. Cassels, *Mussolini's Early Diplomacy*, 373–374.

8. Zanetti, "Italian Refugees," 11.

9. Norman Bentwich, *The Refugees from Germany: April 1933 to December 1935* (London, 1936), 175.

10. Klaus Mann, *The Turning Point: Thirty-Five Years in this Century* (London, 1944), 206.

11. Bentwich, *Refugees from Germany*, 198.

12. Herbert A. Strauss, "Jewish Emigration from Germany: Nazi Policies and

Jewish Response (I)," *Leo Baeck Institute Year Book,* 25 (1980), 317, 326–29; Raul Hilberg, *The Destruction of the European Jews* (Chicago, 1961), 56.

13. S. Adler-Rudel, *Ostjuden in Deutschland 1880–1940: Zugleich eine Geschichte der Organisationen, die sie Betreuten* (Tübingen, 1959), 9–49; Helmut Genschel, *Die Verdrängung der Juden aus der Wirtschaft im Dritten Reich* (Göttingen, 1966), 20, n. 26.

14. Genschel, *Verdrängung der Juden,* 261; Hilberg, *Destruction,* 90–91.

15. McDonald's letter may be found in an appendix to Bentwich, *Refugees from Germany,* 219–28.

16. Simpson, *Refugee Problem,* 142. Cf. Frederick A. Norwood, *Strangers and Exiles: A History of Religious Refugees,* 2 vols. (Nashville, Tenn., 1965), II, 281.

17. Alfred Meusel, "Die Saar Emigranten," RIIA (1937); Dieter Marc Schneider, "Saarpolitik und Exil, 1933–1955," *Vierteljahrshefte für Zeitgeschichte,* 25 (1977), 467–545.

18. Ruth Fabian and Corinna Coulmas, *Die deutsche Emigration in Frankreich nach 1933* (Munich, 1978), 62–64; Simpson, *Refugee Problem,* 155–59.

19. Herbert S. Levine, *Hitler's Free City: A History of the Nazi Party in Danzig, 1925–1939* (Chicago, 1973), 127–30.

20. Bentwich, *Refugees from Germany,* 104; Yehuda Bauer, *My Brother's Keeper: A History of the American Jewish Joint Distribution Committee, 1929–1939* (Philadelphia, 1974), 152.

21. Wolfgang Köllman and Peter Marschalck, "Allemagne: German Overseas Emigration Since 1815," in Commission internationale d'histoire des mouvements sociaux et des structures sociales, *Les Migrations internationales de la fin du XVIIIe siècle à nos jours* (Paris, 1980), 459.

22. Yehuda Bauer, *A History of the Holocaust* (New York, 1982), 123; idem, *My Brother's Keeper,* 176.

23. Karl A. Schleunes, *The Twisted Road to Auschwitz: Nazi Policy Toward German Jews, 1933–1939* (Urbana, Ill., 1970), 116, 187–88.

24. Arthur Ruppin, *The Jews in the Modern World* (London, 1934), 65.

25. Strauss, "Jewish Emigration from Germany (I)," 350–58; Bauer, *History of the Holocaust,* 126–67.

26. Simpson, *Refugee Problem,* 515–16.

27. Siegfried Kraft, *Die russland-deutschen Flüchtlinge des Jahres 1929/30 und ihre Aufnahme in Deutschen Reich* (Halle, 1939), 57–58; Frank H. Epp, *Mennonite Exodus: The Rescue and Resettlement of the Russian Mennonites Since the Communist Revolution* (Altona, Manitoba, 1962), 231; Harvey Dyck, "Collectivization, Depression, and Immigration, 1929–30: A Chance Interplay," in Harvey Dyck and Peter Krosby, eds., *Empire and Nations: Essays in Honour of Frederick H. Soward* (Toronto, 1969), 144–59.

28. Robert Conquest, *The Great Terror: Stalin's Purge of the Thirties* (New York, 1973), 128.

29. Gustav Helger and Alfred G. Meyer, *The Incompatible Allies: A Memoir-History of German-Soviet Relations, 1918–1941* (New York, 1953), 252; Nahum Goldmann, *The Autobiography of Nahum Goldmann: Sixty Years of Jewish Life,* trans. Helen Sebba (New York, 1969), 172–73; Conquest, *Great Terror,* 683.

30. David Pike, *German Writers in Soviet Exile, 1933–1945* (Chapel Hill, N.C., 1982), 57–59.

31. Arieh Tartakower and Kurt B. Grossmann, *The Jewish Refugee* (New York,

1944), 263–64; Margaret Ernestine Burton, "The Assembly of the League of Nations," Ph.D. dissertation, Columbia University, 1941, 201–03.

32. A. J. Sherman, Island Refuge: Britain and Refugees from the Third Reich, 1933–1939 (London, 1973), 206.

33. Margarete Buber, *Under Two Dictators*, trans. Edward Fitzgerald (London, 1949), 155. Cf. Wolfgang Leonhard, *Child of the Revolution*, trans. C. M. Woodhouse (London, 1957).

34. Pike, *German Writers*, 357.

35. Leonhard, *Child of the Revolution*, 12–14; Pike, *German Writers*, 341–42.

36. Joseph B. Schechtman, *Fighter and Prophet: The Vladimir Jabotinsky Story*, Vol. II, *The Last Years* (New York, 1961), 352.

37. Sherman, *Island Refuge*, 94.

38. Arthur Ruppin, *The Jews in the Modern World* (London, 1934), 67.

39. Sherman, *Island Refuge*, 74, 230; Józef Beck, *Dernier rapport: politique polonaise, 1926–1939* (Neuchatel, 1951), 136, 139.

40. Ezra Mendelsohn, *The Jews of East Central Europe Between the World Wars* (Bloomington, Ind., 1983), 7 and passim. Cf. Hugh Seton-Watson, "Government Policies Towards the Jews in Pre-Communist Eastern Europe," *Soviet Jewish Affairs*, 4 (December 1969), 20–25.

41. Eugene M. Kulischer, *Europe on the Move: War and Population Changes, 1917–47* (New York, 1948), 143; Yisrael Gutman, *The Jews of Warsaw, 1939–1943: Ghetto, Underground, Revolt* (Bloomington, Ind., 1982), xvii.

42. Raymond Leslie Buell, *Poland: Key to Europe* (London, 1939), 295–96.

43. Simpson, *Refugee Problem*, 517.

44. Henry L. Feingold, *The Politics of Rescue: The Roosevelt Administration and the Holocaust, 1938–1945* (New Brunswick, New Jersey, 1970), 53.

45. Rita Thalmann, "L'Emigration du IIIᵉ Reich dans la France de 1933 à 1939," *Le Monde Juif*, 96 (October–December 1979), 128; Simpson, *Refugee Problem*, 266.

46. Bentwich, *Refugees from Germany*, 77, 133. See Jean-Charles Bonnet, *Les Pouvoirs publics français et l'immigration dans l'entre-deux-guerres* (Lyon, 1976); Barbara Vormeier, "Dokumentation zur französischen Emigrantenpolitik (1933–1944): ein Beitrag," in Hanna Schramm, *Menschen in Gurs: Erinnerungen an ein französisches Internierungslager (1940–1941)* (Worms, 1977), 157–245.

47. Marcel Livian, *Le Parti Socialiste et l'immigration* (Paris, 1982).

48. Bonnet, *Pouvoirs publics*, 315.

49. Timothy Maga, "Closing the Door: The French Government and Refugee Policy," *French Historical Studies*, 12 (Spring 1982), 436.

50. Thalman, "Emigration," 128; Maga, "Closing the Door," 39; Sir John Hope Simpson, *Refugees: A Review of the Situation Since September 1938* (London, 1939), 52; Kurt R. Grossmann, *Emigration. Geschichte der Hitler-Flüchtlinge, 1933–1945* (Frankfurt am Main, 1969), 161.

51. Norman Bentwich, *My Seventy-Seven Years: An Account of My Life and Times* (London, 1962), 167–68.

52. Sherman, *Island Refuge*, 49.

53. A. J. P. Taylor, *English History, 1914–1945* (Oxford, 1965), 419–20.

54. Sherman, *Island Refuge*, 30, 260, and passim.

55. Simpson, *Refugee Problem*, 340.

56. Richard Griffiths, *Fellow Travellers of the Right: British Enthusiasts for Nazi Germany, 1933–39* (Oxford, 1983), 108.

57. *Palestine, Royal Commission Report, 1937,* Command Paper 5479 of 1937 (London, 1937), 113, 279, and passim.

58. Simpson, *Refugees: A Review of the Situation,* 69.

59. Sherman, *Island Refuge,* 264–65; D. Gurevich et al., eds., *Statistical Handbook of Jewish Palestine* (Jerusalem, 1947), passim. Sherman's evaluation is seconded by Bernard Wasserstein, "The British Government and the German Immigration, 1933–1945," in Gerhard Hirschfeld, ed., *Exile in Great Britain: Refugees from Hitler's Germany* (London, 1984), 79.

60. Carl Ludwig, *La Politique pratiquée par la Suisse à l'égard des réfugiés au cours des années 1933 à 1955* [Annexe au rapport du Conseil fédérale à l'assemblée fédérale sur la politique practiquée par la Suisse à l'égard des réfugiés au cours des années 1933 à nos jours] (Bern, 1957), 3 ff; Hans B. Kunz, *Weltrevolution und Volkerbund: Die schweitzerische Aussenpolitik unter dem Eindruck der bolschewistischen Bedrohung, 1918–1923* (Bern, 1981), 210–11.

61. Hans-Joachim Hoffmann-Nowotny, "European Migration after World War II," in William H. McNeill and Ruth S. Adams, eds., *Human Migration: Patterns and Policies* (Bloomington, Ind., 1978), 94; Alfred A. Häsler, *The Lifeboat Is Full: Switzerland and the Refugees, 1933–1945,* trans. Charles Lam Markmann (New York, 1969), 4–24; Ludwig, 44–47; Jacques Vernant, *The Refugees in the Post-War World* (London, 1953), 328–29.

62. Bauer, *My Brother's Keeper,* 173; Ludwig, *La Politique,* 47, 59; Eliahu Ben Elissar, *La Diplomatie du IIIᵉ Reich et les Juifs* (Paris, 1981), 267–68; Tartakower and Grossmann, *Jewish Refugee,* 286–91.

63. Ben Elissar, *Diplomatie du IIIᵉ Reich,* 270–79; Ludwig, *La Politique,* 90–111; Georg Kreis, "Flüchtlingspolitik und Presselpolitik," *Neue Züricher Zeitung,* 4 May 1979.

64. Ludwig, *La Politique,* 138, 144, 150.

65. Norman Bentwich, *My Seventy-Seven Years,* 126; J. W. Breugel, "The Bernheim Petition: A Challenge to Nazi Germany in 1938," *Patterns of Prejudice,* 17 (1983), 17–25.

66. Bentwich, *Refugees from Germany,* 60–61; Ben Elissar, *Diplomatie du IIIᵉ Reich,* 96–101.

67. Bentwich, *Refugees from Germany,* 156.

68. Haim Genizi, "James G. McDonald: High Commissioner for Refugees, 1933–1935," *Wiener Library Bulletin,* 30:43/44 (1977), 40–52; Bentwich, *My Seventy-Seven Years,* 156.

69. Sherman, *Island Refuge,* 51–52.

70. Ibid., 65.

71. See Simpson, *Refugee Problem,* Appendix VII.

72. Burton, "The Assembly," 201–03; Alexander Dallin, *The Soviet Union at the United Nations: An Enquiry into Soviet Motives and Objectives* (London, 1962), 17–19.

73. Simpson, *Refugee Problem,* 139–40, 515–16.

74. See Abraham Margaliot, "The Problem of the Rescue of German Jewry During the Years 1933–1939: the Reasons for the Delay in Their Emigration from the Third Reich," in Yisrael Gutman and Ephraim Zuroff, eds., *Rescue Attempts During the Holocaust: Proceedings of the Second Yad Vashem International Historical Conference* (Jerusalem, 1977), 262–63.

75. Gerhard Botz, *Wien vom "Anschluss" zum Krieg: Nationalsozialistische Mach-*

tübernahme und politische-soziale Umgestaltung am Beispiel der Stadt Wien 1938/39 (Vienna, 1978), 98–105.

76. Ibid., 253–54; Nora Levin, *The Holocaust: The Destruction of European Jewry, 1933–1945* (New York, 1973), 102; Uwe Dietrich Adam, *Judenpolitik im Dritten Reich* (Düsseldorf, 1972), Chap. 5.

77. Israel Cohen, *Travels in Jewry* (London, 1952), 46.

78. Sherman, *Island Refuge*, 93.

79. William L. Shirer, *Berlin Diary: The Journal of a Foreign Correspondent 1934–1941* (Harmondsworth, England, 1979), 120; Feingold, *Politics of Resuce*, 22–23.

80. Irving Abella and Harold Troper, *None Is Too Many: Canada and the Jews of Europe, 1933–1948* (Toronto, 1982), 28.

81. Botz, *Wien*, 251–52; Schleunes, *Twisted Road*, 232–33; Feingold, *Politics of Rescue*, 36–37.

82. Sherman, *Island Refuge*, 105.

83. Ibid., 230; Mendelsohn, *Jews*, 206–08.

84. Tartakower and Grossmann, *Jewish Refugee*, 37 ff; Simpson, *Refugees: A Review of the Situation*, 53; Ben Elissar, *Diplomatie du IIIe Reich*, 284–90.

85. Levine, *Hitler's Free City*, chap. 7; Joshua B. Stein, "Britain and the Jews of Danzig, 1938–1939," *Wiener Library Bulletin*, 32:49/50 (1979), 29–33.

86. Simpson, *Refugees: Review of the Situation*, 1–30.

87. Marc Vischniac, "Le statut international des apatrides," *Académie de droit international. Recueil des cours*, 43 (1933), 133.

88. Oscar I. Janowsky and M. M. Fagen, *International Aspects of German Racial Policies* (New York, 1937), 158–61; Lawrence Preuss, "La dénationalisation imposée pour des motifs politiques," *Revue international française du droit des gens*, 4 (1937), 241; Yewdall R. Jennings, "Some International Law Aspects of the Refugee Question," *British Yearbook of International Law*, 20 (1939), 98–114.

89. Hannah Arendt, *The Origins of Totalitarianism* (Cleveland, 1958), 266–87. Cf. Edigo Reale, "Passport," *Encyclopaedia of the Social Sciences*, 12 (New York, 1950), 13–16; and idem, "Le problème des passports," *Académie de droit international. Recueil des cours*, 50 (1934), 170–71.

90. Sherman, *Island Refuge*, 169.

91. David Kranzler, "The Jewish Refugee Community of Shanghai, 1938–1945," *Wiener Library Bulletin*, 26:3/4 (1972/73), 28–37.

92. Strauss, "Jewish Emigration from Germany: Nazi Policies and Jewish Responses (II)," *Leo Baeck Institute Year Book*, 26 (1981), 383; Yehuda Bauer, *American Jewry and the Holocaust: The American Jewish Joint Distribution Committee, 1939–1945* (Detroit, 1981), 302–14; Kranzler, "Jewish Refugee Community."

93. Ben Elissar, *Diplomatie du IIIe Reich*, 420–21.

94. Simpson, *Refugee Problem*, 186–89; Tartakower and Grossmann, *Jewish Refugee*, 457.

95. Bauer, *My Brother's Keeper*, chap. 4.

96. Tartakower and Grossmann, *Jewish Refugee*, 496.

97. Bauer, *American Jewry*, 37–40.

98. Ronald Steel, *Walter Lippmann and the American Century* (New York, 1981), 373; Dorothy Thompson, *Refugees: Anarchy or Organization?* (New York, 1938), 80–81.

99. Ben Elissar, *Diplomatie du IIIe Reich*, 404–06.

100. Sherman, *Island Refuge*, 39, 65.

101. Margaliot, "Problem of Rescue," 255–56.

102. See Joseph Schechtman, *Jabotinsky*, I, 338 ff; Pawel Korzec, *Les Juifs en Pologne: la question juive pendant l'entre-deux-guerres* (Paris, 1980), 253–54.

103. Strauss, "Jewish Emigration from Germany (II)," 346. Dalia Ofer, "The Rescue of European Jewry and Illegal Immigration to Palestine in 1940—Prospects and Reality: Berthold Storfer and the Mossad Le'Aliyah Bet," *Modern Judaism*, 4 (1984), 159–81.

104. Ben Elissar, *Diplomatie du IIIe Reich*, 407.

105. Korzec, *Les Juifs en Pologne*, 250; Beck, *Dernier rapport*, 140.

106. Michael R. Marrus and Robert O. Paxton, *Vichy France and the Jews* (New York, 1981), 60–61.

107. J. B. Trent to Consul General, Madagascar, 6 April 1946, Public Record Office [henceforth PRO]: FO 371/57690/WR1045.

108. Sherman, *Island Refuge*, 97, 127.

109. Meir Michaelis, *Mussolini and the Jews: German-Italian Relations and the Jewish Question in Italy, 1922–1945* (Oxford, 1978), 195–200.

110. Feingold, *Politics of Rescue*, 104.

111. Sherman, *Island Refuge*, 209.

112. See Isaiah Bowman, ed., *Limits of Land Settlement: A Report on Present-Day Possibilities* (New York, 1937); Feingold, *Politics of Rescue*, 103; idem, "Roosevelt and the Resettlement Question," in Gutman and Zuroff, *Rescue Attempts*, 123–80.

113. Simpson, *Refugee Problem*, 160–65.

114. Louis Stein, *Beyond Death and Exile: The Spanish Republicans in France, 1939–1955* (Cambridge, Mass., 1979), 33, 40.

115. Simpson, *Refugees: Review of the Situation*, 57; LNA 1919–1946: 20A/38141/27698.

116. Maga, "Closing the Door," 440; Stein, *Beyond Death*, 85–86.

117. Vernant, *Refugees*, 280; Stein, *Beyond Death*, passim.

118. Valentín González and Julian Gorkin, *El Campesino: Life and Death in Soviet Russia* (London, 1952), 39.

119. Ibid.; Raymond Carr, *The Spanish Tragedy: The Civil War in Perspective* (London, 1977), 107.

120. Jésus Hernandez, *La Grande Trahison*, trans. Pierre Berthelin (Paris, 1953), 222–24; Leonhard, *Child of the Revolution*, 152.

121. Alexander Donat (pseud.), *The Holocaust Kingdom: A Memoir* (New York, 1965), 3.

122. Tartakower and Grossmann, *Jewish Refugee*, 271; Ben-Cion Pinchuk, "Jewish Refugees in Soviet Poland, 1939–1941," *Jewish Social Studies*, 40 (Spring 1978), 142–45; Gutman, *Jews of Warsaw*, 17.

123. Buber, *Under Two Dictators*, passim; Pike, *German Writers*, 344–46; Vojtech Mastny, *Russia's Road to the Cold War* (New York, 1979), 27; Boris Levytsky, *The Uses of Terror: The Soviet Secret Service, 1917–1970*, trans. H. A. Piehler (London, 1971), 150.

124. Nicolai Tolstoy, *Stalin's Secret War* (London, 1982), 112; Oscar Halecki, *East-Central Europe Under the Communists: Poland* (New York, 1957), 45–46; Bauer, *American Jewry*, 295; Leon W. Wells, *The Death Brigade (The Janowska Road)* (New York, 1978), 277–80.

125. Tolstoy, *Stalin's Secret War*, 201, 222–23; Levytsky, *Uses of Terror*, 149.

126. See Shimon Redlich, "The Jews in the Soviet-Annexed Territories, 1933–41," *Soviet Jewish Affairs*, 1 (June 1971), 81–90.

127. Yisrael Gutman, "Jews in General Anders' Army in the Soviet Union," *Yad Vashem Studies*, 21 (1977), 234–35.

128. Pinchuk, "Jewish Refugees," 153.

129. Yehuda Bauer, "Rescue Operations Through Vilna," *Yad Vashem Studies*, 9 (1973), 215–23; Yitzhak Arad, "Concentration of Refugees in Vilna on the Eve of the Holocaust," *Yad Vashem Studies*, 9 (1973), 201–14.

130. Joseph B. Schechtman, *European Population Transfers, 1939–1945* (New York, 1946), 388–89.

131. Tolstoy, *Stalin's Secret War*, 157.

132. Malcolm J. Proudfoot, *European Refugees, 1939–52: A Study in Forced Population Movement* (London, [1956]), 41. Schechtman, *Population Transfers*, 392–93.

133. *Report of the International Committee of the Red Cross on its Activities During the Second World War (September 1, 1939–June 30, 1947)*, Vol. II, *Relief Activities* (Geneva, 1948), 416; *Inter Arma Caritas: The Work of the International Committee of the Red Cross During the Second World War* (Geneva, 1947).

134. See J. Vidalenc, *L'Exode de mai-juin 1940* (Paris, 1957), x, passim; W. D. Halls, *The Youth of Vichy France* (Oxford, 1981), 3; Jean-Pierre Azéma, *De Munich à la Libération, 1938–1944* (Paris, 1979), 61–62.

135. Marrus and Paxton, *Vichy France and the Jews*, 11.

136. Ibid., 161–64; Zosa Szajkowski, *Analytical Franco-Jewish Gazetteer, 1939–1945* (New York, 1966), 90; Bauer, *American Jewry*, 44.

137. Ruth Fabian and Corinna Coulmas, *Die deutsche Emigration in Frankreich nach 1933* (Munich, 1938), 79; Peter Gillman and Leni Gillman, *"Collar the Lot!" How Britain Expelled Its Wartime Refugees* (London, 1980); Miriam Kochan, *Britain's Internees in the Second World War* (London, 1983); Tartakower and Grossmann, *Jewish Refugee*, 33; Gilbert Badia, "L'émigration en France: ses conditions et ses problèmes," in Gilbert Badia, ed., *Les Barbelés de l'exil: Etudes sur l'émigration allemande et autrichienne (1938–1940)* (Grenoble, 1979), 88–95, and passim; Barbara Vormeier, *Die Deportierungen deutscher und österreichischer Juden aus Frankreich* (Paris, 1980), 7–8.

138. Gilman and Gilman, *Collar the Lot*, passim; Kochan, *Britain's Internees*, 175; Bernard Wasserstein, *Britain and the Jews of Europe, 1939–1945* (London, 1979), 98–108.

139. Bermuda Conference Report, 29 April 1943, PRO: FO 371/5770/WR 67; Michel Abitbol, *Les Juifs d'Afrique du Nord sous Vichy* (Paris, 1983), 98–99; Wasserstein, *Britain*, 52; Bauer, *American Jewry*, 48, 66; Herbert Rosenkranz, "The Anschluss and the Tragedy of Austrian Jewry, 1938–1945," in Josef Fraenkel, ed., *The Jews of Austria: Essays on their Life, History and Destruction* (London, 1967), 479–545.

140. David S. Wyman, *Paper Walls: America and the Refugee Crisis 1938–1941* (Amherst, Mass., 1968), 176.

141. Monty N. Penkower, "The Bermuda Conference and Its Aftermath: An Allied Quest for 'Refuge' During the Holocaust," *Prologue*, Fall 1981, 170; Eleanor F. Rathbone, *Rescue the Perishing: An Appeal, a Programme and a Challenge* (London, 1943), 10; Shirer, *Berlin Diary*, 602; Marrus and Paxton, *Vichy France and the Jews*, 162.

CHAPTER 4

1. Norman Rich, *Hitler's War Aims*, Vol. I, *Ideology, the Nazi State, and the Course of Expansion* (New York, 1973); Robert L. Koehl, *RKFVD: German Resettlement and Population Policy 1939–1945: A History of the Reich Commission for the Strengthening of Germandom* (Cambridge, Mass., 1957).

2. Eberhard Jäckel, *Hitler's Weltanschauung: A Blueprint for Power*, trans. Herbert Arnold (Middletown, Conn., 1972), 48; Erich Goldhagen, "Weltanschauung und Endlösung: zum Antisemitismus der nationalsozialistischen Führungsschicht, *Vierteljahrshefte für Zeitgeschichte*, 24 (1976), 379–405.

3. David Schoenbaum, *Hitler's Social Revolution: Class and Status in Nazi Germany 1933–1939* (Garden City, N.Y., 1967), 44; Karl A. Schleunes, *The Twisted Road to Auschwitz: Nazi Policy Toward German Jews, 1933–1939* (Urbana, Ill., 1970), 95–96.

4. Arthur Ruppin, *The Jews in the Modern World* (London, 1934), 257.

5. See Schleunes, *Twisted Road to Auschwitz*, 178–88.

6. David Yisraeli, "The Third Reich and the Transfer Agreement," *Journal of Contemporary History*, 6 (1971), 141–42; Gerhard Weinberg, *The Foreign Policy of Hitler's Germany*, Vol. II, *Starting World War II, 1937–1939* (Chicago, 1980), 246.

7. Circular of the state secretary, 8 July 1938, *Documents on German Foreign Policy* [henceforth DGFP], Series D, V, 894–95, and the minister in Hungary to the foreign minister, 11 January 1939, ibid., 360–61; Hans Mommsen, "Die Realisierung des Utopischen: Die 'Endlösung der Judenfrage' im 'Dritten Reich,' " *Geschichte und Gesellschaft*, 9 (1983), 402–3.

8. Schleunes, *Twisted Road to Auschwitz*, 247–48.

9. Emil Schumberg circular, 25 January 1939, DGFP, Series D, V, 926–33. Shlomo Aronson, "Die Dreifache Falle: Hitlers Judenpolitik, die Allierten und die Juden," *Vierteljahrshefte für Zeitgeschichte*, 32 (1984), 29–65.

10. Heydrich to Foreign Ministry, 14 February 1939, ibid., 933–36.

11. Yehuda Bauer, *American Jewry and the Holocaust: The American Jewish Joint Distribution Committee, 1939–1945* (Detroit, 1981), 66; Dalia Ofer, "The Rescue of European Jewry and Illegal Immigration to Palestine in 1940—Prospects and Reality: Berthold Storfer and the Mossad Le'Aliyah Bet," *Modern Judaism*, 4 (1984), 159–81; Yisrael Gutman, *The Jews of Warsaw, 1939–1943: Ghetto, Underground, Revolt*, trans. Ina Friedman (Bloomington, Ind., 1982), 17–18.

12. Uwe Dietrich Adam, *Judenpolitik im Dritten Reich* (Düsseldorf, 1972), 306–10; Michael R. Marrus and Robert O. Paxton, *Vichy France and the Jews* (New York, 1981), 247–48; idem, "The Nazis and the Jews in Occupied Western Europe, 1940–1944," *Journal of Modern History*, 54 (1982), 697.

13. Anthony Komjathy and Rebecca Stockwell, *German Minorities and the Third Reich: Ethnic Germans of East Central Europe between the World Wars* (New York, 1980).

14. Gerald Reitlinger, *The House Built on Sand: The Conflicts of German Policy in Russia* (London, 1960), 154.

15. David Dallin, *Soviet Russia's Foreign Policy, 1939–1942*, trans. Leon Dennen (New Haven, Conn., 1942), 94–99.

16. Joseph B. Schechtman, *European Population Transfers, 1939–1945* (New York, 1946), 195–96.

17. Ibid., 95.

18. Norman Rich, *Hitler's War Aims*, Vol. II, *The Establishment of the New Order* (New York, 1974), 82–83; Schechtman, *European Population Transfers*, 255; G. C. Paikert, *The Danubian Swabians: German Populations in Hungary, Rumania and Yugoslavia and Hitler's Impact on Their Patterns* (The Hague, 1967), 282–83; Koehl, *RKFDV*, 210.

19. Rich, *Hitler's War Aims*, II, 83–84; Koehl, *RKFDV*, 100–10, 182.

20. Norman Davies, *God's Playground: A History of Poland*, Vol. II. *1795 to the Present* (Oxford, 1981), 445.

21. Ibid., 446; Martin Broszat, *Nationalsozialistische Polenpolitik, 1939–1945* (Stuttgart, 1961), 101–2; Raul Hilberg, *The Destruction of the European Jews* (Chicago, 1961), 137–38; Schechtman, *European Population Transfers*, 215.

22. Jan Tomasz Gross, *Polish Society and German Occupation: The Generalgouvernement, 1939–1944* (Princeton, N.J., 1979), 71–73.

23. Yitzhak Arad, Yisrael Gutman, and Abraham Margaliot, eds., *Documents on the Holocaust: Selected Sources on the Destruction of the Jews of Germany, Poland, and the Soviet Union* (Jerusalem, 1981), 198.

24. Hilberg, *Destruction of the European Jews*, 138.

25. See Christopher R. Browning, "Zur Genesis der 'Endlösung': Eine Antwort an Martin Broszat," *Vierteljahrshefte für Zeitgeschichte*, 29 (1981), 97–109.

26. Isaiah Trunk, *Judenrat: The Jewish Councils in Eastern Europe Under the Nazi Occupation* (New York, 1972), 129–35; Gutman, *Jews of Warsaw*, 63; Lucjan Dobroszycki, ed. *The Chronicle of the Lodz Ghetto, 1941–1944*, trans. Richard Lourie, Joachim Neugroschel, et alia (New Haven, 1984), xxxviii, and passim.

27. Raul Hilberg, Stanislaw Staron, and Josef Kermisz, eds., *The Warsaw Diary of Adam Czerniakow*, trans. Stanislaw Staron et al. (New York, 1979), 336–47 and passim.

28. Trunk, *Judenrat*, 135; Hilberg, *Destruction of the European Jews*, 174.

29. Seev Goshen, "Eichmann und die Nisko-Aktion im Oktober 1939: eine Fallstudie zur NS/Judenpolitik in der letzten Etappe vor der 'Endlösung,' " *Vierteljahrshefte für Zeitgeschichte*, 29 (1981), 74–96; Phillip Friedman, "The Lublin Reservation and the Madagascar Plan: Two aspects of Nazi Jewish Policy During the Second World War," in Joshua A. Fishman, ed., *Studies in Modern Jewish History* (New York, 1972), 354–80.

30. Nora Levin, *The Holocaust: The Destruction of European Jewry, 1933–1945* (New York, 1973), 182; Erich Kulka, "The Plight of Jewish Refugees from Czechoslovakia in the U.S.S.R.," *Yad Vashem Studies*, 11 (1976), 298–328.

31. Adam, *Judenpolitik im Dritten Reich*, 303–16; Martin Broszat, "Hitler and the Genesis of the 'Final Solution': an Assessment of David Irving's Theses," *Yad Vashem Studies*, 13 (1979), 73–125. Cf. Kurt Pätzold, "Von der Vertreibung zum Genozid. Zu der Ursachen, Triebkräften und Bedingungen der anijüdischen Politik des faschistischen deutschen Imperialismus," in D. Eichholtz and K. Grossweiler (eds.), *Faschismusforschung. Positionen, Probleme, Polemik* (Berlin, 1980), 181–208.

32. Christopher R. Browning, *The Final Solution and the German Foreign Office: A Study of Referat D III of Abteilung Deutschland 1940–1943* (New York, 1978), 83. For the minutes of the Wannsee meeting, see Lucy S. Dawidowicz, ed., *A Holocaust Reader* (New York, 1976), 73–82.

33. Alexander Dallin, *German Rule in Russia, 1941–1945: A Study of Occupation Policies* (London, 1957), 279, 284–85; Rich, *Hitler's War Aims*, II, 392–93.

34. Rich, *Hitler's War Aims*, II, chap. 30; Reitlinger, *House Built on Sand*, 176. Koehl, *RKFDV*.

35. Hilberg, *Destruction of the European Jews*, 311; Alexander Werth, *Russia at War, 1941–1945*, (London, 1964), 800; Edward L. Homze, *Foreign Labor in Germany* (Princeton, N.J., 1967), passim.

36. Leszek A. Kosinski, "International Migration of Yugoslavs During and Immediately After World War II," *East European Quarterly*, 16 (1982), 185–89; Paikert, *Danubian Swabians*, 283–85.

37. Joseph B. Schechtman, "The Transnistria Reservation," *YIVO Annual of Jewish Social Science*, 8 (1953), 178–96; Martin Broszat, "Das Dritte Reich und die Rumänische Judenpolitik," in *Gutachen des Instituts für Zeitgeschichte* (Munich, 1958), 162.

38. Reitlinger, *House Built on Sand*, 220.

39. Werth, *Russia at War*, 630–31, 863.

40. Frank Epp, *Mennonite Exodus: The Rescue and Resettlement of the Russian Mennonites Since the Communist Revolution* (Altona, Manitoba, 1962), 357.

41. Dallin, *German Rule in Russia*, 292–93; Fred C. Koch, *The Volga Germans in Russia and the Americas from 1763 to the Present* (University Park, Penn., 1977), 286; Reitlinger, *House Built on Sand*, 303.

42. Paikert, *Danubian Swabians*, 189.

43. Nicolai Tolstoy, *Stalin's Secret War* (London, 1982), 244 ff.

44. See Dov Levin, "Estonian Jews in the U.S.S.R. (1941–1945): Research Based on Survivors' Testimony," *Yad Vashem Studies*, 11 (1976), 278 ff.

45. Erich Kulka, "The Plight of Jewish Refugees from Czechoslovakia in the U.S.S.R.," *Yad Vashem Studies*, 11 (1976), 312.

46. Levin, "Estonian Jews," 281.

47. Werth, *Russia at War*, 164, 213.

48. Levin, "Estonian Jews," 279 n. 8; Werth, *Russia at War*, 208–09, 332–33.

49. Koch, *Volga Germans*, 284; Schechtman, *European Population Transfers*, 384; Epp, *Mennonite Exodus*, 353.

50. Yisrael Gutman, "Jews in General Anders' Army in the Soviet Union," *Yad Vashem Studies*, 12 (1977), 284–85; Louise W. Holborn, *The International Refugee Organization: A Specialized Agency of the United Nations: Its History and Work, 1946–1952* (London, 1956), 176.

51. Hilberg, *Destruction of the European Jews*, 192; Helmut Krausnick and Hans-Heinrich Wilhelm, *Die Truppe des Weltanschauungkrieges: Die Einsatzgruppen der Sicherheitspolizei und des SD, 1938–1942* (Stuttgart, 1981).

52. Hilberg, *Destruction of the European Jews*, 192, 196; Yitzhak Arad, "Jewish Family Camps in the Forests—An Original Means of Rescue," in Yisrael Gutman and Efraim Zuroff, eds., *Rescue Attempts During the Holocaust: Proceedings of the Second Yad Vashem International Historical Conference* (Jerusalem, 1977), 333–53.

53. Arieh Tartakower and Kurt R. Grossman, *The Jewish Refugee* (New York, 1944), 268.

54. Valentín González and Julian Gorkin, *El Campesino: Life and Death in Soviet Russia* (London, 1952), 92–93.

55. Aleksandr M. Nekrich, *The Punished Peoples: The Deportation and Fate of Soviet Minorities at the End of the Second World War* (New York, 1978), 113–14.

56. Ibid., passim; Werth, *Russia at War*, 589–91; Boris Levytsky, *The Uses of Terror: The Soviet Secret Service, 1917–1970*, trans. H. A. Piehler (London, 1971), 164–65.

57. Tolstoy, *Stalin's Secret War*, 267 and passim; Tartakower and Grossman, *Jewish Refugee*, 274.

58. Livia Rothkirchen, "Hungary—an Asylum for the Refugees of Europe," *Yad Vashem Studies*, 7 (1968), 127–42; idem, "The Role of Czech and Slovak Jewish Leadership in the Field of Rescue Work," in Gutman and Zuroff, eds., *Rescue Attempts*, 423–34; Randolph L. Braham, *The Politics of Genocide: The Holocaust in Hungary*, 2 vols. (New York, 1981), I, 200.

59. See André Biss, *Der Stopp der Endlösung* (Stuttgart, 1966).

60. Carl Ludwig, *La Politique pratiquée par la Suisse à, l'égard des réfugiés au cours des années 1933 à 1955*, [Annexe au rapport du Conseil fédéral à l'Assemblée fédérale sur la politique pratiquée par la Suisse à l'égard des réfugiés au cours des années 1933 à nos jours] (Bern, 1957), 152; Ladislas Mysyrowicz and Jean-Claude Favez, "Refuge et représentation d'intérêts étrangers," *Revue d'histoire de la deuxième guerre mondiale*, 31 (January 1981), 109–20.

61. Alfred A. Häsler, *The Lifeboat Is Full: Switzerland and the Refugees, 1933–1945*, trans. Charles Lam Markmann (New York, 1969), 175.

62. Tartakower and Grossman, *Jewish Refugee*, 291; Mysyrowicz and Favez, "Refuge et représentation," 112.

63. Mysyrowicz and Favez, "Refuge et représentation," 115; Ludwig, *Politique*, passim.

64. Ian Guest, "The Skeletons Popping out of the Cuckoo Clock," *The Guardian*, 25 April 1979.

65. Georg Kreis, "Flüchtlingspolitik und Pressepolitik," *Neue Züricher Zeitung*, 4 May 1979. Cf. Albert Müller, "Hitlers Judenausrottung im Spiegel der Schweizer Presse," ibid.; and Ludwig, *La Politique*, 225–57.

66. Häsler, *Lifeboat*, 169.

67. Donald A. Lowrie, *The Hunted Children* (New York, 1963), 157–58; Tartakower and Grossman, *Jewish Refugee*, 301.

68. Häsler, *Lifeboat*, 110; Ludwig, *La Politique*, 224–25, n. 1.

69. Ludwig, *La Politique pratiquée par la Suisse*, 255, 281.

70. Ibid., 224.

71. Malcolm J. Proudfoot, *European Refugees, 1939–52: A Study in Forced Population Movement* (London, [1956]), 56; Hugh Thomas, *The Spanish Civil War* (London, 1961), 606–07; Gabriel Jackson, *The Spanish Republic and the Civil War, 1931–1939* (Princeton, 1965), 539.

72. Haim Avni, *Spain, the Jews, and Franco*, trans. Emanuel Shimoni (Philadelphia, 1982), 182; Michel Abitbol, *Les Juifs d'Afrique du Nord sous Vichy* (Paris, 1983), 97.

73. Varian Fry, *Surrender on Demand* (New York, 1945), 14–15; Proudfoot, *European Refugees*, 57.

74. Bauer, *American Jewry*, 206–07; Carlton J. H. Hayes, *Wartime Mission in Spain, 1942–1945* (New York, 1945), 111–28; John P. Willson, "Carlton J. H. Hayes, Spain, and the Refugee Crisis, 1942–1945," *American Jewish Historical Quarterly*, 62 (1972), 99–110.

75. Avni, *Spain, the Jews, and Franco*, chaps. 5, 6.

76. Hugh Kay, *Salazar and Modern Portugal* (London, 1970), chap. 7.

77. Moshe Bejski, "The 'Righteous Among the Nations' and Their Part in the Rescue of the Jews," in Gutman and Zuroff, eds., *Rescue Attempts*, 646; Bauer, *American Jewry*, 44–45.

78. Browning, *Final Solution* 45–46, 68.

79. Bauer, *American Jewry*, 255; Avni, *Spain, the Jews, and Franco*, 91; Martin Gilbert, *Auschwitz and the Allies* (London, 1981), 108; Proudfoot, *European Refugees*, 56; War Cabinet document on the Bermuda Conference, 4 May 1943, Annex No. 5, Public Record Office (henceforth PRO): FO371/5770/WR67.

80. Léon Papeleux, *Les Silences de Pie XII* (Brussels, 1980), 210–16.

81. Hansjacob Stehle, *Eastern Politics of the Vatican, 1917–1979*, trans. Sandra Smith (Athens, Ohio, 1981), 213.

82. Actes et documents du Saint Siège relatifs à la Seconde Guerre Mondiale, Vol. III, *Le Saint Siège et la situation religieuse en Pologne et dans les pays baltes, 1939–1945* (Vatican, 1967), 489–91; John Conway, "Catholicism and the Jewish People During the Nazi Period and Afterwards," *Papers Presented to the International Symposium on Judaism and Christianity Under the Impact of National Socialism (1919–1945)* (Jerusalem, 1982), 352; Papeleux, *Silences de Pie XII*, 149; idem, "Le Vatican et le problème juif (1944–1945)," *Revue d'histoire de la deuxième guerre mondiale*, 31 (1981), 47 and n. 5; Bernard Wasserstein, *Britain and the Jews of Europe, 1939–1945* (London, 1979), 141.

83. Ibid., 365; Papeleux, "Vatican," 48 n. 6, 66; John Morley, *Vatican Diplomacy and the Jews During the Holocaust, 1939–1943* (New York, 1980), passim; Victor Conzemius, "Le Saint-Siège et la deuxième guerre mondiale, deux éditions de sources," *Revue d'histoire de la deuxième guerre mondiale*, 128 (1982), 77–83; Daniel Carpi, "The Rescue of Jews in the Italian Zone of Occupied Croatia," in Gutman and Zuroff, eds., *Resuce Attempts*, 490; John S. Conway, "Records and Documents of the Holy See Relating to the Second World War," *Yad Vashem Studies*, XV (1984), 327–45.

84. Meir Michaelis, *Mussolini and the Jews: German-Italian Relations and the Jewish Question in Italy, 1922–1945* (London, 1978), 364–66; Owen Chadwick, "Weizsäcker, the Vatican, and the Jews of Rome, *Journal of Ecclesiatical History*, 28 (1977), 179–99; Leonidas E. Hill, "History and Rolf Hochhuth's *The Deputy*," in R. G. Collins, ed., *From an Ancient to a Modern Theatre* (Winnipeg, 1972), 145–57.

85. Braham, *Politics of Genocide*, II, 1075.

86. Proudfoot, *European Refugees*, 67.

87. Leni Yahil, "Scandinavian Countries to the Rescue of Concentration Camp Prisoners," *Yad Vashem Studies*, 6 (1967), 186.

88. "Survey of Measures Taken by H. M. G. in Connexion with Refugees," 16 January 1945, PRO: FO371/51112/WR523. Hugo Valentin, "Rescue and Relief Activities in Behalf of Jewish Victims of Nazism in Scandinavia," *YIVO Annual of Jewish Social Science*, 8 (1953), 224–51; Yahil, "Scandinavian Countries to the Rescue," 181–220; Count Folke Bernadotte, *The Curtain Falls: The Last Days of the Third Reich*, trans. Eric Lewenhaupt (New York, 1945); Leonard Gross, *The Last Jews of Berlin* (New York, 1982), 285–86; Hilberg, *Destruction of the European Jews*, 373–74; Peter R. Black, *Ernst Kaltenbrunner: Ideological Soldier of the Third Reich* (Princeton, N.J., 1984), 230–33; Leni Yahil, Raoul Wallenberg—His Mission and His Activities in Hungary," *Yad Vashem Studies*, XV (1984), 7–53. Franklin D. Scott, *Sweden: The Nation's History* (Minneapolis, 1977), 503–09; Frederick E. Werbell and Thurston Clarke, *Lost Hero: The Mystery of Raoul Wallenberg* (New York, 1982), 18–19; Jacques Derogy, *Le Cas Wallenberg* (Paris, 1980), chap. 2.

89. Jacques Vernant, *The Refugee in the Post-War World* (London, 1953), 244; War Cabinet document on the Bermuda Conference, 4 May 1943, Annex No. 5, PRO: FO371/5770/WR67.

90. Dalia Ofer, "The Activities of the Jewish Agency Delegation in Istanbul in 1943," in Gutman and Zuroff, eds., *Rescue Attempts*, 435–50; Ira A. Hirschman, *Life Line to a Promised Land* (New York, 1946), 42.

91. Wasserstein, *Britain and the Jews of Europe*, 143–56.

92. Browning, *Final Solution*, 155–56.

93. Barry Rubin, "Ambassador Laurence A. Steinhardt: The Perils of a Jewish Diplomat, 1940–1945," *American Jewish History*, 70 (1981), 331–46; Henry L. Feingold, *The Politics of Rescue: The Roosevelt Administration and the Holocaust, 1938–1945* (New Brunswick, N.J., 1970), 286–88, 291.

94. Howard Morley Sachar, *A History of Israel: From the Rise of Zionism to Our Time* (New York, 1976), 223–24; Michael J. Cohen, *Palestine: Retreat from the Mandate: The Making of British Policy, 1936–1945* (London, 1978), chap. 5.

95. Bauer, *American Jewry*, 130–34; Walter Laqueur, *A History of Zionism* (London, 1972), 530–31.

96. Proudfoot, *European Refugees*, 69–70; Wasserstein, *Britain and the Jews of Europe*, chap. 2; Nicholas Bethell, *The Palestine Triangle: The Struggle Between the British, the Jews, and the Arabs, 1935–1948* (London, 1980), chap. 3; Bauer, *American Jewry*, 137.

97. Martin Gilbert, *Exile and Return: The Emergence of Jewish Statehood* (London, 1978), 245; Wasserstein, *Britain and the Jews of Europe*, 57–58.

98. Gilbert, *Auschwitz and the Allies*, 109.

99. Wasserstein, *Britain and the Jews of Europe*, 184–85.

100. Tartakower and Grossman, *Jewish Refugee*, 346; Proudfoot, *European Refugees*, 68; Colin Lucas, *The Modern History of Israel* (London, 1974), 198.

101. Carpi, "Rescue of Jews," in Gutman and Zuroff, eds., *Rescue Attempts*, 465–526; idem, "Notes on the History of the Jews in Greece During the Holocaust Period: The Attitude of the Italians (1941–1943)," *Festschrift in Honor of Dr. George S. Wise* (Tel Aviv, 1981), 25–62; idem, "Nuovi documenti per la storia dell'olocausto in Grecia—l'atteggiamento degli Italiani (1941–1943)," in *Michael*, 7 (Tel Aviv, 1981), 119–200; Ivo Herzer, "How Italians Rescued Jews," *Midstream*, 29 (June/July 1983), 35–38.

102. Marrus and Paxton, *Vichy France and the Jews*, 315–21.

103. Browning, *Final Solution*, 135.

104. Renzo De Felice, *Storia degli ebrei italiani sotto il fascismo* (Turin, 1961), 73–75 and passim; Michaelis, *Mussolini and the Jews*, chap. 1 and passim; Gene Bernardini, "The Origins and Development of Racial Anti-Semitism in Facist Italy," *Journal of Modern History*, 49 (1977), 431–53; Juan Linz, "Some Notes Toward a Comparative Study of Fascism in Sociological Historical Perspective," in Walter Laqueur, ed., *Fascism: A Reader's Guide* (Berkeley, 1976), 86.

105. Louis P. Lochner, ed., *The Goebbels Diaries, 1942–1943*, trans. Louis P. Lochner (Garden City, N.Y., 1948), 241.

106. Léon Poliakov and Jacques Sabille, *Jews Under the Italian Occupation* (Paris, 1955), 39–44.

107. See Walter Laqueur, *The Terrible Secret: An Investigation into the Suppression of Information About Hitler's 'Final Solution'* (London, 1980), 224–28; Wasserstein, *Britain and the Jews of Europe*, 173–74; Gilbert, *Auschwitz and the Allies*, 101–04.

108. Feingold, *Politics of Rescue*, 174–78; Wasserstein, *Britain and the Jews of Europe*, 177–88; Eleanor F. Rathbone, *Rescue the Perishing: An Appeal, A Pro-*

gramme and a Challenge (London, 1943); Dina Porat, "Al-domi: Palestinian Intellectuals and the Holocaust," *Studies in Zionism*, 5 (1984), 97–124.

109. Irving Abella and Harold Troper, *None Is Too Many: Canada and the Jews of Europe, 1933–1948* (Toronto, 1982), 130–31.

110. War Cabinet document on the Bermuda Conference, 4 May 1943, Annex No. 5, PRO: FO371/5770/WR67; Monty N. Penkower, "The Bermuda Conference and Its Aftermath: An Allied Quest for 'Refuge' During the Holocaust," *Prologue* (Fall 1982), 145–73; Hirschmann, *Life Line*, 16.

111. Feingold, *Politics of Rescue*, 208; Gilbert, *Auschwitz and the Allies*, 267.

112. Avni, *Spain, the Jews, and Franco*, 117 ff.

113. Wasserstein, *Britain and the Jews of Europe*, 218.

114. Feingold, *Politics of Rescue*, 208; Monty N. Pentkower, "In Dramatic Dissent: The Bergson Boys," *American Jewish History*, 70 (1980–81), 281–309.

115. W. M. Carlgren, *Swedish Foreign Policy During the Second World War*, trans. Arthur Spencer (London, 1977), 158; Valentin, "Rescue and Relief Activities," 235–38.

116. Harold Flender, *Rescue in Denmark* (New York, 1963), 237–38; Wasserstein, *Britain and the Jews of Europe*, 216.

117. Feingold, *Politics of Rescue*, 208.

118. Ibid., 246; Monty N. Pentkower, "Jewish Organizations and the Creation of the U.S. War Refugee Board," *Annals, American Academy of Political and Social Science*, 450 (1980), 122–39. Hirschmann, *Life Line*, 19; Bauer, *American Jewry*, chap. 17; *Final Summary Report of the Executive Director, War Refugee Board* (Washington, D.C., 1945).

119. Herbert Emerson, "Post-War Problems of Refugees," in Research Institute on Peace and Post-War Problems, *Post-War Migrations: Proposals for an International Agency* (New York, 1943), 24; George L. Warren, "Post-War Migration Problems," ibid., 39; Bauer, *American Jewry*, 193; Abella and Troper, *None Is Too Many*, 184; Laqueur, *Terrible Secret*, 14–15.

120. Wasserstein, *Britain and the Jews of Europe*, 179–81; Feingold, *Politics of Rescue*, 182–83.

121. Schechtman, "Transnistria," 193–95; Hirschmann, *Life Line*, chap. 5.

122. John S. Conway, "Between Apprehension and Indifference: Allied Attitudes to the Destruction of Hungarian Jewry," *Wiener Library Bulletin* 27 (1973/74), 37–48; idem, "Der Holocaust in Ungarn: Neue Kontroversen und Überlegungen," *Vierteljahrshefte für Zeitgeschichte*, 32 (1984), 179–212; Yehuda Bauer, *The Holocaust in Historical Perspective* (Seattle, 1978), chap. 4; idem, *The Jewish Emergence from Powerlessness* (Toronto, 1979), 7–25.

123. Hirschmann, *Life Line*, 166; Feingold, *Politics of Rescue*, 246; Yahil, "Scandinavian Countries to the Rescue," 210.

CHAPTER 5

1. Theodor Schieder, ed., *Documents on the Expulsion of the Germans from Eastern-Central-Europe*, 4 vols. (Bonn, 1960–1), I, 1–2.

2. Albert Speer, *Inside the Third Reich*, trans. Richard and Clara Winston (New York, 1971), 557.

3. Eugene M. Kulischer, *Europe on the Move: War and Population Changes, 1917–47* (New York, 1948), 305.

4. Malcolm J. Proudfoot, *European Refugees, 1939–42: A Study in Forced Population Movement* (London, [1956]), 158–59.

5. George L. Warren, "Post-War Migration Problems," in Research Institute on Peace and Post-War Problems, *Post-War Migrations: Proposals for an International Agency* (New York, 1943), 38–39; George Woodbridge, *UNRRA: The History of the United Nations Relief and Rehabilitation Administration*, 3 vols. (New York, 1950), II, 469.

6. Alexander Donat (pseud), *The Holocaust Kingdom: A Memoir* (New York, 1965), 318–19; Proudfoot, *European Refugees*, 158–59; Pascal Ory, *Les Collaborateurs, 1940–1945* (Paris, 1976), 43.

7. Oscar Halecki, *East Central Europe Under the Communists: Poland* (New York, 1957), 49; Daniel Yergin, *Shattered Peace: the Origins of the Cold War and the National Security State* (Harmondsworth, England, 1980), 316.

8. George F. Kennan, *Memoirs 1925–1950* (New York, 1969), 279.

9. *Personnes déplacées* (Paris, 1948), 228.

10. Louise W. Holborn, *The International Refugee Organization: A Specialized Agency of the United Nations: Its History and Work, 1946–1952* (London, 1956), 138: Kulischer, *Europe on the Move*, 293; Anthony T. Bouscaren, *International Migrations Since 1945* (New York, 1963), 50; Kalman Janics, *Czechoslovak Policy and the Hungarian Minority, 1945–1948*, trans. and ed. Stephen Borsody (New York, 1982); Leszek A. Kosinski, "International Migration of Yugoslavs During and Immediately After World War II," *East European Quarterly*, 16 (1982), 189.

11. Bouscaren, *International Migrations*, 25, 73; Jacques Vernant, *The Refugee in the Post-War World* (London, 1953), 183–84.

12. *Report of the International Committee of the Red Cross [ICRC] on Its Activities During the Second World War (September 1, 1939–June 30, 1947)*, Vol. III, *Relief Activities* (Geneva, 1948), 97; Ralph Hewins, *Count Folke Bernadotte: His Life and Work* (Minneapolis, 1950), 164.

13. Janet Flanner, *Paris Journal, 1944–1965* (London, 1966), 26.

14. Vernant, *Refugee*, 367.

15. Max Hastings, *Bomber Command* (New York, 1979), 11, 353; Schieder, *Documents on the Expulsion*, I, 2–3; Bertram Gresh Lattimore, *The Assimilation of German Expellees into the West German Polity and Society Since 1945: A Study of Eutin, Schlesswig-Holstein* (The Hague, 1974).

16. Marcus J. Smith, *The Harrowing of Hell: Dachau* (Albuquerque, N.M., 1972), 68–69; *Report of ICRC*, III, 434.

17. Benedikt Kautsky, *Teufel und Verdammte: Erfahrungen und Erkenntnisse aus sieben Jahren in deutschen Konzentrationslagern* (Vienna, 1948), 77. Report of Brigadier H. L. Glyn Hughes, in Sir Henry Letheby Tidy, ed., *Inter-allied Conferences on War Medicine, 1942–1945* (London, 1947), 457. See also Eberhard Kolb, *Bergen-Belsen: Geschichte des "Aufenhaltslagers" 1943–1945* (Hannover, 1962).

18. Zdenek Lederer, *Ghetto Theresienstadt* (London, 1953), 197.

19. Richard Crossman, *Palestine Mission: A Personal Record* (London, n.d.), 21. Cf. Smith, *Dachau*.

20. Proudfoot, *European Refugees*, 343; Zorach Warhaftig, *Uprooted: Jewish Refugees and Displaced Persons After Liberation* (New York, 1946), 88.

21. "SHAEF Outline Plan for Refugees and Displaced Persons, 4 June 1944,"

4 June 1944, (SHAEF/G-5/9) Archives Nationales (henceforth AN): AJ⁴³76/128/309; Woodbridge, *UNRRA,* II, 477; Proudfoot, *European Refugees,* 147–52.

22. See Lucius D. Clay, *Decision in Germany* (London, 1950), 15–16.

23. Proudfoot, *European Refugees,* 167, chap. 8.

24. Leonard Dinnerstein, *America and the Survivors of the Holocaust* (New York, 1982), 53.

25. Major Greenwood, *Epidemics and Crowd Diseases* (London, 1935), 177–79; Frederick F. Cartwright, *Disease and History* (London, 1972); Friedrich Prinzip, *Epidemics Resulting from Wars* (Oxford, 1916), 163, 177; Ralph H. Major, *War and Disease* (London, n.d.), 142–44.

26. Sir Frederick Morgan, *Peace and War: A Soldier's Life* (London, 1961), 221.

27. Donat, *Holocaust Kingdom,* 75.

28. Smith, *Dachau,* 112, 131.

29. Proudfoot, *European Refugees,* 174; Woodbridge, *UNRRA,* II, 505; Comité international de la Croix-Rouge, *Health Conditions Among the Civilian Population of Certain European Countries Affected by the War, August 1947, June 1948* (Geneva, n.d.), 19–20, passim.

30. Nicolai Tolstoy, *Victims of Yalta* (London, 1979), 51.

31. Ibid., 188 ff.; Nicolas Bethell, *The Last Secret: Forcible Repatriation to Russia, 1944–1947* (London, 1976), 107.

32. Proudfoot, *European Refugees,* 177–78; Bethell, *Last Secret,* 53–54; Mark R. Elliot, *Pawns of Yalta: Soviet Refugees and America's Role in their Repatriation* (Urbana, Ill., 1982), x.

33. Proudfoot, *European Refugees,* 210–12.

34. Gerald Reitlinger, *The House Built on Sand: The Conflicts of German Policy in Russia, 1939–1945* (London, 1960), 284.

35. Tolstoy, *Victims of Yalta,* 405, 477; Bethell, *Last Secret,* chap. 8.

36. Elliot, *Pawns of Yalta,* 243.

37. Dean Acheson, *Present at the Creation: My Years at the State Department* (London, 1969), 69, 78. Cf. Robert H. Johnson, "International Politics and the Structure of International Organization: The Case of UNRRA," *World Politics,* 3 (1950/51), 520–38.

38. SHAEF Administrative Memorandum No. 39 (Revised—16 April 1945), "Displaced Persons and Refugees in Germany," in Proudfoot, *European Refugees,* 461.

39. Proudfoot, *European Refugees,* 136–38; Woodbridge, *UNRRA,* II, 490–91.

40. René Ristelhueber, *Au secours des réfugiés (O.I.R.)* (Paris, 1951), 31; R. L. Coigny, "Service de santé," in *Personnes déplacées,* 840; David Wodlinger, "An UNRRA Field Supervisor Looks Back," *Canadian Welfare,* 24 (15 April 1948), 12–16; Patrick M. Malin, "The Refugee: A Problem for International Organization," *International Organization,* I (1947), 449.

41. Morgan, *Soldier's Life,* 222. Cf. E. F. Penrose, *Economic Planning for the Peace* (Princeton, 1953), 322–23.

42. Proudfoot, *European Refugees,* 300–01.

43. Morgan, *Soldier's Life,* 229, and passim.

44. Dinnerstein, *America,* 17.

45. Harrison report in ibid., 303. Cf. Malin, "The Refugee," 449.

46. Ira A. Hirschmann, *The Embers Still Burn* (New York, 1949), 9; idem, *Caution to the Winds* (New York, 1962), 195 ff.

47. Woodbridge, *UNRRA*, III, 424–25; Proudfoot, *European Refugees*, 276; Ristelhueber, *Au secours des réfugiés*, 87, 189, 248–49; H. B. M. Murphy, "The Camps," in H. B. M. Murphy, ed., *Flight and Resettlement* (Paris, 1955), 58.

48. United Nations, Official Records. Plenary Meetings of the General Assembly. First Session. (1946). Part I, 426–27. Roger Nathan-Chapolot, *Les Nations unies et les réfugiés: le maintien de la paix et le conflit de qualification entre l'Ouest et l'Est* (Paris, 1949).

49. Woodbrige, *UNRRA*, I, 46; Comité internationale, *Health Conditions*, 46.

50. Alfred M. de Zayas, *Nemesis at Potsdam: The Anglo-Americans and the Expulsion of the Germans: Background, Execution, Consequences* (London, 1977), 65; Alexander Werth, *Russia at War, 1941–1945* (London, 1964), 964–65.

51. Schieder, *Documents*, I, 20, 52.

52. Eugen Lemberg and Friedrich Edding, eds., *Die Vertriebenen in West-Deutschland: Ihre Eingliederung und Ihr Einfluss auf Gesellschaft, Wirtschaft, Politik und Geistesleben*, 3 vols. (Kiel, 1959).

53. Stephen Kertesz, "The Expulsion of the Germans from Hungary: A Study in Postwar Diplomacy," *Review of Politics*, 15 (1953), 179–208; Winston S. Churchill, *Triumph and Tragedy* (New York, 1962), chap. 20.

54. De Zayas, *Nemesis at Potsdam*, 81, 109–15; Proudfoot, *European Refugees*, 380–89; Joseph B. Schechtman, *Postwar Population Transfers in Europe, 1945–55* (Philadelphia, 1962), 37.

55. Count Hans von Lehndorff, *East Prussian Diary: Journal of Faith, 1945–47* (London, 1963), 228.

56. United Nations. Official Records. Plenary Meetings of the General Assembly. First Session. (1946). Part I, 424.

57. G. C. Paikert, *The Danubian Swabians: German Populations in Hungary, Rumania and Yugoslavia and Hitler's Impact on their Patterns* (The Hague, 1967); Kertesz, "Expulsion of the Germans"; Schieder, *Documents*, III–V; Janics, *Czechoslovak Policy and the Hungarian Minority*.

58. Clay, *Decision in Germany*, 313.

59. Vernant, *Refugee*, 158; Proudfoot, *European Refugees*, 378–79; Ristelhueber, *Au secours des réfugiés*, 69–70; Louise Holborn, *Refugees: A Problem of Our Time: The Work of the United Nations High Commissioner for Refugees, 1951–1972* (2 vols., Metuchen, N.J., 1975), 338; De Zayas, *Nemesis at Potsdam*, chap. 7.

60. Anna J. Merritt and Richard L. Merrit, eds., *Public Opinion in Occupied Germany: The OMGUS Surveys, 1945–1949* (Urbana, Ill., 1970), 90 ff., 298. See Lemberg and Edding, *Vertriebenen in West-Deutschland*, passim.

61. See Lattimore, *Assimilation of the German Expellees*; De Zayas, *Nemesis at Potsdam*, chap. 7.

62. Wolfgang Jacobmeyer, "Jüdische Uberlebene als 'Displaced Persons': Untersuchungen zur Besatzungspolitik in den deutschen Westzonen und zur Zuwanderung osteuropäischer Juden 1945–1947," *Geschichte und Gesellschaft*, 9 (1983), 421–52; Proudfoot, *European Refugees*, 321; Dinnerstein, *America*, 24.

63. Herman Dicker, *Piety and Perseverance: Jews from the Carpathian Mouhntains* (New York, 1981), 98–9.

64. Dinnerstein, *America*, 28; PRO: FO 371/45380/E7251.

65. Jacobmeyer, "Judische Uberlebene," 421; idem, "Polnische Juden in der Amerikanischen Besatzungzone Deutschlands 1946/7," *Vierteljahrshefte für Zeitgeschichte*, 25 (January 1977), 129–32.

66. Dinnerstein, *America*, 17, and chap. 1, passim; idem, "The U.S. Army

and the Jews: Policies Toward the Displaced Persons After World War II," *American Jewish History,* 68 (1979), 353–66.

67. Harrison report in Dinnerstein, *America,* 291–305.

68. Crossman, *Palestine Mission,* 87–88; PRO: CO537/1765/HM08714.

69. Woodbridge, UNRRA, III, 423, 442; Proudfoot, *European Refugees,* 341; Michael J. Cohen, *Palestine and the Great Powers, 1945–1948* (Princeton, 1982), 57.

70. "Jews in Eastern Europe After World War II: Documents from the British Foreign Office," *Soviet Jewish Affairs,* 10 (1980), 55.

71. Lucjan Dobroszycki, cited in Abraham Brumberg, "The Ghost in Poland," *New York Review of Books,* 2 June 1983, 37; Warhaftig, *Uprooted,* 39; Antony Polonsky, Introduction to "Jews in Eastern Europe," 52.

72. Crossman, *Palestine Mission,* 97.

73. Yehuda Bauer, *Flight and Rescue: Brichah* (New York, 1970).

74. Morgan, *Soldier's Life,* 235, 243–46.

75. Hirschmann, *Embers,* 227; Warhaftig, *Uprooted,* 77 ff. See I. F. Stone, *Underground to Palestine* (New York, 1978).

76. Dinnerstein, *America,* 112–13.

77. Roger Allen to R. M. A. Hankey, 8 July 1946, PRO: F0371/57691/N6029 and N9100.

78. Alec Dickson, "Point de vue anglaise," in *Personnes déplacées,* 142.

79. Proudfoot, *European Refugees,* 355–56.

80. Ibid., 357–59; Howard Morley Sachar, *A History of Israel from the Rise of Zionism to Our Time* (New York, 1976), 395.

81. See Andrei Vyshinsky's speech to the General Assembly, United Nations. Official Records. Plenary Meetings of the General Assembly. First Session (1946), Part I, 414. Roger Nathan-Chapolot, *Les Nations unies,* 79–175.

82. E. F. Penrose, "Negotiating on Refugees and Displaced Persons, 1946," in Raymond Dennett and Joseph Johnson, eds., *Negotiating with the Russians* (Boston, 1951), 139–68.

83. Proudfoot, *Refugees,* 402–04.

84. Holborn, *International Refugee Organization,* 49; Elfan Rees, "The Refugee and the United Nations," *International Conciliation,* 492 (1953), 272: Penrose, "Negotiating on Refugees," 164–67.

85. Ristelhueber, *Au secours des réfugiés,* 66; Holborn, *Refugees,* I, 59; Clay, *Decision in Germany,* 233; John George Stoessinger, *The Refugee and the World Community* (Minneapolis, 1956), chaps. 6, 7; Proudfoot, *European Refugees,* 414–15; interview with J. D. Kingsley, 3 March 1952, IRO Papers, AN: AJ[43]140/35.

86. Ristelhueber, *Au secours des réfugiés,* 77–87; Holborn, *International Refugee Organization,* 195.

87. Stoessinger, *Refugee and the World Community,* chap. 8: Proudfoot, *European Refugees,* 418–19, 424–25; Holborn, *International Refugee Organization,* chap. 20.

88. Pierre Dominique, *Europe of the Heart* (London, 1960), 110.

89. Vernant, *Refugee,* 52; Holborn, *Refugees,* I, 42.

EPILOGUE

1. Louise W. Holborn, *The International Refugee Organization: A Specialized Agency of the United Nations: Its History and Work, 1946–1952* (London, 1956), 180–83; Frederick A. Norwood, *Strangers and Exiles: A History of Religious Refugees,* 2 vols. (Nashville, Tenn., 1965), II, 413; Robert Lee Wolff, *The Balkans in Our Time,* rev. ed. (New York, 1978), 478.

2. E. F. Penrose, "Negotiating on Refugees and Displaced Persons, 1946," in Raymond Dennet and Joseph E. Johnson, eds., *Negotiating with the Russians* (Boston, 1951), 144–47.

3. See John Loftus, *The Belarus Secret* (New York, 1982); Tom Bower, *The Pledge Betrayed: America and Britain and the Denazification of Postwar Germany* (Garden City, N.Y., 1982).

4. Draft Report of Refugee Department, F.O., Public Record Office (henceforth PRO): FO371/72042/WR899; British Legation, Helsinki, to F.O., PRO: FO371/72046/WR 872; Roger Nathan-Chapotot, *Les Nations unies,* 91–92, and passim, Wolff, *The Balkans,* 483.

5. Heinz Höhne, *The Order of Death's Head: The Story of Hitler's SS,* trans. Richard Barry (London, 1972), 289; Robert L. Koehl, *RKFD: German Resettlement and Population Policy, 1939–1945: A History of the Reich Commission for the Strengthening of Germandom* (Cambridge, Mass., 1957), 219–21, and passim; idem, *The Black Corps: The Structure and Power Struggles of the Nazi SS* (Madison, 1983), 190–92.

6. UN General Assembly. Third Session. Part I, Official Records, Plenary Meetings, 23 September 1948, 53; Holborn, *International Refugee Organization,* 493–512.

7. Ibid., 513–14; cf. Koehl, *RKFDV,* 220.

8. Ira A. Hirschmann, *The Embers Still Burn* (New York, 1949), 82; Siegfried Kracauer and Paul L. Berkman, *Satellite Mentality: Political Attitudes and Propaganda Susceptibilities of Non-Communists in Hungary, Poland and Czechoslovakia* (New York, 1956), 157–59.

9. Kracauer and Berkman, *Satellite Mentality,* 10–11.

10. Jacques Vernant, *The Refugee in the Postwar World* (London, 1953), 228.

11. George L. Warren, "The Escapee Program," *Journal of International Affairs,* 7 (1953), 82–83.

12. Paul Tabori, *The Anatomy of Exile: A Semantic and Historical Study* (London, 1972), 260–61.

13. Joseph B. Schechtman, *The Refugee in the World* (New York, 1963), 56–57; Vernant, *Refugee,* 238–41.

14. Anthony T. Bouscaren, *International Migrations Since 1945* (New York, 1963), 15, 27; John George Stoessinger, *The Refugee and the World Community* (Minneapolis, 1956), 176; Vernant, *Refugee,* 53; Warren, "Escapee Program"; Louise Holborn, *Refugees: A Problem of our Time: The Work of the United Nations High Commissioner for Refugees, 1951–1972,* 2 vols. (Metuchen, N.J., 1975), I, 570.

15. Stoessinger, *Refugee,* 163 and 223 n. 19; Vernant, *Refugee,* 38.

16. Holborn, *Refugees,* I, chap. 3; Stoessinger, *Refugee,* 164–65; Robert E. Asher et al., *The United Nations and Economic and Social Cooperation* (Washington, D.C., 1957), 528.

17. United Nations, Statute of the Office of the United Nations High Commissioner for Refugees (HCR/INF/1/Rev. 3).

18. United Nations, Convention and Protocol Relating to the Status of Refugees (HCR/INF/29/Rev. 3).

19. Holborn, *Refugees*, I, 349.

20. Ibid., I, 136; UNHCR, *A Mandate to Protect and Assist Refugees* (Geneva, [1971]).

21. Aristide R. Zolberg, "The Formation of New States as a Refugee-Generating Process," *Annals, American Academy of Political and Social Science*, 467 (May 1983), 26.

22. Holborn, *Refugees*, I, 391; UNHCR, *Mandate to Protect*, 62.

23. Bouscaren, *International Migrations*; James A. Michner, *The Bridge at Andau* (London, 1957), 218–21; Noel H. Moynihan, *The Light in the West* (London, n.d.), 12; Norwood, *Strangers and Exiles*, 321; Tabori, *Anatomy of Exile*, 254. Cf. Egon F. Kunz, "The Refugee in Flight: Kinetic Models and Forms of Displacement," *International Migration Review*, 7 (1973), 125–46.

24. Robert A. Kann, "Looking Across the Iron Curtain," *Journal of International Affairs*, 16 (1962), 71; Manfred Rauchensteiner, *Spätherbst 1956. Die Neutralität auf dem Prüfstand* (Vienna, 1981), 84–89.

25. Report of the Special Committee on the Problem of Hungary. General Assembly. Official Records: Eleventh Session. Supplement No. 18 (A/3592) (New York, 1957), 211–14.

26. Elfan Rees, *We Strangers and Afraid. The Refugee Story Today* (New York, 1969), 52.

27. Holborn, *Refugees*, I, 414–15.

28. Hans-Joachim Hoffmann-Nowotny, "European Migration after World War II," in William H. McNeil and Ruth S. Adams, eds., *Human Migration: Patterns and Policies* (Bloomington, Ind., 1978), 89.

29. Norwood, *Strangers and Exiles*, II, 237–39.

30. Josef Banas, *The Scapegoats: The Exodus of the Remnants of Polish Jewry* (London, 1979); Nicholas Bethell, *Gomulka, His Poland and His Communism* (London, 1969), chap. 16.

31. Holborn, *Refugees*, I, 508–09, 517–18; James L. Carlin, "Significant Refugee Crises Since World War II and the Response of the International Community," in *Transnational Legal Problems of Refugees, 1982. Michigan Yearbook of International Legal Studies* (New York, 1982), 11.

32. Abraham Karlikow, "Soviet Jewish Emigration: The Closing Gate," *World Refugee Survey, 1983*, 34–35.

33. Peter Reddaway, "The Development of Dissent and Opposition," in Archie Brown and Michael Kaser, eds., *The Soviet Union Since the Fall of Krushchev*, 2d ed. (London, 1978); *The New York Times*, 8 November 1981; *The Times* (London), 19 May 1980; Zvi Gitelman, "Exiting from the Soviet Union: Emigrés or Refugees?" in *Transnational Legal Problems of Refugees*.

34. "Israeli Concern at Jews Given Refuge in the US," *The Times* (London), 3 April 1980.

35. Victor Zaslavsky and Robert J. Brym, *Soviet-Jewish Emigration and Soviet Nationality Policy* (New York, 1983); Zvi Gitelman, "What Happened," *Moment*, October 1982, 34–37.

36. Mary C. Spilane, "Europe—An Anxious Year," *World Refugee Survey, 1982* 35–36: "Europe," ibid., *1983*, 67–68.

37. Report on the Activities of the Intergovernmental Committee on European Migration (ICEM), Council of Europe, Parliamentary Assembly, Document 4388, 8 August 1979; ibid., Document 4820, 15 December 1981; ibid., Document 4759, 14 August 1981.

38. Holborn, *Refugees*, II, chaps. 14, 17.

39. Ibid., chaps. 7, 8.

40. David Scott Nance, "The Individual Right to Asylum Under Article 3 of the European Convention on Human Rights," in *Transnational Legal Problems of Refugees*; Guy S. Goodwin-Gill, *International Law and the Movement of Persons Between States* (Oxford, 1978).

41. Pierre Ballet, "Les difficultés d'adaptation en France des réfugiés nationaux d'Afrique du Nord," *AWR Bulletin*, 13 (1975), 217; idem, "Un exemple d'intégration économique: celui des réfugiés d'Algérie," *AWR Bulletin*, 13 (1975), 221; William B. Cohen, "Legacy of Empire: The Algerian Connection," *Journal of Contemporary History*, 15 (1980), 97–123; Schechtman, *Refugee*, 75–86; Alistaire Horne, *A Savage War of Peace: Algeria, 1954–1962* (Harmondsworth, England, 1962), 533; Jane Kramer, *Unsettling Europe* (New York, 1981), 172–80.

42. Schechtman, *Refugee*, 71–72; Claudio Segrè, *Fourth Shore: The Italian Colonization of Libya* (Chicago, 1975), 171–81.

43. Carlin, "Significant Refugee Crises," 13–14; A. B., "Portugal: Open Door Tradition," *Refugees*, May 1982.

44. Leon Gordenker, "The Refugee Explosion: Implications for Humanitarian Organizations," *ICVA News*, 85 (April 1980), 36–38.

45. *World Refugee Survey, 1981*, 41.

Index

Abdul Hamid II (Ottoman sultan), 22, 42, 75
Acheson, Dean, 318, 319
Aga Khan, Sadruddin, 369
Albania, 49, 104, 132
Alexander II (Russian tsar), 24, 26–28
Alexander III (Russian tsar), 25, 29
Algeria, 262, 367–68
Alien Act (1905), 37–38
Alienation, and refugees, 13
Alliance Israélite Universelle, 33, 34, 38, 39, 66, 67
Allies, the, 52, 59, 98, 241, 251, 255, 263, 270, 291, 292, 327
  anti-Bolshevik activities of, 56, 57, 59, 84, 87
  and Armenians, 79, 81
  condemnation of Nazi genocide by, 282, 283
  and rescue efforts for refugees, 284–85
  and Spain, 262, 264–65
  and Turkey, 78, 97, 102, 272, 273
  victory of, in the Second World War, 301
American Committee for Relief in the Near East (ACRNE), 83–84, 101, 111
American Jewish Committee, 290
American Jewish Congress, 159
American Jewish Joint Agricultural Corporation (Agro-Joint), 118–19, 134
American Jewish Joint Distribution Committee, 67, 86, 111, 118, 162, 181–83, 199, 229, 264, 277, 286, 290, 294, 337
American Relief Administration (ARA), 85, 86, 111, 117, 118

Anarchism, 25, 26
Anatolia, 45, 48, 75, 97, 107
Anders, Wladislaw, 246, 278
Anglo-American Committee of Inquiry on Palestine, 334
Ankara, 97, 98, 119, 273, 274, 291
Anschluss, 148, 153, 156, 157, 165–67, 172, 174, 177, 181, 187, 214, 250
Anti-Comintern Pact, 259
Antisemitism, 32, 35, 36, 38, 63, 134, 143, 147. *See also* Pogroms
  among the Allied military, 332–33
  in Hungary, 64, 146
  and Nazism, 122, 138–41, 148, 150–77, 195, 208–19, 227–34, 237
  in Poland, 61, 63, 117, 143–44, 170, 172, 173, 185, 335–36, 362
  in Rumania, 33–35, 144, 170, 173, 237
  in the Soviet Union, 140, 352
  in Spain, 260
  in Switzerland, 257
  in tsarist Russia, 28–31, 54, 61–63, 140
Antonescu, Ion, 237, 291
Arabs, 75, 104, 115, 116, 276
  and war with Israel, 339, 345, 354, 364, 370
  and Zionism, 152, 153, 185, 274
Arendt, Hannah, 4, 70, 180
Armenian Republic, 75, 77–79, 83, 84, 101, 119
  under the Soviet Union, 78, 81, 112, 119, 120
Armenians, 75–76, 78, 83
  as refugees, 10, 12, 52, 74–81, 86, 94, 96, 98, 101, 104, 106, 110, 112, 133,

Armenians (*continued*)
149–50, 159, 167, 178, 289, 363
resettlement of, 119–21
and Turks, 75, 76, 79, 84, 104
Arrow Cross movement, 74, 144, 270, 292
Aryans, 130, 157, 177, 208–9, 211, 213, 234, 235, 239
definition of, 129
Asia Minor, 41, 76, 82, 83, 96, 98, 100–102, 105, 167
Atatürk, Mustafa Kemal, 52, 78, 97, 98, 100
Athens, 100, 103, 104, 262
Atlee, Clement, 332, 334
Auschwitz, 269, 270, 279, 281, 292, 306
Australia, 68, 205, 344, 345, 357, 361
Austria, 19, 27, 35, 41, 48, 69, 124, 370.
See also Anschluss
antisemitism in, 23, 38, 57, 62, 63, 156, 168–69
and Germany, 148, 166–68, 214, 220
and refugees, 66, 74, 87, 130, 132–33, 149, 167–68, 172, 175, 177, 181, 205, 219, 229, 345, 360, 365
Austrians, as refugees, 17, 132–33, 179

Babi Yar, 246
Bakunin, Michael, 14, 20–21
Balfour Declaration (1917), 115
Balkans, 40, 42, 44–47, 50, 82, 83, 106–8
Germans in the, 219
maps, 43, 47, 99
refugees of the, 14, 27, 40–50, 74, 95, 106, 158, 186, 272, 294, 317, 348
Baltic states, 29, 56, 67, 234, 241, 250
Germans of the, 219, 223, 224, 240
and refugees, 55, 57, 71, 95, 139, 197, 198, 249, 271, 323
Barbarossa, 230, 231, 234, 237, 241, 243, 245, 246, 250
Bauer, Yehuda, 135, 206, 218, 290
Bavaria, 66, 129
Beck, Jozef, 142, 186
Begin, Menachem, 363
Belgian Congo, 187, 367
Belgian Relief Commission, 85, 86
Belgium, 71, 201, 282
and refugees, 20, 24, 26, 36, 38, 65, 71, 126, 130, 201, 205, 218, 367
U.S. aid to, 82, 85, 137, 146, 169, 258, 345
Belorussia, 196–98, 234, 238, 246, 247
Ben-Gurion, David, 274, 275
Bentwich, Norman, 128, 129, 134, 147, 160, 161, 163, 182

Bergson, Peter H., 283, 286, 288
Berlin, 25, 87, 128, 129, 131, 162, 167, 172, 180, 196, 204, 216–18, 230, 231, 235, 249, 251, 257, 271
and Berlin Wall, 362
blockade of (1948), 344, 347, 349
as hole in the Iron Curtain, 361–62
riots in East (1953), 359
Treaty of (1878), 33, 34
Bermuda Conference (1943), 264, 283–85, 288, 291
Bern, 155, 157, 255
Bernadotte, Folke, 271
Bessarabia, 64, 65, 200, 237, 291
Germans in, 223, 224
Bevin, Ernest, 312, 338
Bey, Talaat, 77
Birobidzhan, 118–19, 338
Bismarck, Otto von, 22, 23, 26
Blum, Léon, 147, 190, 192
Bohemia, 23, 36, 175, 219, 230
Bolsheviks, 55–61, 64, 78, 86, 167, 178, 258
Bolshevism, 56, 57, 59, 82, 84, 93
Boundary changes in Europe after the First World War, *map*, 73
Brand, Joel, 292, 293
Brătianu, Ion, 64
Brazil, 253, 266
Bremen, 35, 36, 65
Brest-Litovsk, Treaty of (1918), 55, 62
Brichah, 336–37, 348
British Union of Fascists, 151
Brussels, 25, 67, 82
Bucharest, 49, 64, 65, 72, 143, 200, 290
Budapest, 39, 72, 74, 251, 262, 268, 270, 272, 359
Bukovina, 27, 74, 200, 223, 224, 237
Bulgaria, 41, 42, 44–46, 48, 49, 106, 200, 236
Communist control in, 348
and Greece, 106–9
and refugees, 48, 49, 52, 59, 103, 104, 106–8, 268, 291, 295, 350
and Turkey, 46, 48, 270, 353
Bulgarians, as refugees, 41, 44, 46, 106, 107, 110
Burckhardt, Carl, 134
Burgos, 192
Burtsev, Vladimir, 26

Canada, 68, 205, 284, 345, 361, 368
Carnegie Commission, 45, 46, 48
Carol II (Rumanian king), 143, 144, 173, 200

Carr, Raymond, 193
Casablanca, 203, 204, 206, 259
Castro, Fidel, 369
Catalonia, 190, 191
Catherine the Great, 7, 20, 24
Catholic Church. *See* Vatican
Caucasus, 54, 76–79, 81, 119, 120, 239, 243, 249, 282
  aid to, 82–84
  Germans in the, 223, 234
  map, 80
Cecil of Chelwood, Robert, 151, 160, 163
Central Asia, 54, 242, 243, 245, 249
Central Europe, 16, 36, 67, 123, 145, 300
  Jews of, 32, 209–19
  refugees of, 10, 23, 71–72, 74, 91, 149, 154, 169, 172, 185, 186, 258, 264
Central Office for Reich Security (RSHA), 209, 215, 221
Central Powers, 48, 62, 63, 78, 87, 97, 106
Chamberlain, Neville, 274
Chautemps, Camille, 146
Chicherin, Georgi, 85
China, and refugees, 60, 180–82
Christians
  as refugees, 6–9, 41, 44, 75, 154
  sympathy of, for Ottoman victims, 50
Churchill, Winston, 58, 84, 100, 116, 195, 262, 274, 276, 286, 293, 327, 334
Cold war, 348–54
  origins of, 348
  and refugees, 11, 351, 358–65
Communists. *See also* Bolsheviks; Bolshevism
  conditions under, in Soviet bloc, 352, 358
  control of Eastern Europe by, after the Second World War, 348–49, 358
  of Germany, 130, 139–41, 193, 196
  of Spain, 193
Concentration camps, 129, 162, 168, 196, 233, 304, 306–8, 312, 331, 337
Congress of Berlin (1878), 42, 44, 46
Constantine I (Greek king), 98
Constantinople (later Istanbul), 10, 42, 44, 97
  and Armenians, 75, 77
  and refugees, 59, 60, 88, 94, 96, 100, 101, 110, 111, 113, 199, 204, 270, 272, 292
  and relief aid, 83–85
Convention of Adrianople (1913), 46
Convention on the Status of Refugees (1951), 356, 357

Copenhagen, 271, 287
Council of Europe (1949), 5, 367
Counter-Reformation, 154
Crimea, 56, 59, 60, 84, 118, 234, 248, 249
Croatia, 225, 236, 268, 278, 279, 281
  Germans in, 236
Crossman, Richard, 307, 334, 335
Cuba, 177, 253
  refugees from, 362, 369–70
Curzon line, 57, 328
Cyprus, 187, 272, 278, 338, 339, 370
Czechoslovakia, 57, 67, 69, 70, 145, 175, 282
  Communist coup in (1948), 344, 349, 353
  Germans in, 219, 220, 330
  and Germany, 174–77, 214
  and Hungary, 174, 302
  and population shifts, 327, 329
  and refugees, 130, 149, 153, 168, 174–76, 181, 212, 242, 250
  and the Soviet Union, 249, 302, 362
Czechs, as refugees, 17, 179, 217, 249

Daladier, Edouard, 148, 192
Dallin, Alexander, 235, 239
Danzig, Free City of, 65, 71, 160, 326
  and Germany, 133–34, 173, 177, 221
De Sousa Mendes, Aristides, 263–64
Decolonization, and refugees, 367–69
Decree on Eliminating Jews from German Economic Life, 213
Decree of Public Safety, 126
Denikin, Anton, 56, 84
Denmark, 71, 90, 270
  and Germany, 201, 271, 304, 305
  rescue of Jews of, 271, 283, 287, 288
Disease control, after Second World War, 311–13
Displaced persons (DPs), after the Second World War, 299–300, 302, 304, 309–13, 336, 351
  definition of, 342
  Germans as, 325–31
  and International Refugee Organization (IRO), 340–41, 344–45
  Jews as, 322–23, 332, 334, 335, 338
  and UNRRA, 319–23
Displaced Persons Act (1948), 344
Dobruja, 49, 200, 223, 224
Dollfuss, Engelbert, 132
Donat, Alexander, 194, 312
Dubček, Alexander, 362
Duma, 62

East Africa, 115, 246
East Germany, refugees from, 353, 361
Eastern Europe, 35, 82, 145, 278
  anti-Jewish pattern of, 64, 142, 166, 335
  empires dominating, 52–74
  exodus of Jews from, 27–39, 61–68, 123,
      130, 137, 141–45, 153, 200, 338
  after the First World War, 53, 132
  and Germany, 208–9, 223, 241, 326, 330
  map (1947), 303
  non-Jewish emigration from, 154, 200,
      225–26
  resettlement of Jews in, 117–19
  after the Second World War, 297–98,
      300, 314, 323, 326, 335, 338, 340, 341,
      344, 349–51, 355, 358, 365
Eastern Question, 21, 40, 50
Eden, Anthony, 284–85
Egypt, 77, 81, 272
Eichmann, Adolf, 168, 215, 217, 229, 233,
    281
Eisenhower, Dwight D., 310, 314, 332, 336
El Alamein, 282
Emergency Rescue Committee, 260
Emerson, Herbert, 166, 217, 286
Emigdirect (United Committee for Jewish
    Emigration), 67–68.
Empires, crumbling of, and refugees, 9–
    10, 70–72, 74
Enabling Act (1933), 129
Engels, Friedrich, 17
England, 34, 87, 109, 143, 151, 194, 263
  and Armenians, 75, 81, 150
  decolonization and, 367
  fascists in, 151
  and Germany, 149, 152, 189, 201, 275,
      282
  and Greece, 41, 102, 105
  and Madagascar, 186–87, 232
  and Palestine and Zionism, 81, 115,
      116, 152–54, 170, 184–88, 212, 216–17,
      273–78, 285, 323, 332, 334–39
  and refugees, 17, 20–22, 24, 26, 29, 31,
      32, 36–38, 49, 60, 85, 90, 126, 140,
      149–54, 165, 169–71, 175, 181, 204,
      205, 217, 246, 261, 264, 275–78, 282–
      86
  and Russia, 42, 56, 317
  and Turkey, 97, 100, 102
Erevan, 78, 81, 112
Escape routes for refugees, during the
    Second World War, 240–82
  Eastern Europe (1941–44), 250–52
  Italy, 278–81

Palestine, 274–78
Portugal, 258–59, 262–65
Soviet Union (1941–44), 241–50
Spain, 241, 258–65
Sweden, 270–72
Switzerland, 241, 252–58
Turkey, 270, 272–74
Vatican, 265–70
Estonia, 71, 197, 221, 235, 242, 243
Europe under Nazi occupation, map, 244
European Convention of Human Rights
    (1951), 367
Evian Conference (1938), 165, 166, 170–72,
    214, 216
Exiles, vs. refugees, 15, 17–22
Exodus affair, 338–39

Far East, and refugees, 60, 61, 138, 180–82
Fascism, 90, 136, 190. See also Terrorism
  and creation of refugees, 122–28, 190–
      94
  definition of, 122
  in Italy, 122–28
"Final Solution of the Jewish Question,"
    187, 209, 215, 218, 219, 227–33, 254,
    256, 279, 287, 289, 290
Finland, 231
  and refugees, 55–57, 199–200, 270,
      350
  and Soviet Union, 199–200, 271, 349
First World War, 4, 9, 10, 14, 23, 26, 32,
    33, 37, 40, 56, 85, 110, 189, 219
  demise of empires after, 52
  end of, 81
  and nationalism, 51
  outbreak of, in 1914, 41, 45, 48, 53
  refugee crisis after, 52–81, 106, 119, 120,
      128, 149, 167, 178
  refugee crisis before, 76, 106
  refugee crisis during, 76, 201
  relief aid after, 83
  and Swiss neutrality, 154–55
  and Zionism, 115, 116
France, 7, 9, 23, 71, 75, 81, 82, 93, 97, 98,
    109, 123, 127, 147, 154, 194, 195, 261,
    263. See also Vichy
  decolonization and, 367–68
  and Germany, 23, 145, 148, 189, 192,
      202–3, 230, 258, 330
  liberation of, 255
  and refugees, 15–17, 20, 21, 23, 26, 29,
      36, 49, 60, 65, 85, 90, 96, 111–14,
      125–27, 130, 133, 134, 137, 145–50,

156, 165, 171, 172, 175, 181, 184, 191, 200–205, 217, 232, 257, 258, 260, 261, 264, 279, 281, 286, 345, 370
  and Russia, 56, 316, 317
Franco, Francisco, 123, 149, 190–92, 259
Frank, Hans, 221, 225–27, 230, 231
Frankfurt, 39, 321, 337
  Treaty of (1871), 23
Frederic, Harold, 36, 38
French Revolution, 9, 15, 18, 191
Freud, Sigmund, 168
Fry, Varian, 260

Galicia, 25, 57, 63
  Germans in, 223
  Jewish condition in, 32, 36, 63
  Jewish exodus from, 27, 32–33, 35, 62, 66
  and refugees, 53, 56, 65
General Plan East (1942), 234
Generalgouvernement, 221, 223–26, 228–30
Geneva, 109, 155, 354
  and refugees, 17–19, 22, 25, 26, 64, 72, 94, 111, 113, 137, 148, 164–65, 256
Geneva Convention on Refugees Coming from Germany (1938), 165
George V (British monarch), 59
Georgia (U.S.S.R.), 78, 79, 84
German Empire, 23, 36
German unification, wars of (1864–71), 23
Germans
  as refugees, 10–11, 16–17, 19, 23, 24, 52–54, 56, 71, 74, 110, 324–31, 363
  in Russia, 29, 219, 240, 241, 245
Germany, 34, 57, 69, 71, 81, 109, 113, 263, 270, 278. *See also* Antisemitism; East Germany; Hitler, Adolf; Nazism; Refugees; West Germany
  antisemitism in, 36, 66, 163, 176–77, 208–19, 227–34
  Auslandsdeutsche, 219–20, 225
  and Austria, 148, 157, 214
  and communism, 130
  and Czechoslovakia, 174–76, 214
  and Danzig, 133, 134
  defeat of, 296, 299, 302, 326
  and England, 201, 212, 275, 282, 292–93
  and France, 145, 148, 189, 192, 200, 201, 258
  and Hungary, 144, 236, 251–52, 269
  and Italy, 236, 279–81

  and League of Nations, 109, 158, 161, 164, 171
  and Nazism, 122–23, 128–35, 152, 157, 208–95
  and Poland, 133, 134, 173, 189, 194–96, 198, 209, 218, 237, 250
  post-World War II division of, 347
  post-World War II recovery of, 330–31, 357
  and refugees, 16, 17, 20, 24, 25, 36, 60, 61, 63, 65, 66, 71–72, 87, 95, 154
  and Russia, 55, 61–63, 139–40, 189, 193, 195, 196, 199, 219, 223, 232, 235, 238, 241, 242, 247, 249–51, 259, 262, 282, 287, 297
  and Spain, 258–60, 262
  and Switzerland, 156–58, 256
  and Turkey, 231, 270, 272, 273
Gestapo, 129, 130, 134, 177, 196, 217, 219, 269, 276, 279
Goebbels, Joseph, 280–81
Goedhart, G. J. van Heuven, 357–58
Goga, Octavian, 144
Gömbös, Gyula, 144, 174
Gomulka, Wladyslaw, 196, 359, 362
González, Valentín (El Campesino), 193, 194, 248
Göring, Hermann, 213, 215–17, 219, 232
Grabski, Wladyslaw, 117
Great Britain. *See* England
Great Depression, effect of, on refugees, 108, 118, 123, 131, 134–41, 143–44, 146, 150, 155, 179, 270
Great Purges of the U.S.S.R., 139, 141, 193
Greco-Bulgarian Exchange, 106–9
Greco-Bulgarian Mixed Commission, 106–7
Greco-Turkish War (1919–23), 97–100
Greece, 45–46, 49, 70, 106, 111, 279, 282
  and Bulgaria, 106–9
  civil war in, 302, 355
  independence for, 41
  and refugees, 41, 48, 52, 81, 101, 107, 108, 278, 281
  and Turkey, 41, 44–46, 48, 52, 97–108
  and the United States, 98, 100
Greek Refugee Settlement Commission, 104, 105
Greeks, as refugees, 41, 48, 52, 98, 106, 107, 110, 113, 167, 272, 278, 279, 355
Grossmann, Kurt, 256–57
Grosz, Bandi, 292
Grynszpan, Herschl, 176

Ha'avara (Transfer) Agreement, 212–13, 217
Hamburg, 34–36, 38, 65, 177, 266, 305, 339
Hapsburg empire, 9, 20, 36, 40, 44–45, 52, 60, 72, 74
Harbin, and refugees, 60
Harrison, Earl G., 322, 333–34, 336
Havana, 177, 369
Hayes, Carlton J. H., 261, 262, 286, 288
Hebrew Sheltering and Immigrant Aid Society (HIAS), 68
Herzen, Alexander, 18–22, 24
Herzl, Theodor, 115
Heydrich, Reinhard, 215–17, 221, 232, 233
HICEM, 68, 182–83, 203, 206, 259, 264
High Commission for Refugees (1939), 165–66
High Commission for Refugees, Jewish and Other (1933), 161–66
High Commission for Refugees of the League of Nations (associated with Nansen), 10, 67, 82, 88, 89–91, 94–96, 101, 109–13, 128, 147, 158, 166, 356, 357
Hilberg, Raul, 229, 247
Himmler, Heinrich, 209, 219, 221, 224–26, 232, 239, 271
Hindenburg, Paul von, 210
Hirschmann, Ira, 274, 285, 289, 291, 295, 321, 322, 351
Hitler, Adolf, 12, 55, 122, 123, 146, 147, 184, 186, 221, 265, 270, 275, 298. See also Antisemitism; Germany; Nazism
    and Czechoslovakia, 166, 174, 217
    and Danzig, 133, 134
    and Denmark, 287
    Germany under, 128–35, 142, 149, 152, 171, 204, 206, 213–14
    ideology of, 208–11, 215, 234
    and Jewish emigration, 211–13, 215, 218–19, 230–33, 259, 261, 264
    punishment of Germany by, 297
    and repression, 129, 139, 159, 177
    and the Second World War, 189, 246
    and the Soviet Union, 193, 195
    and Spain, 190, 259
    success of, as mass murderer, 331
Hoare, Samuel, 60, 84, 169
Holland. See Netherlands
Holocaust, 182, 283, 335. See also "Final Solution of the Jewish Question"
Hoover, Herbert, relief activities of, 85, 86, 88, 91, 117

Hope Simpson, John, 61, 95, 105, 113, 114, 138, 145, 149, 151, 153, 166, 167, 177, 178, 182, 191
Horthy, Miklós, 64, 269, 270, 292, 293
Hull, Cordell, 285, 288
Hungarians, as refugees, 17, 21, 52, 72, 74, 110, 132
Hungary, 27, 50, 72
    antisemitism in, 144, 174, 335
    Communist control of (1948), 344, 348
    and Czechoslovakia, 174, 302
    fascist sympathies of, 144, 174
    Germans in, 219, 240, 330
    and Germany, 144, 236, 251–52
    pogroms in, 61, 64
    and population shifts, 327, 329
    and refugees, 48, 130, 141–42, 144, 156, 168, 169, 173, 174, 195, 250–51, 262, 268, 269, 272, 292–94, 359–61
    and Rumania, 200, 301–2
    and the Soviet Union, 348, 359–61
    uprising in (1956), 359

Imrédy, Béla, 174
India, 246, 354, 355
Indochina, refugees from, 365, 367, 370
Indonesia, 367
Inter-Allied Committee on Post-War Requirements, 298–99
Inter-Governmental Committee for European Migration (ICEM), 361, 365, 369
Inter-Governmental Committee on Refugees (IGCR; 1938), 163, 171, 182, 216, 218, 285, 286, 340, 343
International Labor Organization (ILO), 49, 61, 112, 119, 159
International Refugee Organization (IRO), 5, 11, 340–45, 353, 355, 356, 365
    criticism of, 349, 351, 355
    and definition of refugees, 341–42
    and divisions between East and West, 340–41, 344
    Preparatory Commission of, 342, 343
    role of, 342–45, 354
    successes of, 345
    and UNRRA, 340, 343
Iran. See Persia
Ireland, 16, 231, 266
Iron Guard, 173
Israel, 162, 368
    opening of, to Jews, 339, 345
    and war with the Arabs, 339, 345, 354, 364, 370
Istanbul. See Constantinople

Italians, as refugees, 16, 17, 21, 26, 27, 122–28, 146, 150, 178–79
Italy, 16, 17, 69, 109, 115, 145. *See also* Fascism
  antisemitism in, 280
  and Austria, 132, 259
  decolonization and, 368
  Fascism in, 122–28, 280
  and Germany, 236, 279–81
  and Greece, 98, 279
  under Mussolini, 90, 122–28
  and refugees, 49, 122–23, 156, 169, 178, 205, 217, 218, 241, 257, 266, 278–81, 302, 319, 338, 365
  surrender of, 255, 262, 269, 281, 304
  and Turkey, 7, 98
  and Vichy, 279

Jabotinsky, Vladimir (Ze'ev), 142, 185
Japan, 259, 319
  and refugees, 181, 199
Jewish Agency, 164, 167, 175, 182, 186, 206, 272–76, 335, 337, 363
Jewish Colonization Association (ICA), 67, 68, 85, 182
Jewish relief agencies, 62, 65–68, 82, 182–83, 206
Jews. *See also* Antisemitism; Palestine; Pogroms; Zionism
  aid to, by Jews, 38–39, 62, 65–68, 82, 182–83
  and England, 115–16, 152–54, 274–78, 294
  fascism and, 10, 26–32, 362–64
  Final Solution for, 209, 215, 218, 219, 227–33, 251, 254, 256, 279, 287, 289, 290
  and Nazism, 10, 123, 129–83, 195–219, 241–47, 251, 257, 259, 260, 269, 271–74, 278, 280, 283–94
  as refugees, 5–6, 7–10, 14, 27–39, 52, 53, 61–68, 91, 94, 96, 110, 137, 141–45, 153
  resettlement plans for, 183–88
  return to Germany of, 134–35, 137, 156
  in Russia, 9, 28–32, 53, 118–19, 196
  Sephardic, 262, 368, 370
  survivors among, after the Second World War, 300, 304, 306–8, 322–23, 325, 331–39, 343
  and Switzerland, 155–58
  Western sympathy for, 28–29, 159
Johnson Act (1924), 65, 112

Joint Emergency Committee on European Jewish Affairs, 283, 286
Jones, Ernest, 168

Kádár, János, 361
Kállay, Miklós, 351, 352
Karlsbad Decrees (1819), 16
Katyn massacre, 250
Kennan, George, 301
Kenya, 188, 367, 369
Keynes, John Maynard, 136
Khrushchev, Nikita, 249
Kiev, 54–57, 62, 246, 291
Koestler, Arthur, 277
Kolchak, Alexander, 57, 84
Korea, 354–55
Kossuth, Lajos, 19, 20
Kot, Stanislaw, 198, 246
Kovno, 62, 65
Krasnov, Peter, 314, 316, 317
Kristallnacht, 149, 176–77, 181, 213, 214
Kuibyshev, 243
Kun, Béla, 64, 72

La Guardia, Fiorello, 322–23
Latin America, 67, 68, 113, 177, 206, 218, 260, 344
Latvia, 71, 197, 221
Lausanne Convention (1923), 102–4
Laval, Pierre, 202, 279
League of Nations, 88, 93, 110, 113, 259
  and Danzig, 133, 177
  effectiveness of, 52, 66, 91, 105–6, 108–9, 111–12, 133, 259
  and Germany, 109, 158, 161, 164, 171
  ineffectiveness of, 158–66, 180
  and minority rights, 70, 160
  and Palestine Mandate, 115–16
  problems of, 110–11, 161, 165
  and refugees, 10, 11, 52, 60, 64, 66, 67, 72, 75, 81, 89–90, 96, 101, 107, 108, 123, 128, 140, 147, 158–66, 171, 192
  resettlement/repatriation programs of, 85, 89, 95–96, 102–4, 108, 119, 120
  and the Soviet Union, 88, 110, 165
Lebensborn program, 350–51
Lebensraum, 208, 219–27, 234–40
Lehman, Herbert, 317, 318, 322
Lenin, Vladimir Ilyich, 55, 56
Leningrad, 199, 234, 243
Lippmann, Walter, 184
Lisbon, 203, 204, 206, 207, 258, 259, 263, 264

Lithuania, 57, 71, 195, 197, 198, 221, 235
  and Jews, 62, 198–99
Litvinov, Maxim, 140
Locarno, 109
Lockerwood, Oliver, 184
Lodz, 63, 227, 228
London, 25, 39, 82
  and refugees, 14–15, 17–19, 21, 22, 24,
    41, 90, 150, 168, 171, 174, 217, 218,
    250
Long, Breckinridge, 284, 288
Louis XIV (French king), 6
Louis XVI (French king), 9
Louis Philippe (French king), 16
Lowrie, Donald, 256
Lublin, 228–30
Luxembourg, 155, 201, 230, 258, 282

Macartney, C. A., 10, 104, 105
McClellend, Rosell, 294
McDonald, James G., 132, 161–65, 184
Macedonia, 10, 42, 44–46, 48, 106, 279
  and Bulgaria, 106–8, 236
  and Greece, 104, 106–8
Macedonians, as refugees, 44–46, 104,
    106–8, 113, 236
Madagascar, 186–87, 230, 232, 266
Madrid, 190, 191, 258, 261, 262, 268, 287,
    369
Malcolm, Neill, 164–65, 175, 187, 283
Manchuria, 60, 181
Mann, Thomas, 128–29, 141
Marseilles, 19, 41, 114, 203, 206, 253, 259
Marshall Plan, 330, 345, 349, 352
Marx, Karl, 14, 17, 18
Marxism, 265
Masaryk, Thomas, 74
Matteoti, Giacomo, 125
Mauco, Georges, 113
Mauritius, 276
Mazzini, Giuseppe, 14, 17, 19, 20–22
Metternich, Klemens von, 16
Mexico, 187, 192, 203, 246
Middle East, 60, 75, 93, 110, 138, 246
Mongolia, 60
Montenegro, 41, 42
Moravia, 36, 175, 219, 230
Morgan, Frederick, 320–22, 337
Morgenthau, Henry, Sr., 83, 100, 104–5
Morocco, 262, 367
Moscow, 29, 30, 196, 250
  and refugees, 54, 130, 139, 362–63

siege of, 232
  Treaty of (1921), 78
  Treaty of (1940), 199
Mosley, Oswald, 151
Munich Accord, 174
Murphy, Robert, 328
Muslims
  as refugees, 12, 40–42, 44, 45, 104, 286
  in the Soviet Union, 249
Mussolini, Benito, 187, 218, 280
  and anti-Fascists, 124–26, 146, 178
  collapse of, 287
  and Fascism, 122, 124–26
  and France, 127, 255
  and Spain, 190
  and the Vatican, 265, 266
Mutual Security Act (1951), 354

Nagy, Imre, 359
Nansen, Fridtjof, 10, 59, 60, 64, 85, 86
  career of, 86–88
  death of, 112, 114
  difficulties of, 90
  as high commissioner for refugees of
    the League of Nations, 52, 88–91, 96,
    109, 110, 112, 119–21, 271
  as mediator in Greco-Turkish war, 102,
    104
  policies of, 90–91, 94, 95
  and population exchanges, 101–4
  relief aid of, 87–88, 91, 101, 104–5, 111,
    119–21
  successes of, 120–21
Nansen Help, 87–88
Nansen passport, 81, 94–96, 133, 178, 180,
    316
Nantes, Edict of, 7, 9, 154
Napoleon I (French emperor), 16, 23, 311,
    314
Napoleon III (French emperor), 21
Nationalism, 33, 38, 51, 52, 63, 116–17
  definition of, 49–50
  role of, in refugee problems, 23–26, 33,
    49–52, 69–72, 74
Nazism, 68, 162. See also Antisemitism;
    Concentration camps; Refugees; Ter-
    rorism
  Allied condemnation of, 282
  in Austria, 168
  and creation of refugees, 122–23, 162,
    178, 179, 297
  and Danzig, 133–34

in Germany, 122, 123, 128–41, 148, 156, 160, 185, 208–9
goals of, 208–9, 215, 220
and Jews, 10, 123, 128–46, 148, 150–83, 208–19, 227–34, 272
policies under (1933–44), 209–40
and police network, 129
and Saar Valley, 133
victims of, 122–23
Nazi-Soviet Pact, 193, 195, 200
Netherlands, 282
and Germany, 201
and refugees, 29, 36, 38, 65, 130, 134–35, 137, 154, 160–61, 169, 171, 205, 206, 218, 258, 288, 367
Neuilly, Treaty of (1919), 106, 107
Neurath, Konstantin von, 161
Nice, 125–26, 279
Nicholas I (Russian tsar), 15, 26
Nicholas II (Russian tsar), 53
Nicolson, Harold, 68
NKVD, 141, 194, 196, 197, 241
North Africa, 59, 259, 261, 262, 264, 281, 285, 286, 319
Norway, 87, 90, 163–65, 270, 282
and Germany, 88, 201, 231, 271
Nuremberg Laws (1935), 130, 131, 134, 157, 177, 211

Odessa, 56, 57, 199, 237, 243
Orsini, Felice, 21–22
Ottoman Empire, 42, 52. *See also* Turkey
and Armenians, 75, 95
cruelty of, 41, 42, 44, 50, 262
decline of, 40, 41, 44, 45, 97
passport requirement of, 92
and refugees, 9–10, 12, 20, 40–42, 44, 50, 96
Overpopulation, and refugees, 115, 365

Pacelli, Eugenio, 265. See Pius XII
Pale of Settlement, 27, 29, 30, 55, 61
Palestine. *See also* England
colonization of, 115–17, 145
and Jewish refugees, 60, 116, 117, 137, 152–53, 163, 169, 175, 177, 183–88, 199, 206, 212, 213, 217, 218, 251, 269, 272–74
as National Home for Jews, 115–16, 152, 184, 274
opening of, for Jews, 339

partition of, 152, 339
relief aid to, 83
Papen, Franz von, 272
Paraguay, 133, 157
Paris, 24, 25, 82, 93, 186, 201
and peacemaking after the First World War, 68–69, 75, 79
and refugees, 17–19, 22, 24, 26, 35, 52, 60, 68, 88, 90, 125, 133, 146, 150, 176, 192, 279
Paris Peace Conference, 115, 132
Passports, 92–94
Nansen 94–95
Patton, George S., 321, 333
Pavelić, Ante, 279
Peace settlements, following the First World War, 68–72, 74–82
Peel Commission (1937), 152, 185
Permanent Court of International Justice, 134
Persia, 75, 76, 83, 247, 340
Pétain, Philippe, 202, 279
Peter the Great, 7, 24
Pilsudski, Józef, 143
Pius XI (pope), 265
Pius XII (pope), 265, 267
Pogroms, 28, 29, 31, 39, 61, 63, 64, 115, 143, 144, 176–77, 335–36
Poland, 15–17, 21, 24, 29, 57, 63, 64, 69–71, 175, 282, 300
antisemitism in, 61, 63, 117, 143–44, 170, 172, 173, 185, 335–36, 362
Communist control in, 348, 350
and Danzig, 133, 173
Germans in, 219–21, 224, 240, 305, 330
and Germany, 71, 130, 133, 134, 143, 173, 189, 209, 218, 223, 227, 228, 231, 237, 239, 250, 275
independence for, 57
Jews in, 27, 67
under Nazi occupation, 221, 222 *(map)*, 234
pogroms in, 61, 63, 143
and population shifts, 327
refugees and, 53, 55–58, 60, 63–66, 88, 94–96, 130, 135, 139, 141–44, 146, 152, 168, 172, 173, 181, 183, 218, 227, 228, 251, 262, 292
and the Second World War, 153, 189, 192, 194–98, 200, 271
and the Soviet Union, 71, 88, 189, 194–99, 241, 245–46, 249–50, 282, 302, 328, 359, 364

Poland (continued)
  uprising of (1863), 27
  and the Vatican, 267
Poles, as refugees, 12, 15–16, 19–21, 23,
  24, 52, 57–58, 71, 189, 195, 196, 225,
  240–41, 250–51, 253, 264, 278, 323,
  348, 364
Polish Corridor, 71
Popular Front, 147, 186, 190, 192
Population exchange, 46, 101–2, 106–9,
  152, 198, 302, 328–29
Population shifts
  under Nazi domination, 221, 223–25,
    241–43, 245–50
  before Nazism, 5–7, 10, 12
  after the Second World War, 299–302,
    305, 326, 327, 329–31
Portugal, 122, 132, 263
  neutrality of, 258, 263
  and refugees, 203, 206, 258, 259, 262–
    65, 368
Portuguese, as refugees, 16, 132
Potsdam agreement (1945), 327–29, 334
Prague, 23, 65, 67, 153, 160, 174, 328
Proudfoot, Malcolm, 3–4, 313, 134, 320,
  344
Prussia, 23, 24, 128

Qaddafi, Muammar, 368
Quakers, and relief aid, 82, 182
Quisling, Vidkun, 88, 91, 120

Rapallo, Treaty of, 139–40
Rathbone, Eleanor, 283–84, 286
Rebellion/revolution, and refugees, 15–26,
  52, 55–57, 60–61
Red Cross, 271
  and refugees, 11, 89, 111, 159, 192, 201,
    237, 250, 272, 304, 306, 307, 325, 360
  and relief aid, 82, 83, 85, 86, 111, 263
Reestablishment of the Professional Civil
  Service Law (1933), 129
Reformation, 154
Refugee Act (1953), 354
Refugees. See also Displaced persons;
  Great Depression; Jews; High Com-
  mission for Refugees; International
  Refugee Organization; League of Na-
  tions; United Nations and its organizations; and specific countries
  characteristics of, 4–12, 65, 352
  cold war and, 348
  contemporary situation regarding, 347–
    51

decolonization and, 367–71
definitions of, 3–5, 8–12, 52–53, 356–59,
  363–66
from Eastern Europe after the Second
  World War, 358–64
era of the, 3, 13, 347
escape routes during the Second World
  War for, 240–82
vs. exiles, 15, 17–22
fascism and, 10, 90, 96, 122–207, 132–
  35, 145–46, 150
after the First World War, 52–74
hostility to, 48–50, 53–66
lesson of history regarding, 371
nationalism and, 23–26, 49–52, 69–72,
  94
Nazism and, 123, 128–41, 148, 150–83,
  186–87, 196–207, 209, 216–19, 225–40,
  242, 243, 245–95, 345
in the nineteenth century, 14–39, 50
problems of, 123, 136–37, 165, 352–53
and rebellion/revolution, 15–26, 52, 55–
  57, 60, 357
Soviet Union and, 196–200, 310, 313–
  19, 352
and stateless persons, 70, 91–96
of the Third World, 347, 354–55
war and, 6, 7, 23, 52–56, 95–109, 189–
  95, 197–203, 205–6, 231, 238–40, 295,
  354–55
as worldwide problem, 365
Reich Central Office for Jewish Emigration, 215, 217
Reich Commission for the Consolidation
  of Germandom (RKFDV), 209, 221,
  224
Relief agencies, 82–84, 111
Resettlement/repatriation programs, 85,
  87, 89–91, 95–96, 102–3, 106–9, 111–
  21, 138, 183–88, 218, 221, 223, 225,
  235, 305, 328–29, 358, 366, 370–71
  problems of, 315, 334, 337, 342, 344
  of Soviet nationals, 310, 313–19
  and UNRRA, 320, 323
Revolution of 1905, 24, 25, 28, 36
Ribbentrop, Joachim von, 186, 214, 216,
  280
Riga, Treaty of (1921), 57, 71
Rome, 75, 269
  Treaty of (1967), 366–67
Roosevelt, Eleanor, 329, 340
Roosevelt, Franklin D., 137, 145, 168, 170,
  171, 187, 217, 285, 286, 291, 315, 318
Rosenberg, Alfred, 229, 234

Rothmund, Heinrich, 157, 254, 257
Rublee, George, 171, 172, 216, 217
Rumania, 33, 49, 52, 70, 74, 111, 200, 237, 272
  antisemitism in, 33–35, 144, 170, 173, 237
  Germans in, 219, 240
  and Germany, 144, 223
  and Hungary, 200, 301–2
  and Jews, 2, 27, 33–35, 67
  and refugees, 60, 64–66, 72, 94, 95, 106, 139, 141–45, 168, 172, 174, 195, 250, 273, 286, 290–92, 295
  and the Second World War, 200, 237
  and the Soviet Union, 200, 238, 240, 278, 329
Rumanians, as refugees, 19, 49, 104, 132
Rumbold, Horace, 59, 107
Ruppin, Arthur, 30, 32, 65, 142, 167, 184, 211
Russell Bill (1940), 206
Russia, 40, 41, 57, 98, 113. *See also* Soviet Union
  antisemitism in, 28–31, 61–62
  Armenians and, 12, 81, 119
  and the Balkans, 41, 42
  Bolshevik Revolution in, 52, 55
  civil war in, 55–57, 59, 84, 85, 95, 111
  economy of, under communism, 136
  famine in, 85, 86, 89, 95
  and Germany, 55, 61–62
  Jews and, 9, 12, 26–28, 30, 35, 54–55, 63, 65, 67
  Napoleon's invasion of, 23
  pogroms in, 28, 61, 63
  and Poland, 15–16, 27
  and refugees, 54–57, 82, 88, 90, 91, 94, 95, 106, 110
  and Turkey, 42, 44, 75, 76, 78, 97, 98, 119
Russians, as refugees, 19, 20, 24, 25, 52, 59–61, 82, 84, 96, 104, 110–12, 114, 120, 133, 138–39, 149, 154, 158, 159, 167, 178
Russo-Japanese War, 28, 30

Saar Valley, 133, 147, 159
Saint-Germain, Treaty of, 69
Salazar, António de Oliveira, 258, 263, 368
Salengro, Roger, 147–48
Salonika, 45, 100, 103, 104, 262, 279, 281
San Stefano settlement, 42
Sauckel, Fritz, 236, 238
Schacht, Hjalmar, 131, 213, 216, 217

Schechtman, Joseph, 4, 69, 199, 223, 298
Schnyder, Felix, 369
Second World War, 3–5, 10, 144, 258. *See also* Nazism; Refugees
  casualties of, 193
  conflicts in Europe after, 301–2
  end of, 296
  outbreak of, 189, 267
  problems following, 296–345
  refugees during, 194, 195, 197–200
  refugees immediately before, 123, 131–35
Serbia, 41, 42, 45–46, 48–49, 236
Serbs, as refugees, 6, 104, 236, 279
Sèvres, Treaty of (1920), 97
Shanghai, and refugees, 60, 180–82, 218
Shirer, William L., 170–71, 206
Siberia, 54, 57, 64, 181, 196, 197, 234, 243, 245, 247, 249
Sicily, 282, 287
Simon, John, 164
Slovakia, 195, 251, 268, 294, 335
Slovenians, as refugees, 237
Smolensk, 53, 250
Smyrna (Izmir), 81, 98, 100
Sofia, 94, 106, 113, 262, 272, 353
South Africa, 68, 184
Soviet bloc. *See also* Russia; Soviet Union
  and refugees, 11, 348
  unrest in, 352
Soviet Union. *See also* Nazism; Russia
  antisemitism in, 140, 352
  and Armenians, 78, 81, 120
  and Bulgaria, 295
  collectivization of agriculture in, 138–39, 249
  confrontation of, with the West, 349, 350
  and Czechoslovakia, 248
  and denationalization, 178
  and Finland, 199–200, 271
  and foreign Communists, 130, 139–41, 193, 219, 221, 223
  Germans in, 219, 240, 241, 245
  and Germany, 232, 233, 238, 241, 242, 247, 249, 250, 262, 282, 287, 296, 297
  harsh conditions after the Second World War in, 352
  and Hungary, 348, 359–61
  Jews in, 118–19, 336, 362–64
  and League of Nations, 110, 165
  and Poland, 71, 88, 189, 194–99, 241, 245–46, 249–50, 282, 328, 359, 364
  recognition of, 96

Soviet Union (*continued*)
  and refugees, 12, 61, 65, 67–68, 89, 93,
    96, 130, 138–41, 158, 189, 193–94,
    196, 197, 282, 286, 290, 302
  and relief aid, 85, 86, 88, 91, 111
  suspicions of, 85, 88
  and Turkey, 272
  and the United States, 315, 317, 323,
    340, 342–43, 347
  and UNRRA, 318, 319, 323, 324
Spain, 260, 262
  Civil War in, 127, 138, 148, 149, 177,
    189–94, 258–59
  fascism in, 122
  and Germany, 258–60
  neutrality of, 258
  and refugees, 126, 145, 167, 205, 206,
    241, 258–65, 285, 362, 368, 370
Spaniards, as refugees, 16, 17, 123, 138,
    145, 190–94, 212–13, 258, 341
Special Committee on Refugees and Dis-
    placed Persons (1946), 340
SS, 209, 211, 215, 224, 225, 230, 232, 236,
    239, 246, 264, 269, 292
Stabilization, of refugee problem (1924–
    30), 109–21
Stalin, Joseph, 138, 139, 141
  death of, 359
  and Germany, 193, 195
  and the Second World War, 241, 243,
    315, 316
  and Soviet bloc unrest, 352
  and the United States, 318
Stalingrad, 282
Stateless persons, 70–71, 91–96, 178–80,
    187, 262
  problems of, 93–94, 253
Steiger, Edouard von, 252–53
Steinhardt, Laurence, 274
Stockholm, 21, 93, 270, 271, 288
Sudetenland, 153, 166, 174, 175, 177, 329
Supreme Headquarters Allied Expedition-
    ary Force (SHAEF), 309–10, 316, 318,
    331
  and UNRRA, 319
Sweden, 87
  and Finland, 270–71
  and Germany, 270
  and Hitler, 270
  neutrality of, 271
  and refugees, 20–21, 90, 270–72, 282,
    287–88, 294, 304
  and the Soviet Union, 271
Switzerland, 154, 157, 254, 257

and Germany, 156–58, 256
neutrality of, 154–55, 252–53
and refugees, 17–21, 24, 26, 49, 64, 126,
    130, 154–58, 169, 170, 206, 217, 241,
    252–58, 294, 304, 370
Syria, 60, 77, 83, 101, 272, 278
Szálasi, Ferenc, 74, 292

Taylor, A. J. P., 150
Taylor, Myron C., 170
Terrorism, 24, 26, 57, 72, 124–26, 128–31,
    235, 246
Third World, 4, 347, 354–55, 365, 369–71
Thomas, Albert, 112, 115
Thompson, Dorothy, 3, 105, 184
Thrace, 41, 44, 45, 48, 98, 100–102, 106,
    107, 110, 279
Thurstan, Violetta, 54, 64
Tito, Josip Broz, 281, 302, 360
Tolstoy, Nicolai, 197, 199, 241
Toynbee, Arnold, 45, 50
Transcaucasia, 10, 12, 74, 78, 86, 96, 167,
    245, 249
Transnistria, 237, 238, 291
Transylvania, 72, 144, 200, 301
Travel, 92–96
Trianon, Treaty of (1920), 72, 144
Trieste, 41, 348
Trotsky, Leon, 56, 139
Truman, Harry S., 311, 332, 334, 338, 349,
    354
Tsarist empire, 9, 20, 21, 24, 25, 33, 36,
    39, 78, 96
Tunisia, 367
Turkey, 21, 45, 81, 115. *See also* Ottoman
    Empire
  and the Allies, 78, 97, 102, 272, 273
  and Armenians, 75–77, 79, 81, 83, 84,
    96, 178, 289
  and Bulgaria, 46, 270, 353
  and the Central Powers, 76, 78
  and England, 97, 100, 102
  and Germany, 231, 270, 272, 273
  and Greece, 41, 44–46, 48, 52, 97–108
  neutrality of, 272
  and reforms, 52, 78, 106
  and refugees, 43–45, 52, 108, 242, 251,
    275, 277, 286, 291, 293–95, 370
  and Russia, 42, 60, 78, 97, 119, 139, 270,
    272
Turks, as refugees, 45, 46, 106, 353

Uganda, 367, 369
Ukraine, 57

and Bolshevik Revolution of 1917, 56
famine in, 89, 91
Germans in, 221, 223, 235, 245
and Germany, 234, 239
Jews in, 30, 62
pogroms in, 61, 63
and Rumania, 237, 291
Ukrainians, 25, 29, 196–98
as refugees, 52, 56, 249, 323
United Committee for Jewish Emigration
(Emigdirect), 67–68
United Kingdom. *See* England
United Nations. *See also* United Nations
High Commission for Refugees;
United Nations Relief and Rehabilitation Administration
and the Palestine question, 339
and refugees, 10–12, 340, 342, 345
United Nations High Commission for Refugees (UNHCR), 11, 342, 345, 354–
58, 365, 370
and cold war refugees, 361, 364
limited budget, staff, and authority of,
357, 366
and resettlement, 366
and the Third World, 368, 369
United Nations Korean Reconstruction
Agency (UNKRA), 355
United Nations Refugee Fund (UNREF),
366
United Nations Relief and Rehabilitation
Administration (UNRRA), 11, 309–12,
317–24, 334, 337, 356
criticism of, 320–22, 324, 343
financing of, 320, 324
origins of, 317–18
problems of, 318, 323, 324
role of, 319–24
successes of, 320, 324
United Nations Relief and Works Agency
for Palestine Refugees (UNRWA), 355
United States, 34, 92, 98, 100, 181, 261
and Armenians, 75, 79, 81, 83
and Germany, 215–17
immigration to, 28, 30–32, 34–36, 65,
68, 112, 114, 117, 137, 138, 163, 206,
253, 323, 334, 338, 344, 354, 361
and relief aid, 82, 83, 85, 86, 183, 301
and Russia, 56, 315, 317, 318, 323, 340,
343, 347, 348
and UNRRA, 318
United States Escape Program (USEP), 354
Upper Silesia, 38, 71, 160, 221
Uzbekistan, 194, 248

Vatican, 267. *See also* Pius XI, Pius XII.
and Fascism and Nazism, 265–67, 269
neutrality of, 265–67
opposition of, to Nazi racial doctrines,
265–66
and refugees, 265–270, 281, 291
and Zionism, 269
Venizelos, Eleutherios, 97–98, 102
Versailles, Treaty of, 69, 71, 109, 133, 160
Vichy France, 202–4, 219, 254, 259, 261,
262, 278, 279
Vienna, 32, 35, 36, 40, 74, 133, 168, 169
and refugees, 62, 180, 217, 363
Vilna, 57, 63, 199
Volksdeutsche, 220, 221, 223–25, 235, 239,
240, 301, 325, 330
Vyshinsky, Andrei, 324, 340

Wallenberg, Raoul, 272, 294
War, 92
and refugees, 6, 7, 23, 52–56, 95–109,
189–95, 197–203, 205–6, 231, 238–40,
296
War Refugee Board (WRB), 272, 274, 283,
289–91, 293, 294
Warsaw, 57, 63, 134, 172, 195, 329
ghetto uprising in, 286, 300
and refugees, 62, 65, 66
Warsaw Pact, 362
Washington, D.C., 82, 216–18
Wavell, Archibald, 276
Weimar Republic, 129, 131, 212, 219
and the Great Depression, 136
and refugees, 71, 74, 139
Weizmann, Chaim, 117, 185, 275, 290
Weizsäcker, Ernst von, 172, 214, 257, 269
Welles, Sumner, 186
Werth, Alexander, 238, 243
West Germany, 11, 361–62
West Prussia, 221
Western Europe
Jews of, 32, 66–67, 209–19
post-World War II prosperity of, 347,
352
refugees from Nazism in, 135–41, 159
White Paper (1922), 116
White Paper (1939), 152–53, 217, 274–76,
278, 285
White Russia, 56, 57, 62, 63
Wilhelminian empire, 9, 52
Wilson, Woodrow, 64, 85
Wise, Stephen, 285
Wohlthat, Helmut, 217
World Council of Churches, 283

World Jewish Congress, 164, 199, 274, 283, 290, 291, 294
World Refugee Year, 365
World Zionist Organization, 117, 185, 218
Wrangel, Peter, 57, 59, 84

Yalta agreement (1945), 315–17, 328
Young Turks, 22, 44, 46, 76, 106
Yudenich, Nikolai, 57, 84
Yugoslavia, 49, 70, 106, 282, 302
    Communist control in, 348
    Germans in, 219, 224, 240

and Germany, 236
and population shifts, 329
and refugees, 59, 60, 72, 74, 108, 130, 132, 156, 169, 219, 249, 251, 277, 360–61
and the Soviet Union, 352

Zionism, 115–17, 152–53, 162, 171, 184–86, 212, 274, 275, 304, 323
    and the Vatican, 269
Zogu, Ahmed, 132
Zurich, and refugees, 19, 154